Endorsements for *Sustainable Energy Systems Engineering*

Sustainable Energy Systems Engineering soars with practical advice. This book is a must read for everyone who has any role in renewable energy-from architects, installers, purchasers, governmental and regulatory agency staff to lenders and all citizens who care about the global environment and sustainable technology. This book is the leading edge of the biggest crisis facing the human race—energy production—and the common sense solutions to that crisis. Dr. Gevorkian has once again produced an informative and practical book that is chock full of powerful examples, charts and diagrams.

GENE BECK, CLP
President
EnviroTech Financial, Inc.

Where Dr. Gevorkian's first book, *Sustainable Energy Systems in Architectural Design: A Blueprint for Green Building*, was the manual of Sustainable Energy Systems, his latest tome *Sustainable Energy Systems Engineering*, is the Encyclopedia of Green Energy. I keep this book near my desk, and it is my sole source Green Design Manual, whether I am comparing systems for selection, or am delving a bit deeper to fully understand a particular system's application. Dr. Gevorkian's work concisely covers a library of information as well as hours of research at each glance.

MARK GANGI, AIA
GANGI Architects

Thanks, Dr. Gevorkian, for giving me a chance to go over the introduction manuscript.

This book is for everyone who is interested or in the renewable energy field. If you are allowed to have only one book on the renewable energy topics, this is the book to have.

The book gives you an overall assessment of the immediate crisis and possible future damage to the earth environment due to human being expansion and development. It also suggests possible solutions—well-rounded information and knowledge of renewable technologies available today. The green technology development in chronologic order is also valuable information.

MS. YING WANG, REGISTERED ARCHITECT
LEED Accredited Program Manager CHPS/Energy Program, School Planning & Design
Los Angeles Unified School District Facilities Services Division

Dr. Gevorkian, I would like to thank you for the opportunity to read your book's manuscript. The book is an excellent source of information. I will be first in line to get my copy. I found the colorful diagrams and photos to be just enough to bring system designs to life.

WADE WEBB
Engineering Manager
Integrated Solar Power Technologies

Dr. Gevorkian's description of the basic system components and all of the variables associated with alternative energy, in particular wind energy systems, is very informative. I found the examples on wind farms and their impacts to our environment very descriptive. Wind farms have a great potential on otherwise under-developed or non-developable lands, especially in the Mid-West. Thank you for sharing your information on these incredible technologies.

JEFF ZOOK, AIA
Coastal Architects
Oxnard, CA

The book is a comprehensive presentation of alternative sources of energy from the physics of solar cells to the detailed study of many technologies, their environmental impact as well as economics. Throughout the book, the author retains a simple approach in presenting the subject matter. Abundance of informative illustrations in the book provide ample help for readers to understand otherwise complicated technologies.

Highly recommended, especially at a time when the negative impact of conventional power generation on many aspects of environment and human life is becoming so obvious and alternative approaches are so critically needed.

RAZMIK MATHEVOSIAN EE, PE

Sustainable Energy Systems Engineering is a very useful reference book for those of us in the environmental energy field. The systems which you have so clearly outlined in your book, illustrate environmental impact and great potential of waste matters which are used for generating electrical energy.

In particular, the section on generation of energy from capture of methane gas from sanitary landfills is particularly illuminating.

I believe that your book provides an excellent practical advice and references necessary to implement sustainable energy strategies, especially excellent graphics are in particular very illuminating.

KEN HEKIMIAN PH.D., RCE, REA
President, HVN Environmental Service Corporation, Inc.

SUSTAINABLE ENERGY SYSTEMS ENGINEERING

About the Authors

Dr. Peter Gevorkian, Ph.D. is president of Vector Delta Design Group, Inc., an electrical engineering and solar power design consulting organization. Dr. Gevorkian has been an active member of the Canadian and California Board of Professional Engineers. In addition, he has taught computer science and automation control, and has authored numerous technical papers on sustainable energy systems.

SUSTAINABLE ENERGY SYSTEMS ENGINEERING

THE COMPLETE GREEN BUILDING DESIGN RESOURCE

PETER GEVORKIAN

New York   Chicago   San Francisco   Lisbon   London   Madrid
Mexico City   Milan   New Delhi   San Juan   Seoul
Singapore   Sydney   Toronto

The **McGraw·Hill** Companies

Library of Congress Cataloging-in-Publication Data.

Gevorkian, Peter.
  Sustainable energy systems engineering : the complete green
building design resource / Peter Gevorkian.
      p. cm.
  Includes index.
  ISBN-13: 978-0-07-147359-0
  ISBN-10: 0-07-147359-9 (alk. paper)
1. Renewable energy sources. 2. Sustainable buildings. I. Title.
  TJ808.G48 2006
  696—dc22

                                    2006020280

1 2 3 4 5 6 7 8 9 0   DOC/DOC   0 1 3 2 1 0 9 8 7 6

ISBN-13: 978-0-07-147359-0 ►
ISBN-10: 0-07-147359-9

The sponsoring editor for this book was Cary Sullivan, the editing supervisor was David E. Fogarty, and the production supervisor was Richard C. Ruzycka. It was set in Times by International Typesetting and Composition. The art director for the cover was Anthony Landi.

Printed and bound by RR Donnelley.

This book was printed on acid-free paper.

McGraw-Hill books are available at special quantity discounts to use as premiums and sales promotions, or for use in corporate training programs. For more information, please write to the Director of Special Sales, McGraw-Hill Professional, Two Penn Plaza, New York, NY 10121-2298. Or contact your local bookstore.

CONTENTS

FOREWORD

Pursuing green is the most important endeavor in the world today. This book is an excellent resource for understanding the state of green technology today. Each technology is set into its historical context with engaging prose and lucid scientific and engineering rigor. It is an encyclopedic source for a wide range of green technologies. Numerous technology case studies throughout the book emulate the process of designing the green systems.

This is a professional's book. Architects, engineers, environmentalists, developers, governmental agencies, and others will be enabled in their pursuit of green design. It serves the purposes of architects by showcasing the broad range of options and the conceptual underpinnings available to them when they begin designing their own green projects. It serves the purposes of engineers by providing the scientific bases for each system. It serves the purposes of the entire building team by discussing the economics, strategies, typical scenarios, and documentation required to implement green projects.

MICHAEL B. LEHRER, FAIA
Adjunct Associate Professor
University of Southern California

INTRODUCTION

Since the dawn of agriculture and civilization, human beings have hastened deforestation, impacting climatic and ecological conditions. Deforestation and the use of fossil fuel energy diminish the natural recycling of carbon dioxide gases. This accelerates and increases the inversion layer that traps the reflected energy of the sun. The augmented inversion layer has an elevated atmospheric temperature, giving rise to global warming, which in turn has caused melting of the polar ice, substantial changes to climatic conditions, and depletion of the ozone layer.

Within a couple of centuries, the unchecked effects of global warming will not only change the makeup of the global land mass but will affect the life style of humans on the planet.

Continued melting of the polar ice caps will increase seawater levels and will gradually cover some habitable areas of global shorelines. It will also result in unpredictable climatic changes, such as unusual precipitation, floods, hurricanes, and tornados.

In view of the rapid expansion of the world's economies, particularly those of developing countries with large populations, such as China and India, demand for fossil fuel and construction materials will become severe. Within the next few decades, if continued at the present projected pace, the excessive demand for fossil fuel energy resources, such as crude oil, natural gas, and coal, will result in the demise of the ecology of our planet and, if not mitigated, may be irreversible. Today China's enormous demand for energy and construction materials has resulted in considerable cost escalations of crude oil, construction steel, and lumber, all of which require the expenditure of fossil fuel energy.

Developing countries are the most efficient consumers of energy, since every scrap of material, paper, plastic, metal cans, rubber, and even common trash, is recycled and reused. However, when the 2.3 billion combined populations of China and India attain a higher margin of families with middle-class incomes, the new demand for electricity, manufacturing, and millions of automobiles will undoubtedly change the balance of ecological and social stability to a level beyond imagination.

The United States is the richest country in the world. With 5 percent of the world's population, the country consumes 25 percent of the global aggregate energy. As a result of its economic power, the United States enjoys one of the highest standards of life with the best medical care and human longevity. The relative affluence of the country as a whole has resulted in the cheapest cost of energy and its wastage.

Most consumption of fossil fuel energy is a result of inefficient and wasteful transportation and electric power generation technologies. Because of a lack of comprehensive energy control policies and lobbying efforts of special-interest groups, research-and-development funds to accelerate sustainable and renewable energy technologies have been neglected.

In order to curb the waste of fossil fuel energy, it is imperative that our nation, as a whole, from politicians and educators to the general public, be made aware of the dire consequences of our nation's energy policies and make every effort to promote use of all available renewable energy technologies so that we can reduce the demand for nonrenewable energy and safeguard the environment for future generations.

As scientists, engineers, and architects, we have throughout the last few centuries been responsible for the elevation of human living standards and contributed to the advancement of technology. We have succeeded in putting a man on the moon, while ignoring the devastating side effects to the global ecology. In the process of creating betterment and comforts of life, we have tapped into the most precious nonrenewable energy resources, miraculously created over the life span of our planet, and have been misusing them in a wasteful manner to satisfy our most rudimentary energy needs.

The deterioration of our planet's ecosystem and atmosphere cannot be ignored or considered a matter that is not of immediate concern. Our planet's ozone layer according to scientists has been depleted by about 40 percent over the past century and greenhouse gases have altered meteorological conditions. Unfortunately, the collective social consciousness of the educated masses of our society has not concerned itself with the disaster awaiting our future generations and continues to ignore the seriousness of the situation.

PRESS CLIPPINGS

As principal agents of human welfare, we scientists, engineers, and architects must individually and collectively assume responsibility for correcting the course of future technological development. To accentuate the seriousness of the misuse of the nonrenewable energy resources, I would like the reader to take notice of the following news articles published in various periodicals.

Detroit Free Press, Michigan—04-23-2005

Emissions of "greenhouse gasses," believed to cause global warming, have risen in Michigan. University of Michigan researchers released findings from the first statewide inventory of such emissions. The report said that nine percent more emissions were released in 2002 than in 1990.

Reuters South Africa, Minnesota—02-20-2005

Global warming cloud stifle cleansing summer winds across parts of northern United States over the next 50 years and worsen air pollution. Further warming of the atmosphere, as is happening now, would block cold fronts bringing cooler, cleaner air from Canada and allow stagnant air and ozone pollution to build up over the cities in the Northeast and Midwest. "The air just cooks" said Loretta Mickley of Harvard University's Division of Engineering and Applied Sciences.

Vallejo Times-Herald, California—03-01-2005

The geologists, who believe in global warming, said the Carquinez Strait's future would depend on how much the trend plays out. "If the sea level continues to rise, it would continue to flood the San Francisco Bay and the Delta. There would be several tens of meters of rise in world's oceans, then it would be Venice, then it would be under water."

Environment News Service, New Zealand—03-03-2005

A major study of Arctic lake sediments provides new evidence of human-induced climate change and concludes it may soon be impossible to find pristine Arctic environment untouched by climate warming. Arctic lakes have undergone dramatic ecological change in the past 150 years, and the timing of these changes mirrors the warming trends that commenced when humans began the widespread burning of fossil fuels. The findings were published in the *Proceedings of the National Academy of Sciences*.

ABS CBN News, Philippines—12-01-2004

Singapore—The weather predictions for Asia in 2050 read like a script from a doomsday movie. Except many climatologists and green groups fear they will come true unless there is a concerted global effort to rein in greenhouse gas emissions. In the decades to come, Asia, home to more than 6.3 billion people, will lurch from one climate extreme to another, with impoverished farmers battering droughts, floods, diseases, food shortages, and rising sea levels.

CTV, Canada—03-12-2005

Acid rain is causing forest decline in much of Eastern Canada, with losses to the forest industry estimated at hundreds of millions of dollars annually in the Atlantic region alone, says an Environmental Canada report. Even though acid pollution in Canada has been cut in half over the past 20 years, an additional 75 percent cut is needed says the Canadian Acid Deposition Sciences Assessment.

GLOBAL WARMING AND CLIMATE CHANGE

Ever since the Industrial Revolution, human activities have constantly changed the natural composition of Earth's atmosphere. Concentrations of trace atmospheric gases, nowadays termed "greenhouse gases," are increasing at an alarming rate. There is conclusive evidence that consumption of fossil fuels, conversion of forests to agricultural land, and the emission of industrial chemicals are the principal contributing factors to air pollution.

According to the National Academy of Sciences, Earth's surface temperature has risen by about 1°F in the past century, with accelerated warming occurring in the past three decades. According to statistical reviews of the atmospheric and climatic records, there is substantial evidence that global warming over the past 50 years is directly attributable to human activities.

Under normal atmospheric conditions, energy from the sun controls Earth's weather and climate patterns. Heating of Earth's surface resulting from the sun radiates energy back into space. Atmospheric greenhouse gases, including carbon dioxide (CO_2), methane (CH_4), nitrous oxide (N_2O), tropospheric ozone (O_3), and water vapor (H_2O) trap some of this outgoing energy, retaining it in the form of heat, somewhat like a glass dome. This is referred to as the *greenhouse effect*.

Without the greenhouse effect, surface temperatures on Earth would be roughly 30°C (54°F) colder than they are today—too cold to support life. Reducing greenhouse gas emissions depends on reducing the amount of fossil fuel fired energy that we produce and consume.

Fossil fuels include coal, petroleum, and natural gas, all of which are used to fuel electric power generation and transportation. Substantial increases in the use of nonrenewable fuels have been principal factors in the rapid increase in global greenhouse gas emissions.

Use of renewable fuels can be extended to power industrial, commercial, residential, and transportation applications to substantially reduce air pollution.

Examples of zero-emission, renewable fuels include solar, wind, geothermal, and renewably powered fuel cells. These fuel types, in combination with advances in energy-efficient equipment design and sophisticated energy management techniques, can reduce the risk of climate change and the resulting harmful effects on the ecology. It should be kept in mind that natural greenhouse gases are a necessary part of sustaining life on Earth. It is the anthropogenic or human-caused increase in greenhouse gases that is of concern to the international scientific community and governments around the world.

Since the beginning of the modern industrial revolution, atmospheric concentrations of carbon dioxide have increased nearly 30 percent, methane concentrations have more than doubled, and nitrous oxide concentrations have also risen by about 15 percent. These increases in greenhouse gas emissions have enhanced the heat-trapping capability of Earth's atmosphere.

Fossil fuels burned to operate electric power plants, run cars and trucks, and heat homes and businesses are responsible for about 98 percent of U.S. carbon dioxide emissions, 24 percent of U.S. methane emissions, and 18 percent of U.S. nitrous oxide emissions. Increased deforestation, landfills, large agricultural production, industrial production, and mining also contribute a significant share of emissions. In 2000, the United States produced about 25 percent of total global greenhouse gas emissions, the largest contributing country in the world.

Estimating future emissions depends on demographic, economic, technological policy, and institutional developments. Several emissions scenarios have been developed based on differing projections of these underlying factors. It is estimated that by the year 2100, in the absence of emission control policies, carbon dioxide concentrations will be about 30 to 150 percent higher than today's levels.

Increasing concentrations of greenhouse gases are expected to accelerate global climate change. Scientists expect that the average global surface temperatures could raise an additional 1 to 4.5°F within the next 50 years and 2.2 to 10°F over the next century, with significant regional variation. Records show that the 10 warmest years of the twentieth century all occurred in the last 15 years of that century. The expected impacts of the weather warming trend include the following.

Water resources: Warming-induced decrease in mountain snowpack storage will increase winter stream flows (and flooding) and decrease summer flows. This along with an increased evapotranspiration rate is likely to cause a decrease in water deliveries.

Agriculture. The agricultural industry will be adversely affected by lower water supplies and increased weather variability, including extreme heat and drought.

Forestry. An increase in summer heat and dryness is likely to result in forest fires, insect population increases, and disease.

Electric energy. Increased summer heat is likely to cause an increase in the demand for electricity due to increased reliance on air conditioning. Reduced snowpack is likely to decrease the availability of hydroelectric supplies.

Regional air quality and human health. Higher temperatures may worsen existing air quality problems, particularly if there is a greater reliance on fossil fuel generated electricity. Higher heat would also increase health risks for segments of the population.

Rising ocean levels. Thermal expansion of the ocean and glacial melting are likely to cause a 0.5- to 1.5-m (2 to 4 ft) rise in the ocean level by 2100.

Natural habitat. Rising ocean levels and reduced summer river flow are likely to reduce coastal and wetland habitats. These changes could also adversely affect spawning fish populations. The general increase in temperatures and accompanying increase in summer dryness could also adversely affect wild land plant and animal species.

Scientists calculate that without considering feedback mechanisms a doubling of carbon dioxide would lead to a global temperature increase of 1.2°C (2.2°F). But, the net effect of positive and negative feedback patterns appear to cause substantially more warming than would the change in greenhouse gases alone.

POLLUTION ABATEMENT CONSIDERATION

According to a 1999 study report by the U.S. Department of Energy (DOE), 1 kW of energy produced by a coal-fired electric power generating plant requires about 5 lb of coal. Likewise generation of 1.5 kWh of electric energy per year requires about 7400 lb of coal that in turn produces 10,000 lb of carbon dioxide (CO_2). Commercial energy pollution is further illustrated in Figure I.4

Roughly speaking, the calculated projection of the power demand for the project totals to about 2500 to 3000 kWh. This will require between 12 million and 15 million lb of coal, thereby producing about 16 million to 200 million lb of carbon dioxide. Solar power, if implemented as previously discussed, will substantially minimize the air pollution index. The Environmental Protection Agency (EPA) will soon be instituting an air pollution indexing system that will be factored into all future construction permits. All major industrial projects will be required to meet and adhere to the air pollution standards and offset excess energy consumption by means of solar or renewable energy resources.

ENERGY ESCALATION COST PROJECTION

According to the Energy Information Administration data source published in 1999, California consumes just as much energy as Brazil or the United Kingdom. The entire global crude oil reserves are estimated to last about 30 to 80 years, and over 50 percent of the nation's energy is imported from abroad. It is inevitable that energy costs will surpass historical cost escalation averaging projections. Growth of fossil fuel consumption is illustrated in Figure I.2. It is estimated that the cost of nonrenewable energy will, within the next decade, increase by approximately 4 to 5 percent by producers.

When compounded with a general inflation rate of 3 percent, the average energy cost increase, over the next decade, could be expected to rise at a rate of about 7 percent per year. This cost increase does not take into account other inflation factors, such as regional conflicts, embargos, and natural catastrophes.

Figure I.1   **The collection of renewable energy.** *Photo courtesy of Shell Solar.*

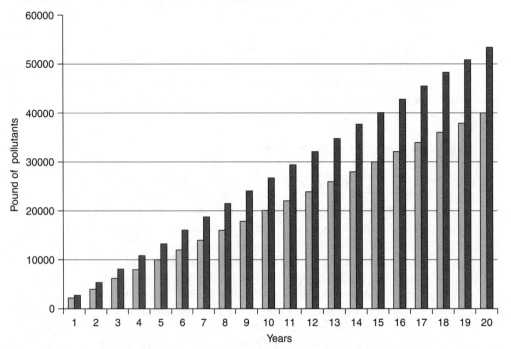

Figure I.2   **Growth in fossil fuel consumption courtesy of DOE.**

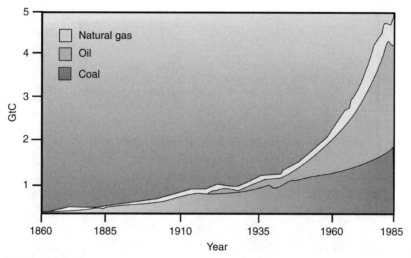

Figure I.3 **Growth in fossil fuel consumption.** *Courtesy of Geothermal Education Office.*

Solar power cogeneration systems require nearly zero maintenance and are more reliable than any human-made power generation devices. The systems have an actual life span of 35 to 40 years and are guaranteed by the manufacturers for a period of 25 years. It is my opinion that in a near-perfect geographic setting, the integration of the systems into the mainstream of architectural design will not only enhance the design aesthetics but also will generate considerable savings and mitigate adverse effects on the ecology and global warming. Figure I.3 is graph which depicts growth of fossil fuel consumption.

SOCIAL AND ENVIRONMENTAL CONCERNS

Nowadays, we do not think twice about leaving lights on or turning off the television or computers, which run for hours. Most people believe that energy seems infinite, but in fact, that is not the case. World consumption of fossil fuels, which supply us with most of our energy, is steadily rising. In 1999, it was found that out of 97 quads of energy used (a quad is 3×10^{11} kWh), 80 quads came from coal, oil, and natural gas. As we know, sources of fossil fuels will undoubtedly run out within a few generations and the world has to be ready with alternative and new sources of energy. In reality, as early as 2020, we could be having some serious energy deficiencies. Therefore, interest in renewable fuels such as wind, solar, hydropower, and others is a hot topic among many people.

Renewable fuels are not a new phenomenon; although, they may seem so. In fact, the Industrial Revolution was launched with renewable fuels. The United States and the world has, for a long time, been using energy without serious concern, until the 1973 and 1974 energy conferences, when the energy conservation issues were brought to the attention of the industrialized world. Ever since, we were forced to realize that the supply of fossil fuels would one day run out, and we had to find alternate sources of energy.

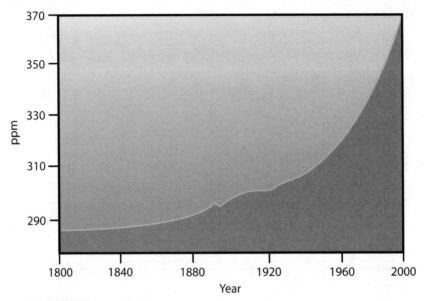

Figure I.4   **Growth of carbon dioxide in the atmosphere.** *Courtesy of Geothermal Education Office.*

In 1999, the U.S. Department of Energy published a large report in which it was disclosed that by the year 2020 there will be a 60 percent increase in carbon dioxide emissions which will create a serious strain on the environment, as it will further aggravate the dilemma with greenhouse gases. Figure I.4 shows the growth of carbon dioxide in the atmosphere.

A simple solution may seem to be to reduce energy consumption; however, it would not be feasible. It has been found that there is a correlation between high electricity consumption (4000 kWh per capita) and a high Human Development Index (HDI), which measures quality of life. In other words there is a direct correlation between quality of life and amount of energy used.

This is one of the reasons that our standard of living in the industrialized countries is better than in third-world countries, where there is very little access to electricity. In 1999, the United States had 5 percent of the world's population and produced 30 percent of the gross world product. We also consumed 25 percent of the world's energy and emitted 25 percent of the carbon dioxide. It is not hard to imagine what countries, such as China and India, with increasing population and economic growth could do to the state of the global ecology.

The most significant feature of solar energy is that it does not harm the environment. It is clean energy. Using solar power does not emit any of the extremely harmful greenhouse gases that contribute to global warming. There is a small amount of pollution when the solar panels are produced, but it is miniscule in comparison to fossil fuels. The sun is also a free source of energy. As technology advances, solar energy will become increasingly economically feasible because the price of the photovoltaic modules will go down. The only concern with solar power is that it is not energy on demand and that it only works during the day and when it is very sunny. The only way to overcome this problem is to build storage facilities to save up some of the energy in batteries; however, that adds more to the cost of solar energy.

CONCLUSION

Even though a solar power cogeneration system requires an initial investment, the long-term financial and ecological advantages are so significant that deployment in the existing project should be given special consideration.

MY REASON FOR WRITING THIS BOOK

During years of practice as a research and design engineer, I have come to realize that the best way to promote the use of sustainable energy design is to properly educate key professionals, such as architects and engineers, whose opinions direct project development.

I have found that even though solar power at present is a relatively mature technology, its use and application in the building industry is hampered due to lack of exposure and education. Regardless of present federal and state incentive programs, sustainable design by use of renewable energy will not be possible without a fundamental change in the way we educate our architects, engineers, and decision makers.

As engineers, scientists, architects, and public policy makers, we are collectively and individually responsible for weighing in on every aspect of the renewable energy design and making maximum use of resources and technologies available to us today. Development of renewable energy technologies requires substantial national investment and commitment to our research and development effort, which inevitably will define the future economic well-being of our nation, which is as significant as the national defense and social health and welfare.

PETER GEVORKIAN, PH.D., P.E.

ACKNOWLEDGMENTS

I would like to thank my colleagues and individuals who have encouraged and assisted me in writing this book. I am especially grateful to all agencies and organizations that provided photographs and allowed use of some textual material and my colleagues who read the manuscript and provided valuable insight.

Michael Lehrer FAIA; Joe DesMeres, PhD.; Ken Hekimian PhD; Mark Gangi, AIA; Mary Olson Khanian, environmentalist; Brian Schwagerl of Hearst Corporation; A. Tom Scarola of Tishman Speyer; Gene Beck of EnviroTech Financial, Inc.; Joel Davidson of Solar Integrated Technologies; Alice Barsoumian of Capstone MicroTurbine, Bob Russel of Ducon Industries, John Zinner of Zinner Consultants, Heather smith of Atomic Energy of Canada Ltd., Wade Webb of Solar Integrated Technologies, Mike Linenberger of NREL, Tracy Bea of UMA/HELICOL, Joseph Morrissey of Atlantis Energy, Tina Nickerson of Shell Solar, Bruce Grant of Grant electric, Joel Glodblatt, Steve Chaduma of Energy Innovations, Inc., Mark Skowroski of Solargenix, Mark Robinson of NEXTEK Sara Eva Winkler for the graphics, ; and to the following individuals and organizations for providing photographs and technical support materials.

Special thanks to Katrina Lyons of Murdoch University of Research Institute for Sustainable Energy providing textual material on tidal energy.

American Wind Energy Manufacturers Association

Ballard Power Systems, 4343 North Fraser Way, Burnaby, BC, Canada V5J 5J9

California Energy Commission, 1516, 9th St, MS-45 Sacramento, CA 95814-5512

California Green Design, Inc., 18025 Rancho Street, Suite 200, Encino, CA 91316

Capstone Turbine Corporation, 21211 Nordhoff Street, Chatsworth Street, CA 91311

Danish Wind Turbine Manufacturers Association, Vester Voldgade 106, Copenhagen, Denmark

EnviroTech Financial, Inc., 333 City Blvd. West 17th, Orange, CA 92868

J.A. Consulting Tidal Turbine Specialists, 76 Dules Ave., London, England W4 2AF

Marine Current Turbines Ltd., The Court, The Green Stoke Gifford, Bristol, BS 34 8PD, UK

National Renewable Energy Laboratories Airplus Engineering Consultants, LLC 10850 Riverside Dr, Suite 509, Toluca Lake, CA 91602

Nemzer, Marilyn, Executive Director Geothermal Education Office, 664 Hilary Dr, Tiburon, CA 94920

Office of Industrial Technologies Energy Efficiency and Renewable Energy, U.S. Department of Energy, Washington, DC 20585

Sandia National Laboratories P.O.BOX 5800, Albuquerque, NM 87185-0753

Solar Electric Company, 2500 Townsgate Rd., Suite J, Westlake Village, CA 91361

Solargenix Energy, 3501 Jamboree Road, Suite 606, Newport Beach, CA 92660

Southwest Technology Development Institute, New Mexico State University, 1505 Payne St, Las Cruces, NM 88003

Tidal Stream Turbine Specialists, 26 Dukes Avenue, London, England

UMA/Heliocol, 13620 49th St Clearwater, FL 33762

U.S. Department of Energy, Office of Scientific and Technical Information, P.O. Box 62, Oakridge, TN 37831

DISCLAIMER NOTE

This book examines solar power generation and renewable energy sources, with the sole intent to familiarize the reader with the existing technologies in order to encourage policy makers, architects, and engineers to use available energy conservation options in their designs.

The principal objective of the book is to emphsize solar power cogeneration design, application, and economics. This book also covers passive solar, wind energy, hydrogen fuel cell, ocean tidal and ocean stream energy renewable technologies, which have in the past few decades, undergone notable improvements throughout the industrialized world.

Neither the author, individuals, organizations, or manufacturers referenced or credited in this book make any warranties, express or implied, or assume any legal liability or responsibility for the accuracy, completeness, or usefulness of any information, products, or processes disclosed or presented.

Reference to any specific commercial product, manufacturer, or organization does not constitute or imply endorsement or recommendation by the author.

SOLAR POWER TECHNOLOGY

Introduction

Solar or photovoltaic (PV) solar cells are electronic devices that essentially convert the solar energy of sunlight into electric energy or electricity. The physics of solar cells is based on the same semiconductor principles as diodes and transistors, which form the building blocks of the entire world of electronics.

Solar cells convert energy as long as there is sunlight. In the evenings and during cloudy conditions, the conversion process diminishes. It stops completely at dusk and resumes at dawn. Solar cells do not store electricity, but batteries can be used to store the energy.

One of the most fascinating aspects of solar cells is their ability to convert the most abundant and free form of energy into electricity, without moving parts or components and without producing any adverse forms of pollution that affect the ecology, as is associated with most known forms of nonrenewable energy production methods, such as fossil fuel, hydroelectric, or nuclear energy generating plants.

In this chapter we will review the overall solar energy conversion process, system configurations, and the economics associated with the technology. We will also briefly look into the mechanism of hydrogen fuel cells.

In Chapter 2 of this book we will review the fundamentals of solar power cogeneration design and explore a number of applications including an actual design of a 500-kilowatt (kW) solar power installation project, which also includes a detailed analysis of all system design parameters.

A Brief History of the Photoelectric Phenomenon

In the later part of the 19th century, physicists discovered a phenomenon: when light is incident on liquids and metal cell surfaces, electrons are released. However, no one

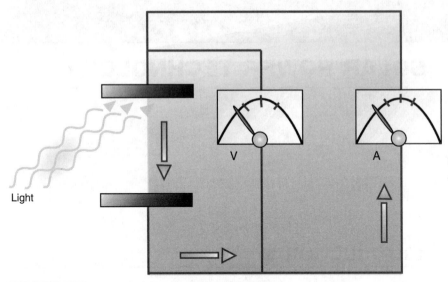

Figure 1.1 The photoelectric effect experiment.

had an explanation for this bizarre occurrence. At the turn of the 20th century, Albert Einstein provided a theory for this—for which he won the Nobel Prize in physics—which laid the groundwork for the theory of the *photoelectric effect*. Figure 1.1 shows the photoelectric effect experiment. When light is shone on metal, electrons are released, which are attracted toward a positively charged plate, thereby giving rise to a photoelectric current.

Einstein explained the observed phenomenon by a contemporary theory of *quantized energy levels,* which was previously developed by Max Planck. The theory described light as being made out of miniscule bundles of energy called *photons*. Photons impinging on metals or semiconductors knock electrons off atoms.

In the 1930s, these theorems led to a new discipline in physics called *quantum mechanics,* which consequently led to the discovery of transistors in the 1950s and to the development of semiconductor electronics.

Solar Cell Physics

Most solar cells are constructed from semiconductor material, such as silicon (the fourteenth element in the Mendeleyev table of elements). Silicon is a semiconductor that has the combined properties of a conductor and an insulator.

Metals such as gold, copper, and iron are conductors; they have loosely bound electrons in the outer shell or orbit of their atomic configuration. These electrons can be detached when subjected to an electric voltage or current. On the contrary, atoms of insulators, such as glass, have very strongly bonded electrons in the atomic configuration and do not allow the flow of electrons even under the severest application of voltage or current. Semiconductor materials, on the other hand, bind electrons midway between that of metals and insulators.

Semiconducting elements used in electronics are constructed by fusing two adjacently doped silicon wafer elements. Doping implies impregnation of silicon by positive and negative agents, such as phosphor and boron. Phosphor creates a free electron that produces so-called N-type material. Boron creates a "hole," or a shortage of an electron, that produces so-called P-type material. Impregnation is accomplished by depositing the previously referenced "dopants" on the surface of silicon using a certain heating or chemical process. The N-type material has a propensity to lose electrons and gain holes, so it acquires a positive charge. The P-type material has a propensity to lose holes and gain electrons, so it acquires a negative charge.

When N-type and P-type doped silicon wafers are fused together, they form a *PN junction*. The negative charge on P-type material prevents electrons from crossing the junction, and the positive charge on the N-type material prevents holes from crossing the junction. A space created by the P and N, or PN, wafers creates a potential barrier across the junction.

This PN junction, which forms the basic block of most electronic components, such as diodes and transistors, has the following specific operational uses when applied in electronics:

In *diodes*, a PN device allows for the flow of electrons and, therefore, current in one direction. For example, a battery, with direct current (dc), connected across a diode allows the flow of current from positive to negative leads. When an alternating sinusoidal current (ac) is connected across the device, only the positive portion of the waveform is allowed to pass through. The negative portion of the waveform is blocked.

In *transistors*, a wire secured in a sandwich of a PNP-junction device (formed by three doped junctions), when properly polarized or biased, controls the amount of direct current from the positive lead to the negative, thus forming the basis for current control, switching, and amplification, as shown in Figure 1.2. In *light-emitting diodes* (LEDs), a controlled amount and type of doping material in a PN-type device connected across a dc voltage source converts the electric energy to visible light with differing frequencies and colors, such as white, red, blue, amber, and green.

In *solar cells*, when a PN junction is exposed to sunshine, the device converts the stream of photons (packets of quanta) that form the visible light into electrons (the reverse of the LED function), making the device behave like a minute battery with a unique characteristic voltage and current, which is dependent on the material dopants and PN-junction physics. This is shown in Figure 1.3.

The bundles of photons that penetrate the PN junction randomly strike silicon atoms and give energy to the outer electrons. The acquired energy allows the outer electrons to break free from the atom. Thus, the photons in the process are converted to electron movement or electric energy.

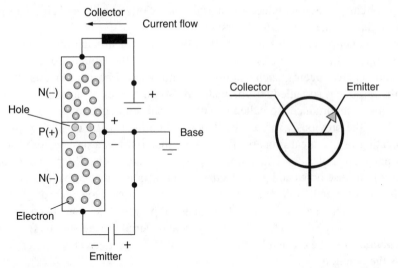

Figure 1.2 NPN junction showing holes and electron flow in an NPN transistor.

It should be noted that the photovoltaic energy conversion efficiency is dependent on the wavelength of the impinging light. Red light, which has a lower frequency, produces insufficient energy, whereas blue light, which has more energy than needed to break the electrons, is wasted and dissipates as heat.

Solar Cell Electronics

An electrostatic field is produced at a PN junction of a solar cell by impinging photons that create 0.5 volt (V) of potential energy, which is characteristic of most PN junctions and all solar cells. This miniscule potential resembles in function a small battery with positive and negative leads. These are then connected front to back in series to achieve higher voltages.

For example, 48 solar cell modules connected in series will result in 24 V of output. An increase in the number of solar cells within the solar cell bank will result in higher voltage. This voltage is employed to operate inverters (discussed later), which convert the dc power into a more suitable ac form of electricity.

In addition to the previously discussed PN-junction device, solar cells contain construction components, for mechanical assembly purposes, that are laid over a rigid or flexible holding platform or a substrate, such as a glass or a flexible film, and are interconnected by micron-thin, highly conductive metals. A typical solar panel used in photovoltaic power generation is constructed from a glass supportive plate that houses solar PV modules, each formed from several hundreds of interconnected PN devices. Depending on the requirements of a specific application, most solar panels

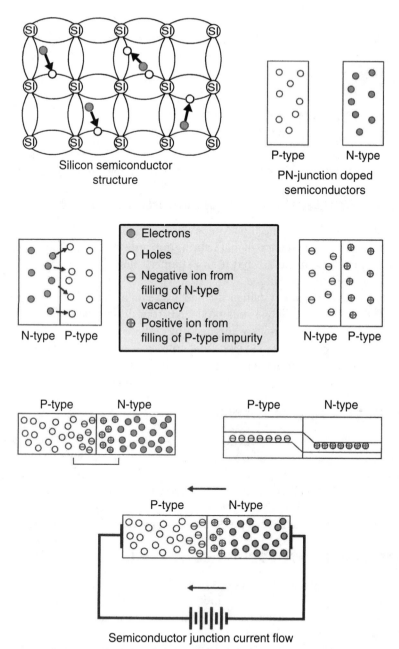

Figure 1.3 Semiconductor depletion region formations.

manufactured today produce an output of 6, 12, 24, or 48 V dc. The amount of power produced by a solar panel, expressed in watts, represents an aggregate power output of all solar PN devices. For example, a manufacturer will express various panel characteristics by voltage, wattage, and surface area.

Types of Solar Cells

Solar cell technologies at present fall into three main categories: monocrystalline (single-crystal construction), polycrystalline (semicrystalline), and amorphous silicone.

MONOCRYSTALLINE PHOTOVOLTAIC SOLAR CELLS

Monocrystalline cells have the highest conversion efficiency. They currently have a well-established semiconductor manufacturing process that dates back over several decades. The cells are manufactured from extremely pure silicon by the *Czochralsky* process or *floating zone technique.*

In these processes, monocrystalline silicone grows on a seed, which is pulled slowly out of the silicon melt. Silicone rods produced from these processes are sliced into 0.2- to 0.4-millimeter (mm)-thick wafer disks by carbide thread saws. The wafers produced undergo several production steps that consist of grinding, polishing, cleaning, doping, and antireflective and coating processing. The manufacturing process for monocrystalline silicon solar cells is highly intensive and expensive.

POLYCRYSTALLINE PHOTOVOLTAIC SOLAR CELLS

In the polycrystalline process, the silicon melt is cooled very slowly, under controlled conditions. The silicon ingot produced in this process has crystalline regions, which are separated by grain boundaries. After solar cell production, the gaps in the grain boundaries cause this type of cell to have a lower efficiency compared to that of the monocrystalline process just described. Despite the efficiency disadvantage, a number of manufacturers favor polycrystalline PV cell production because of the lower manufacturing cost.

AMORPHOUS PHOTOVOLTAIC SOLAR CELLS

In this process, a thin wafer of silicon is deposited on a carrier material and doped in several process steps. An amorphous silicon film is produced, by a method similar to the monocrystalline manufacturing process, and is sandwiched between glass plates, which form the basic PV solar panel module.

Even though the process yields relatively inexpensive solar panel technology, it has the following disadvantages:

- Larger installation surface
- Lower conversion efficiency
- Inherent degradation during the initial months of operation, which continues over the life span of the PV panels

The main advantages of this technology are

■ Relatively simple manufacturing process
■ Lower manufacturing cost
■ Lower production energy consumption

Other Technologies

There are other prevalent production processes that are currently being researched and will be serious contenders in the future of solar power production technology. We discuss these here.

Thin-film cell technology. In this process, thin crystalline layers of cadmium telluride (CdTe, of about 15 percent efficiency) or copper indium diselenide (CuInSe$_2$, of about 19 percent efficiency) are deposited on the surface of a carrier base. This process uses very little energy and is very economical. It has simple manufacturing processes and relatively high conversion efficiencies.

Gallium-arsenide cell technology. This manufacturing process yields a highly efficient PV cell. But as a result of the rarity of gallium deposits and the poisonous qualities of arsenic, the process is very expensive. The main feature of gallium-arsenide (GaAs) cells, in addition to their high efficiency, is that their output is relatively independent of the operating temperature and is primarily used in space programs.

Tandem or multijunction cell technology. This process employs two layers of solar cells, such as silicon (Si) and GaAs components, one on top of another, to convert solar power with higher efficiency.

Concentrators

Concentrators are lenses or reflectors that focus sunlight onto the solar cell modules. Fresnel lenses, which have concentration ratios of 10 to 500 times, are mostly made of inexpensive plastic materials engineered with refracting features that direct the sunlight onto the small narrow PN-junction area of the cells. Module efficiencies of GaAs single-crystalline PV cells, which normally range from 10 to 14 percent, can be augmented in excess of 30 percent.

Reflectors are used to increase power output, increase the intensity of light on the module, or extend the time that sunlight falls on the modules. The main disadvantage of concentrators is their inability to focus scattered light, which limits their use to areas such as deserts.

Solar Panel Arrays

Serial or parallel interconnections in solar panels are called *solar panel arrays* (SPA). Generally, a series of solar panel arrays are configured to produce a specific voltage potential and collective power production capacity to meet the demand requirements of a project.

Depending on the size of the mounting surface, solar panels are secured on tilted structures called *stanchions*. Solar panels installed in the northern hemisphere are mounted facing south with stanchions tilted to a specific degree angle. In the southern hemisphere solar panels are installed facing north.

Solar panel arrays feature a series of interconnected positive (+) and negative (−) outputs of solar panels in a serial or parallel arrangement that provides a required dc voltage to an inverter (described later). Figure 1.4 shows the internal wiring of a solar power cell.

The average daily output of solar power systems is entirely dependent on the amount of exposure to sunlight. This exposure is dependent on the following factors. An accurate north-south orientation of solar panels (facing the sun), as referenced earlier, has a significant effect on the efficiency of power output. Even slight shadowing will affect a module's daily output. Other natural phenomena that affect solar production include diurnal variations (due to the rotation of Earth about its axis), seasonal variation (due to the tilt of Earth's axis), annual variation (due to the elliptical orbit of Earth around the sun), solar flares, solar sunspots, atmospheric pollution, dust, and haze.

Photovoltaic solar array installation in the vicinity of trees and elevated structures, which may cast a shadow on the panels, should be avoided. The geographic location of the project site and seasonal changes are also significant factors that must be taken into consideration.

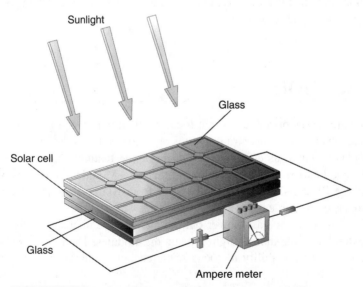

Figure 1.4 **Photovoltaic module operational diagrams.**

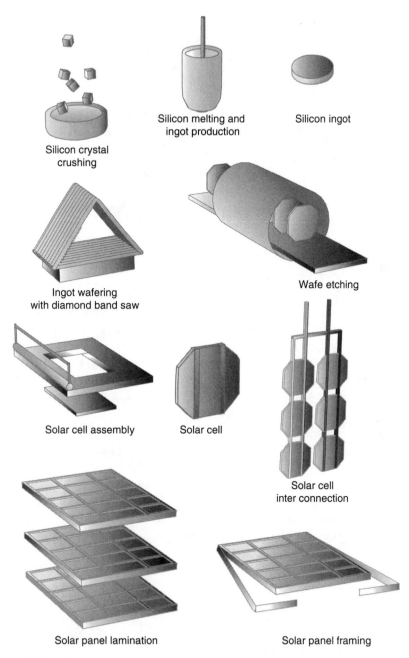

Silicon melting and
ingot production

Silicon ingot

Silicon crystal
crushing

Ingot wafering
with diamond band saw

Wafe etching

Solar cell assembly

Solar cell

Solar cell
inter connection

Solar panel lamination

Solar panel framing

Figure 1.5A Monocrystalline solar panel manufacturing process.

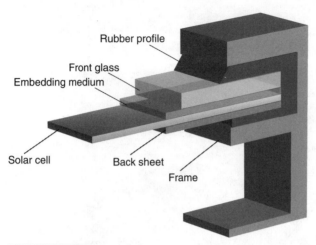

Figure 1.5B Solar power frame assembly.

In order to account for the average daily solar exposure time, design engineers refer to world sunlight exposure maps, shown in Figure 1.8. Each area is assigned an "area exposure time factor," which depending on the location may vary from 2 to 6 hours.

A typical example for calculating daily watt-hours (Wh) for a solar panel array consisting of 10 modules with a power rating of 75 W in an area located with a multiplier of 5 will be $(10 \times 75 \text{ W}) \times 5 \text{ h} = 3750 \text{ Wh}$ of average daily power.

Solar Power System Configurations

Photovoltaic modules only represent the basic element of a solar power system. They work only in conjunction with complementary components, such as batteries, inverters, and transformers. Power distribution panels and metering complete the energy conversion process. Figure 1.6 represent essential components of a solar power system used in a wide variety of applications.

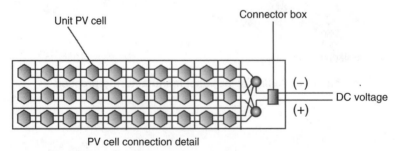

Figure 1.6 Solar power panel internal wiring diagram.

Figure 1.7 Photovoltaic panel.

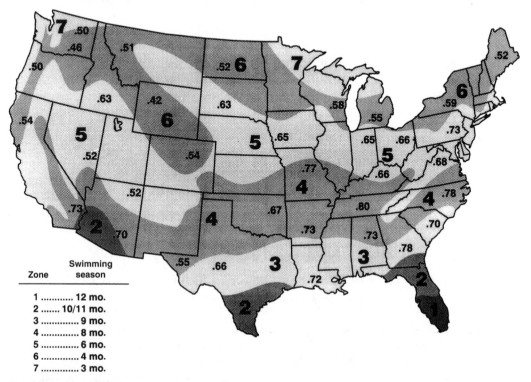

Figure 1.8 Solar insolation map of the United States. *Courtesy of Vector Delta Design Group, Inc.*

STORAGE BATTERIES

As mentioned previously, solar cells are devices that merely convert solar energy into a dc voltage. Solar cells do not store energy. To store energy beyond daylight, the dc voltage is used to charge an appropriate set of batteries.

The reserve capacity of batteries is referred to as *system autonomy*. This varies according to the requirements of specific application. Batteries in applications that require autonomy form a critical component of a solar power system. Battery banks in photovoltaic applications are designed to operate at deep-cycle discharge rates and are generally maintenance-free.

The amount of required autonomy time depends on the specific application. Circuit loads, such as telecommunication and remote telemetry stations, may require two weeks of autonomy, whereas a residential unit may require no more than 12 hours. Batteries must be properly selected to store sufficient energy for the daily demand. When calculating battery ampere-hours and storage capacity, additional derating factors, such as cloudy and sunless conditions, must be taken into consideration. A detailed discussion of battery selection is provided in Chapter 2.

CHARGE REGULATORS

Charge regulators are electronic devices designed to protect batteries from overcharging. They are installed between the solar array termination boxes and batteries.

INVERTERS

As described earlier, photovoltaic panels generate direct current, which can only be used by a limited number of devices. Most residential, commercial, and industrial devices and appliances are designed to work with alternating current. Inverters are devices that convert direct current to alternating current. Although inverters are usually designed for specific application requirements, the basic conversion principles remain the same. Essentially, the inversion process consists of the following.

Wave formation process   Direct current, characterized by a continuous potential of positive and negative references (bias), is essentially chopped into equidistant segments, which are then processed through circuitry that alternately eliminates positive and negative portions of the chopped pattern, resulting in a waveform pattern called a *square wave*.

Wave shaping or filtration process   A square wave, when analyzed mathematically (by Fourier series analysis), consists of a combination of a very large number of sinusoidal (alternating) wave patterns called harmonics. Each wave harmonic has a distinct number of cycles (rise-and-fall pattern within a time period).

An electronic device referred to as a choke (magnetic coils) or filter discriminate passes through 60-cycle harmonics, which form the basis of sinusoidal current. Solid-state inverters use a highly efficient conversion technique known as envelope construction. Direct current is sliced into fine sections, which are then converted into a progressive rising (positive) and falling (negative) sinusoidal 60-cycle waveform pattern. This chopped sinusoidal wave is passed through a series of electronic filters that produce an output current, which has a smooth sinusoidal curvature.

Protective relaying systems   In general, most inverters used in photovoltaic applications are built from sensitive solid-state electronic devices that are very susceptible to external stray spikes, load short circuits, and overload voltage and currents. To protect the equipment from harm, inverters incorporate a number of electronic circuitry:

- Synchronization relay
- Undervoltage relay
- Overcurrent relay
- Ground trip or overcurrent relay
- Overvoltage relay
- Overfrequency relay
- Underfrequency relay

Most inverters designed for photovoltaic applications are designed to allow simultaneous paralleling of multiple units. For instance, to support a 60-kW load, outputs of three 20-kW inverters may be connected in parallel. Depending on the power system requirements, inverters can produce single- or three-phase power at any required voltage or current capacity. Standard outputs available are single-phase 120 V ac and three-phase 120/208 and 277/480 V ac. In some instances step-up transformers are used to convert the output of 120/208 V ac inverters to higher voltages.

Input and output power distribution systems   To protect inverters from stray spikes resulting from lightning or high-energy spikes, dc inputs from PV arrays are protected by fuses, housed at a junction box located in close proximity to the inverters. Additionally, inverter dc input ports are protected by various types of semiconducting devices that clip excessively high voltage spikes resulting from lightning activity.

To prevent damage resulting from voltage reversal, each positive (+) output lead within a PV cell is connected to a rectifier, a unidirectional (forward-biased) element. Alternating-current output power from inverters is connected to the loads by means of electronic or magnetic-type circuit breakers. These serve to protect the unit from external overcurrent and short circuits.

Solar Power System Applications

The simplest solar power applications require output from a PV cell or an array to supply power to dc motors or circuit devices requiring dc power. An example of a rudimentary application is a pump driven by a dc motor that provides a continuous stream of water from a well. This application is quite prevalent in remotely located cattle ranches. Other examples of simple PV cell applications include remotely located communications transmitters, repeater stations, highway signs, roadside emergency telephones, telephone booths, landscape fixtures, surveillance cameras, and a large variety of commercial equipment and appliances.

To prevent damage from power surges resulting from strong magnetic field interference and lightning, most of this equipment incorporates appropriate power input surge protection, lightning arresters, and ground-fault protection devices. Equipment that requires specific autonomy time is provided with battery backup systems that provide adequate power sustenance. In applications requiring ac power, the dc output of a solar array is connected to an inverter system. Likewise, depending on the application, PV systems can be equipped with battery-backed systems and a wide variety of distribution and voltage transformation equipment.

POWER DEMAND CALCULATIONS

To properly size and specify solar power system components, which consist of PV cells, batteries, and inverters, transformer and power distribution systems are determined by the following dc and ac power calculation steps.

Daily dc power requirements

1 List each dc load in watts.
2 Multiply the loads by daily hours used.
3 Sum the loads to establish the average daily dc load.
4 Add a 20 to 30 percent compensation for losses resulting from voltage drops, batteries, and inverters.

Daily ac power requirements

1 List each ac load.
2 Multiply each load by average daily hours used.
3 Sum the loads to find the average daily ac load.
4 Add a 40 percent compensation for losses resulting from voltage drops, batteries, and inverters.

Battery storage capacity sizing

1 Use daily power requirements calculated as specified before (sum the ac and dc power if applicable).
2 Multiply the daily power requirement (as calculated in step 1) by the number of days that the batteries must sustain the load without solar charge (autonomy time). Add a safety factor of 30 percent, for reserve capacity.

Solar module requirements

1 Establish the total solar power requirements as specified before.
2 Divide the load wattage (including compensation losses) by the PV panel wattage to arrive at the number of solar panels required.

3 Multiply the surface area of the PV cells by the number of modules.

4 Form fit the PV cell arrays and stanchion configuration within the available space.

Commercial Projects Best Suited for Solar Power Installation

When planning for photovoltaic power systems, special considerations such as physical space requirements, geographic location, seasonal weather conditions, and initial investment costs must be taken into consideration. As will be discussed later, at present, the initial investment for PV systems constitutes the most significant decision-making parameter.

In remote telemetry or communication system installations, where there are no conventional means of power generation, solar power, regardless of economics, becomes the only viable alternative.

In some industrial and commercial applications, where peak power penalties represent a significant cost of energy, solar power systems could provide a significant peak power shaving, which could result in considerable expenditure reduction. In view of the fact that office building operational hours coincide with the peak power production time of solar power systems, installation of a PV system could, under suitable conditions, result in a significant increase in peak power shaving.

Some important factors when deciding to install solar power systems are the longevity of the PV cells, the absolute minimal maintenance of the equipment, and the fact that, unlike fuel cells and microturbines, PV cells are fuel-free. Most manufacturers warrantee PV cells for a period of 25 years. As will be discussed, system financing, assuming an average investment amortization payoff period of 8 to 10 years, will provide the owner a substantial amount of cost savings for, at least, 15 years.

Another significant factor favoring photovoltaic system power production is that most local municipalities and gas and electric utilities within southern California have power buyback metering systems, as shown in Figure 1.9. Excess power is fed back into the utility grid and results in an energy credit that can be quite significant for reducing overhead costs.

Solar Power Application in Residential Installations

Photovoltaic panels installed atop roofs facing southward have been gaining significant popularity among homeowners and developers throughout Europe and the United States. In addition to added architectural aesthetics, as shown in Figure 1.10, residential solar energy systems can provide a significant reduction of power per day. See the daily residential power usage in Figure 1.11. If each of the 150,000 new homes built in

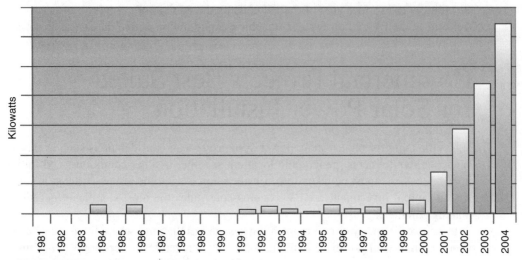

Kilowatts

1981 1982 1983 1984 1985 1986 1987 1988 1989 1990 1991 1992 1993 1994 1995 1996 1997 1998 1999 2000 2001 2002 2003 2004

Figure 1.9 Grid-connected PV capacity installation in California by year.

California had a modest 2-kW PV system installed, it would eliminate 300 megawatts (MW) per hour and 1650 MW of energy per day. Building solar panels into new homes has an intrinsic advantage over form-fitting panels on existing roofs. In a newly constructed single residential unit, the solar panel system can be designed into the house, as shown being installed in Figure 1.12, and significantly reduce the installation cost.

Figure 1.10 Residential solar power installation diagram.
Courtesy of Sharp Solar.

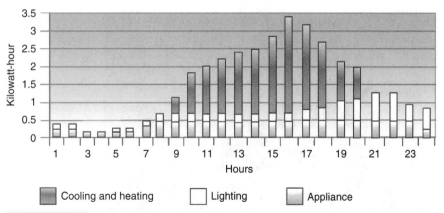

Figure 1.11 **Typical daily residential power usages.** *DOE (US Department of Energy)*

Even though the installation cost of a typical 2-kW residential solar power system may be as high as $15,000 to $20,000, with buyback programs offered by state governments, municipalities, and gas and power utilities, the actual cost to the owner can be reduced to around $6000. Most significantly, cost roll-over into a long-term mortgage (15 to 30 years) will nearly pencil out the extra cost burden. The most significant aspect of PV solar power in residential installations is the added sale and resale value of the home.

Figure 1.12 **Example of a transparent solar power panel installation.** *Courtesy of Atlantis Energy.*

Figure 1.13 Example of a transparent solar power panel application. *Courtesy of Atlantis Energy and Mr. J.Goldbat.*

A 2-kW solar power system can readily provide about one-half of the power used by a house's daylight illumination (300 W), TV (100 W), stereo (60 W), microwave oven (450–750 W), refrigerator (90–150 W), evaporating cooler (200–300 W), and general receptacles (about 400 W).

Design Considerations for Commercial and Residential Projects

When conceptualizing the overall design of a project intended to include a solar power system as part of the integral design, the architect, as prime coordinator, must review

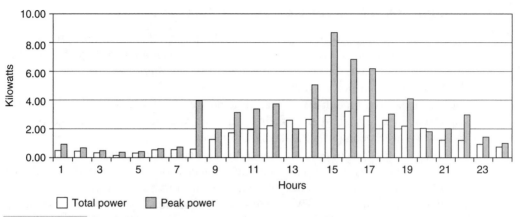

Figure 1.14 Total daily power uses versus peak power demand.

the overall details of the PV system structure and space requirements with the electrical design engineer. When PV cells are to be located on roofs, serious consideration must be given to stanchion locations, stanchion spacing, air-conditioning platforms, solar power system conduit installation, conduit shafts, additional roof loading, wind gust conditions, downspout positions, and roof pitches and slopes. In addition to architectural and structural specifics, the mechanical engineer must take into account the resulting solar panel shade contribution to roof insulation when calculating air conditioning. This may amount to as much as a 10 percent increase in insulation value.

Figure 1.15   **Example of a solar carport design.**
Courtesy of Atlantis Energy.

Design Guidelines for Residential Solar Projects

When planning single or multiple residential units, the following design guideline will result in substantial economical energy use savings if applied properly. Now that utilization has become a practical component of an *energy-efficient home design*, dismissal of a solar power system as a superfluous and unnecessary requirement is shortsighted. Even though PV power systems are at the start of the economic bell curve, within a very short time their application will be as prevalent and basic as microwave ovens, computers, and cellular phones.

To dismiss PV system requirements at the conceptual or planning stages of design will be proven to be a costly error, since undoubtedly, solar power energy contribution in most future residences will be considered a mandatory requirement. Moreover, residential housing units with PV systems will unquestionably have far better resale value than regular ones. The following design guidelines outlined are based on practical construction practices and numerous published research articles.

Orientation   The basic architectural design principle dictates that the longest wall of the home should face south, since the winter sun rises from the southeast and sets in the southwest. Therefore, placing large glass windows on the south wall will ensure that the home will receive optimal solar energy.

Declination of magnetic south varies across the country. In the northern hemisphere photovoltaic cells work best when facing *true south*.

Photovoltaic panel tilt angle   Stanchions or roof tilts, for best efficiency, must be tilted at an angle to provide maximum insolation.

Solar exposure   To ensure complete solar exposure to the sun, PV cells should be installed in locations that avoid the shade of buildings and trees.

Windows   The amount of window surface on the south wall is optimally about 7 percent of a home's total square footage. For example, a 3000-square-foot house, excluding the window trims, should have about 210 square feet of glass. This design criteria applies to conventional home constructions that use wall-to-wall carpeting

The amount of window surface should not exceed 7 percent since it will create unnecessary overheating. To increase the window surface beyond this suggested design limit, an additional increase must be made in thermal mass (concrete, floor tiles, and so on). East wall and north wall window surfaces should be limited to a maximum of 4 percent of the total square footage. West window glass surface areas must be limited to 2 percent of the total square footage. Additional design considerations include structural slab insulation (a sole plate), wall insulation, and attic insulation (R-30 blown insulation for moderate climates, R-40 and R-50 for colder). Use of fluorescent lamps, ceiling fans, and natural ventilation from windows can also drastically reduce the running time of air conditioners.

Electrical engineering system design   To incorporate a solar power generation system as an integral part of an existing or a new project, the electrical engineer must have the required expertise and qualification to evaluate the entire project from both a technical and a costing point of view.

One of the most important steps, when considering a solar system design, is an initial feasibility study, where the electrical engineer or designer undertakes a detailed study of all aspects of the design. These include site evaluation, systems engineering configuration, equipment integration, service demand transfer switching requirements, and all issues that concern system installation and integration. The feasibility study should, in addition to technical issues, include the economics of engineering, material, and installation cost. A typical cost evaluation matrix that outlines cost and equipment performance parameters from various competitors must be included as part of the report.

A thorough understanding of all technical issues concerning particular performance characteristics of existing PV cell technologies and specific system integration requirements are essential for designing an efficient solar power system. The electrical engineer or designer should assume primary responsibility to coordinate all technical aspects of the project with all design disciplines.

Because of the specific performance characteristics of PV cells and conversion equipment, the designer should exercise diligence to minimize power losses. The equipment selected should be reliable, efficient, and durable, requiring a minimal amount of maintenance and care.

Advances in the Design and Development of Photovoltaic Technology

In view of the increasing shortage of nonrenewable energy resources and the increasing cost of conventional energy production in recent years, there has been an accelerated international research and product development effort to produce efficient and inexpensive photovoltaic power production technology.

A large portion of the research-and-development funds for PV cell development is subsidized by national funds. Subsidies in the United States exceed $100 million annually. Sandia National Laboratories in Albuquerque, New Mexico, is a major recipient of research funds from the U.S. Department of Energy, and, in addition to technology research, evaluation, and product development, has a Photovoltaic System Assistance Center, which ensures technology transfer to private industry. Similar entities in Canada, Japan, Germany, and the United Kingdom provide research-and-development assistance to private enterprises.

The impact of photovoltaic technology in view of depleted nonrenewable energy resources is enormous and will represent a significant upward impact on the global economies.

Some of the recent research and development noteworthy of reference are as follow:

Japan Energy Corporation. This company has developed and fabricated a photovoltaic module named Record One-Sun Cell. The unit has 30.28 percent efficiency and is made of InGa/GaAs, a two-junction device. Marketing and mass-fabrication dates are currently unknown.

NREL USA. The lab has recently announced the successful development of a module called Record CIGS Device. This experimental PV cell has a surface area of 1 square centimeter (cm^2) and has an efficiency of 17.7 percent. The device is made from $Cu(InGa)Se_2$ (CIGS) junctions. The date for the production model is currently unknown.

Georgia Institute of Technology. This institute has developed a multicrystalline silicon PV cell that has an efficiency of 18.6 percent. The PV module is named HEM Record Cell. The date for the production model is currently unknown.

University of New South Wales, Australia. This university has reported successful test completion of a PV cell module named Record Module, which has a 22.3 percent efficiency. The date for the production module is currently unknown.

Texas Instruments. This company has come up with other interesting PV technologies including a spherical solar cell and very thin photovoltaic films that are recently being used as curtain walls in high-rise buildings.

U.S. Government Annual Research Expenditure

In the recent past, the U.S. federal government had allocated about $90 million for solar energy research, as shown in Figure 1.16. Unfortunately, the solar power fund, at the expense of coal, oil, and fusion energy research, has been reduced by 50 percent.

U.S. Energy use 1996 (94 Quadrillion Btu)

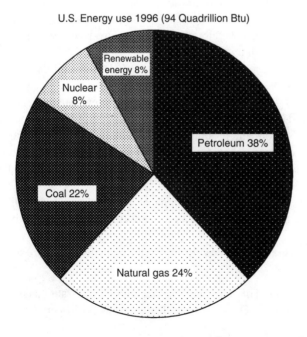

Figure 1.16 U.S. energy consumption in 1996.

TABLE 1.1 TYPICAL SOLAR MODULE AND CELL CONVERSION EFFICIENCIES UNDER THE STANDARD TEST

	CONDITION		
TYPE	TYPICAL MODULE EFFICIENCY (%)	MAXIMUM EFFICIENCY (%)	LABORATORY EFFICIENCY (%)
Single-crystal silicone	12	22	24.7
Multicrystalline silicone	11	15.3	19.8
Amorphous silicone	5	—	12.7
Cadmium telluride	—	10.5	16.0
	—	—	
Gallium diselenide	—	12.1	18.2

Evaluation of Photovoltaic Systems

When evaluating PV cells or panels, system performance and installation costs, which are different for each project, must be the principal guiding factors. For instance, the requirements of a project located in a desert area, which poses no installation space limitations, could tilt the designer's choice to select less expensive, low-efficiency solar panels. In applications such as rooftop installations, design constraints will dictate the use of costlier but highly efficient PV panels, which occupy the least amount of space. Table 1.1 provides the comparative characteristic features of various types of PV technologies.

2

SOLAR POWER GENERATION DESIGN

Introduction

This section is intended to acquaint the reader with the basic design concepts of solar power applications. The typical solar power applications that will be reviewed include stand-alone systems with battery backup, commonly used in remote telemetry; vehicle charging stations; communication repeater stations; and numerous installations where the installation cost of regular electrical service becomes prohibitive. An extended design application of stand-alone systems also includes the integration of an emergency power generator system.

Grid-connected solar power systems, which form a large majority of residential and industrial applications, are reviewed in detail. To familiarize the reader with the prevailing state and federal assistance rebate programs, a special section is devoted to reviewing the salient aspects of existing rebates.

Solar power design essentially consists of electronics and power systems engineering, which requires a thorough understanding of the electrical engineering disciplines and the prevailing standards outlined in Article 690 of the National Electrical Code (NEC).

The solar power design presented, in addition to reviewing the various electrical design methodologies, provides detailed insight into photovoltaic modules, inverters, charge controllers, lightning protection, power storage, battery sizing, and critical wiring requirements. To assist the reader with the economic issues of solar power cogeneration, a detailed analysis of a typical project, including system planning, photovoltaic power system cogeneration estimates, economic cost projection, and payback analysis, are covered later in this chapter.

Solar Power System Components and Materials

As described later in this chapter ("Ground-Mount Photovoltaic Module Installation and Support Hardware"), solar power photovoltaic (PV) modules are constructed from a series of cross-welded solar cells, each typically producing a specific wattage with an output of 0.5 V.

Effectively, each solar cell could be considered as a 0.5-V battery that produces current under adequate solar ray conditions. To obtain a desired voltage output from a PV panel assembly, the cells, similar to batteries, are connected in series to obtain a required output.

For instance, to obtain a 12-V output, 24 cell modules in an assembly are connected in tandem. Likewise, for a 24-V output, 48 modules in an assembly are connected in series. To obtain a desired wattage, a group of several series-connected solar cells are connected in parallel.

The output power of a unit solar cell or its efficiency is dependent on a number of factors such as crystalline silicon, polycrystalline silicon, and amorphous silicon materials, which have specific physical and chemical properties, details of which were discussed in Chapter 1.

Commercially available solar panel assemblies mostly employ proprietary cell manufacturing technologies and lamination techniques, which include cell soldering. Soldered groups of solar cells are in general sandwiched between two tempered-glass panels, which are offered in framed or frameless assemblies.

Solar Power System Configuration and Classifications

There are four types of solar power systems:

1 Directly connected dc solar power system
2 Stand-alone dc solar power system with battery backup
3 Stand-alone hybrid solar power system with generator and battery backup
4 Grid-connected solar power system

DIRECTLY CONNECTED dc SOLAR POWER SYSTEM

As shown in Figure 2.1, the solar system configuration consists of a required number of solar photovoltaic cells, commonly referred to as PV modules, connected in series or in parallel to attain the required voltage output. Figure 2.2 shows four PV modules that have been connected in parallel.

The positive output of each module is protected by an appropriate overcurrent device, such as a fuse. Paralleled output of the solar array is in turn connected to a dc motor via a two-pole single throw switch. In some instances, each individual PV module is also

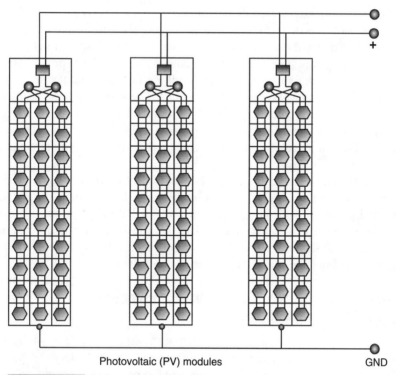

Photovoltaic (PV) modules   GND

Figure 2.1 A three-panel solar array diagram.

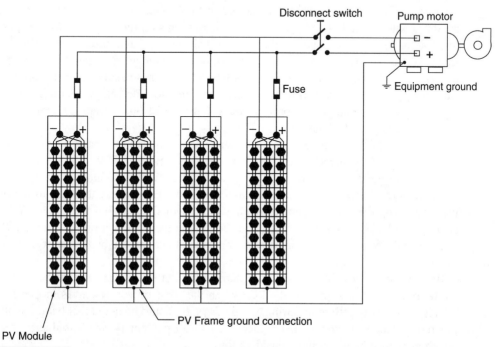

Figure 2.2 A directly connected solar power dc pump diagram.

protected with a forward-biased diode connected to the positive output of individual solar panels (not shown in Figure 2.2). An appropriate surge protector connected between the positive and negative supply provides protection against lightning surges, which could damage the solar array system components. In order to provide equipment-grounding bias, the chassis or enclosures of all PV modules and the dc motor pump are tied together by means of grounding clamps. The system ground is in turn connected to an appropriate grounding rod. All PV interconnecting wires are sized and the proper type selected to prevent power losses caused by a number of factors, such as exposure to the sun, excessive wire resistance, and additional requirements that are mandated by the NEC.

The photovoltaic solar system described is typically used as an agricultural application, where either regular electrical service is unavailable or the cost is prohibitive. A floating or submersible dc pump connected to a dc PV array can provide a constant stream of well water that can be accumulated in a reservoir for farm or agricultural use. In subsequent sections we will discuss the specifications and use of all system components used in solar power cogeneration applications.

STAND-ALONE dc SOLAR POWER SYSTEM WITH BATTERY BACKUP

The solar power photovoltaic array configuration shown in Figure 2.3, a dc system with battery backup, is essentially the same as the one without the battery except that there are a few additional components that are required to provide battery charge stability.

Stand-alone PV system arrays are connected in series to obtain the desired dc voltage, such as 12, 24, or 48 V, outputs of that are in turn connected to a dc collector panel equipped with specially rated overcurrent devices, such as ceramic-type fuses. The positive lead of each PV array conductor is connected to a dedicated fuse, and the negative lead is connected to a common neutral bus. All fuses as well are connected to a common positive bus. The output of the dc collector bus, which represents the collective amperes and voltages of the overall array group, are connected to a dc charge controller, which regulates the current output and prevents the voltage level from exceeding the maximum needed for charging the batteries.

The output of the charge controller is connected to the battery bank by means of a dual dc cutoff disconnect. As depicted in Figure 2.3, the cutoff switch, when turned off for safety measures, disconnects the load and the PV arrays simultaneously.

Under normal operation, during the daytime when there is adequate solar insolation, the load is supplied with dc power while simultaneously charging the battery. When sizing the solar power system, take into account that the dc power output from the PV arrays should be adequate to sustain the connected load and the battery trickle charge requirements.

Battery storage sizing depends on a number of factors, such as the duration of an uninterrupted power supply to the load when the solar power system is inoperative, which occurs at nighttime or during cloudy days. It should be noted that battery banks inherently, when in operation, produce a 20 to 30 percent power loss due to heat, which also must be taken into consideration.

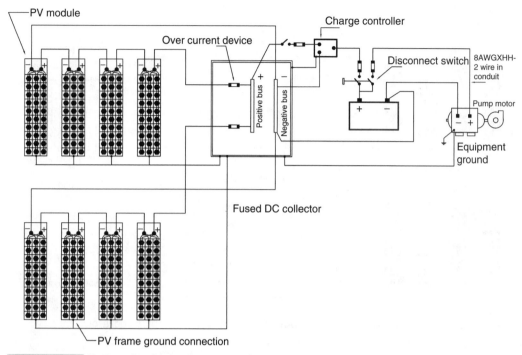

Figure 2.3 **Battery-backed solar power-driven dc pump.**

When designing a solar power system with a battery backup, the designer must determine the appropriate location for the battery racks and room ventilation, to allow for dissipation of the hydrogen gas generated during the charging process. Sealed-type batteries do not require special ventilation.

All dc wiring calculations discussed take into consideration losses resulting from solar exposure, battery cable current derating, and equipment current resistance requirements, as stipulated in NEC 690 articles.

STAND-ALONE HYBRID ac SOLAR SYSTEM WITH STANDBY GENERATOR

A stand-alone hybrid solar power configuration is essentially identical to the dc solar power system just discussed, except that it incorporates two additional components, as shown in Figure 2.4. The first component is an inverter. Inverters are electronic power equipment that are designed to convert direct current into alternating current. The second component is a standby emergency dc generator, which will be discussed later.

Alternating-current inverters   The principal mechanisms of dc-to-ac conversion consists of chopping or segmenting the dc current into specific portions, referred to as square waves, which are filtered and shaped into sinusoidal ac waveforms.

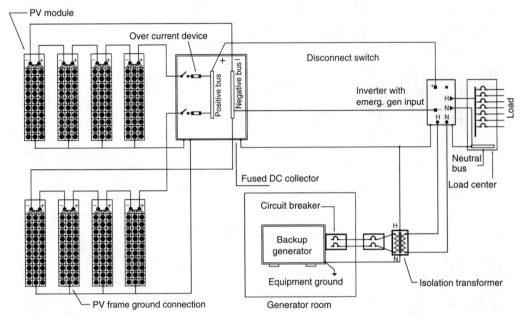

Figure 2.4 Stand-alone hybrid solar power system with standby generator.

Any power waveform, when analyzed from a mathematical point of view, essentially consists of the superimposition of many sinusoidal waveforms, referred to as harmonics. The first harmonic represents a pure sinusoidal waveform, which has a unit base wavelength, amplitude, and frequency of repetition over a unit of time called a cycle. Additional waveforms with higher cycles, when superimposed on the base waveform, add or subtract from the amplitude of the base sinusoidal waveform.

The resulting combined base waveform and higher harmonics produce a distorted wave shape that resembles a distorted sinusoidal wave. The higher the harmonic content, the squarer the wave shape becomes.

Chopped dc output, derived from the solar power, is considered to be a numerous superimposition of odd and even numbers of harmonics. To obtain a relatively clean sinusoidal output, most inverters employ electronic circuitry to filter a large number of harmonics. Filter circuits consist of specially designed inductive and capacitor circuits that trap or block certain unwanted harmonics, the energy of which is dissipated as heat. Some types of inverters, mainly of earlier design technology, make use of inductor coils to produce sinusoidal waveshapes.

In general, dc-to-ac inverters are intricate electronic power conversion equipment designed to convert direct current to a single- or three-phase current that replicates the regular electrical services provided by utilities. Special electronics within inverters, in addition to converting direct current to alternating current, are designed to regulate the output voltage, frequency, and current under specified load conditions. As discussed in the following sections, inverters also incorporate special electronics that allow them to automatically synchronize with other inverters when connected in parallel. Most

Figure 2.5 **Three phase solar power inverter.** *Courtesy of Vector Delta Design Group, Inc.*

inverters, in addition to PV module input power, accept auxiliary input power to form a standby generator, used to provide power when battery voltage is dropped to a minimum level.

A special type of inverter, referred to as the *grid-connected* type, incorporates synchronization circuitry that allows the production of sinusoidal waveforms in unison with the electrical service grid. When the inverter is connected to the electrical service grid, it can effectively act as an ac power generation source. Grid-type inverters used in grid-connected solar power systems are strictly regulated by utility agencies that provide net metering.

Some inverters incorporate an internal ac transfer switch that is capable of accepting an output from an ac-type standby generator. In such designs, the inverters include special electronics that transfer power from the generator to the load.

Standby generators   A standby generator consists of an engine-driven generator that is used to provide auxiliary power during solar blackouts or when the battery power discharge reaches a minimum level. The output of the generator is connected to the auxiliary input of the inverter.

Engines that drive the motors operate with gasoline, diesel, natural gas, propane, or any type of fuel. Fuel tank sizes vary with the operational requirements. Most emergency generators incorporate underchassis fuel tanks with sufficient fuel storage capacity to operate the generator up to 48 hours. Detached tanks could also be designed to hold much larger fuel reserves, which are usually located outside the engine room. In general, large fuel tanks include special fuel-level monitoring and filtration systems.

As an option, the generators can be equipped with remote monitoring and annunciation panels that indicate power generation data and log and monitor the functional and dynamic parameters of the engine, such as coolant temperature, oil pressure, and malfunctions.

Engines also incorporate special electronic circuitry to regulate the generator output frequency, voltage, and power under specified load conditions.

Hybrid system operation   As previously discussed, the dc output generated from the PV arrays and the output of the generator can be simultaneously connected to an inverter. The ac output of the inverter is in turn connected to an ac load distribution panel, which provides power to various loads by means of ac-type overcurrent protection devices.

In all instances, solar power design engineers must ensure that all chassis of equipment and PV arrays, including stanchions and pedestals, are connected together via appropriate grounding conductors that are connected to a single-point service ground bus bar, usually located within the vicinity of the main electrical service switchgear.

In grid-connected systems, switching of ac power from the standby generator and the inverter to the service bus or the connected load is accomplished by internal or external automatic transfer switches.

Standby power generators must always comply with the National Electrical Code requirements outlined in the following articles:

- Electrical Service Requirement, NEC 230
- General Grounding Requirements, NEC 250
- Generator Installation Requirements, NEC 445
- Emergency Power System Safety Installation and Maintenance Requirements, NEC 700

GRID-CONNECTED SOLAR POWER COGENERATION SYSTEM

With reference to Figure 2.6, a connected solar power system diagram, the power cogeneration system configuration is similar to the hybrid system just described. The essence of a grid-connected system is *net metering*. Standard service meters are odometer-type counting wheels that record power consumption at a service point by means of a rotating disc, which is connected to the counting mechanism. The rotating discs operate by an electrophysical principle called *eddy current*, which consists of voltage and current measurement sensing coils that generate a proportional power measurement.

New electric meters make use of digital electronic technology that registers power measurement by solid-state current- and voltage-sensing devices that convert analog measured values into binary values that are displayed on the meter bezels by liquid-crystal display (LCD) readouts.

In general, conventional meters only display power consumption; that is, the meter counting mechanism is unidirectional.

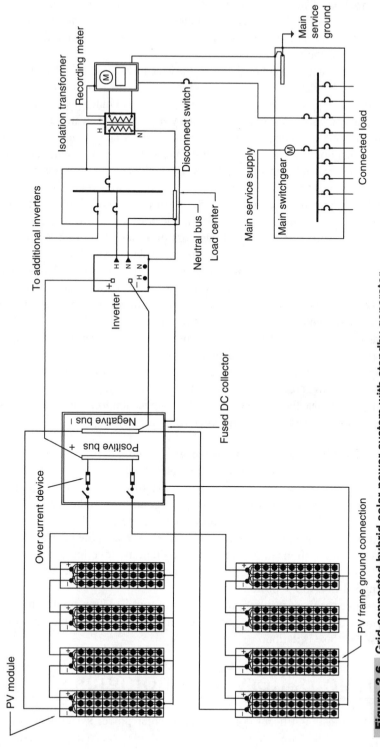

Figure 2.6 Grid-connected hybrid solar power system with standby generator.

Net metering   The essential difference between a grid-connected system and a stand-alone system is that inverters, which are connected to the main electrical service, must have an inherent line frequency synchronization capability to deliver the excess power to the grid.

Net meters, unlike conventional meters, have a capability to record consumed or generated power in an exclusive summation format; that is, the recorded power registration is the net amount of power consumed—the total power used minus the amount of power that is produced by the solar power cogeneration system. Net meters are supplied and installed by utility companies that provide grid-connection service systems. Net-metered solar power cogenerators are subject to specific contractual agreements and are subsidized by state and municipal governmental agencies. The major agencies that undertake distribution of the state of California's renewable energy rebate funds for various projects are the California Energy Commission (CEC), Southern California Edison, Southern California Gas (Sempra Power), and San Diego Gas and Electric (SG&E), as well as principal municipalities, such as the Los Angeles Department of Water and Power. When designing net metering solar power cogeneration systems, the solar power designers and their clients must familiarize themselves with the CEC rebate fund requirements. Essential to any solar power implementation is the preliminary design and economic feasibility study needed for project cost justification and return-on-investment analysis. The first step of the study usually entails close coordination with the architect in charge and the electrical engineering consultant. A preliminary PV array layout and a computer-aided shading study are essential for providing the required foundation for the design. Based on the preceding study, the solar power engineer must undertake an econometrics study to verify the validity and justification of the investment. Upon completion of the study, the solar engineer must assist the client to complete the required CEC rebate application forms and submit it to the service agency responsible for the energy cogeneration program.

Grid-connection isolation transformer   In order to prevent spurious noise transfer from the grid to the solar power system electronics, a delta-y isolation transformer is placed between the main service switchgear disconnects the inverters. The delta winding of the isolation transformer, which is connected to the service bus, circulates noise harmonics in the winding and dissipates the energy as heat.

Isolation transformers are also used to convert or match the inverter output voltages to the grid. Most often, in commercial installations, inverter output voltages range from 208 to 230 V (three phase), which must be connected to an electric service grid that supplies 277/480 V power.

Some inverter manufacturers incorporate output isolation transformers as an integral part of the inverter system, which eliminates the use of external transformation and ensures noise isolation.

Storage Battery Technologies

One of the most significant components of solar power systems consist of battery backup systems that are frequently used to store electric energy harvested from solar photovoltaic systems for use during the absence of sunlight, such as at night and during cloudy conditions. Because of the significance of storage battery systems it is important for design engineers to have a full understanding of the technology since this system component represents a notable portion of the overall installation cost. More importantly, the designer must be mindful of the hazards associated with handling, installation, and maintenance. To provide an in-depth knowledge about the battery technology, this section covers the physical and chemical principles, manufacturing, design application, and maintenance procedures of the storage battery. In this section we will also attempt to analyze and discuss the advantages and disadvantages of different types of commercially available solar power batteries and their specific performance characteristics.

HISTORY

In 1936, while excavating the ruins of a 2000-year-old village near Baghdad, called Khujut Rabu, workers discovered a mysterious small jar identified as a Sumerian artifact dated to 250 BC. This jar, which was identified as the earliest battery, was a 6-in-high pot of bright yellow clay that included a copper-enveloped iron rod capped with an asphalt-like stopper. The edge of the copper cylinder was soldered with a lead-tin alloy comparable to today's solder. The bottom of the cylinder was capped with a crimped-in copper disk and sealed with bitumen or asphalt. Another insulating layer of asphalt sealed the top and also held in place the iron rod that was suspended into the center of the copper cylinder. The rod showed evidence of having been corroded with an agent. The jar when filled with vinegar produces about 1.1 V of electric potential.

A German archaeologist, Wilhelm Konig, who examined the object (see Figure 2.7 and 2.8), came to the surprising conclusion that the clay pot was nothing less than an

Figure 2.7 The Baghdad battery.

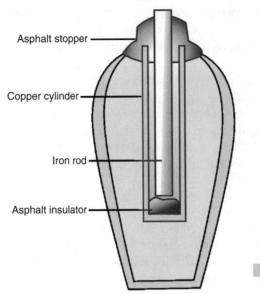

Asphalt stopper

Copper cylinder

Iron rod

Asphalt insulator

Figure 2.8 The Baghdad battery elements.

ancient electric battery. It is stipulated that the Sumerians made use of the battery for electroplating inexpensive metals such as copper with silver or gold.

Subsequent to the discovery of this first battery, several other batteries were unearthed in Iraq, all of which dated from the Parthian occupation between 248 BCE and 226 CE.

In the 1970s, German Egyptologist Arne Eggebrecht built a replica of the Baghdad battery and filled it with grape juice, which he deduced ancient Sumerians might have used as an electrolyte. The replica generated 0.87 V of electric potential. Current generated from the battery was then used to electroplate a silver statuette with gold.

However, the invention of batteries is associated with the Italian scientist Luigi Galvani, an anatomist who, in 1791, published works on animal electricity. In his experiments, Galvani noticed that the leg of a dead frog began to twitch when it came in contact with two different metals. From this phenomenon he concluded that there is a connection between electricity and muscle activity. Alessandro Conte Volta, an Italian physicist, in 1800, reported the invention of his electric battery or "pile." The battery was made by piling up layers of silver, paper or cloth soaked in salt, and zinc (see Figure 2.9). Many triple layers were assembled into a tall pile, without paper or cloth between the zinc and silver, until the desired voltage was reached. Even today the French word for battery is *pile,* pronounced "peel" in English. Volta also developed the concept of the electrochemical series, which ranks the potential produced when various metals come in contact with an electrolyte.

The battery is an electric energy storage device that in physics terminology can be described as a device or mechanism that can hold kinetic or static energy for future use. For example, a rotating flywheel can store dynamic rotational energy in its wheel, which releases the energy when the primary mover such as a motor no longer engages the connecting rod. Similarly, a weight held at a high elevation stores static energy

Figure 2.9 Alessandro Volta's pile.

embodied in the object mass, which can release its static energy when dropped. Both of these are examples of energy storage devices or batteries.

Energy storage devices can take a wide variety of forms, such as chemical reactors and kinetic and thermal energy storage devices. It should be noted that each energy storage device is referred to by a specific name; the word *battery,* however, is solely used for electrochemical devices that convert chemical energy into electricity by a process referred to as galvanic interaction. A galvanic cell is a device that consists of two electrodes, referred to as the anode and the cathode, and an electrolyte solution. Batteries consist of one or more galvanic cells.

It should be noted that a battery is an electrical storage reservoir and not an electricity-generating device. Electric charge generation in a battery is a result of chemical interaction, a process that promotes electric charge flow between the anode and the cathode in the presence of an electrolyte. The electrogalvanic process that eventually results in depletion of the anode and cathode plates is resurrected by a recharging process that can be repeated numerous times. In general, batteries when delivering stored energy incur energy losses as heat when discharging or during chemical reactions when charging.

The Daniell cell   The Voltaic pile was not good for delivering currents over long periods of time. This restriction was overcome in 1820 with the Daniell cell. British researcher John Frederich Daniell developed an arrangement where a copper plate was located at the bottom of a wide-mouthed jar. A cast-zinc piece commonly referred to as a crowfoot, because of its shape, was located at the top of the plate, hanging on the rim of the jar. Two electrolytes, or conducting liquids, were employed. A saturated copper-sulfate solution covered the copper plate and extended halfway up the remaining distance toward the zinc piece. Then, a zinc-sulfate solution, which is a less dense liquid, was carefully poured over a structure that floated above the copper sulfate and immersed zinc.

In a similar experiment, instead of zinc sulfate, magnesium sulfate or dilute sulfuric acid was used. The Daniell cell was the also one of the first batteries that incorporated mercury, which was amalgamated with the zinc anode to reduce corrosion when the batteries were not in use. The Daniell battery, which produced about 1.1 V, was extensively used to power telegraphs, telephones, and even to ring doorbells in homes for over a century.

Plante's battery   In 1859 Raymond Plante invented a battery that used a cell by rolling up two strips of lead sheet separated by pieces of flannel material. The entire assembly when immersed in diluted sulfuric acid produced an increased current that was subsequently improved upon by insertion of separators between the sheets.

The carbon-zinc battery   In 1866, Georges Leclanché, in France developed the first cell battery. The battery, instead of using liquid electrolyte, was constructed from moist ammonium chloride paste and a carbon and zinc anode and cathode. It was sealed and sold as the first dry battery. The battery was rugged and easy to manufacture and had a good shelf life. Carbon-zinc batteries were in use over the next century until they were replaced by alkaline-manganese batteries.

Lead-acid battery suitable for autos   In 1881 Camille Faure produced the first modern lead-acid battery, which he constructed from cast-lead grids that were packed with lead oxide paste instead of lead sheets. The battery had a larger current producing capacity. Its performance was further improved by the insertion of separators between the positive and negative plates to prevent particles falling from these plates, which could short out the positive and negative plates from the conductive sediment.

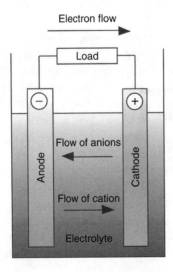

Figure 2.10 Lead acid battery current flow diagram.

The Edison battery   Between the years 1898 and 1908, Thomas Edison developed an alkaline cell with iron as the anode material (–) and nickel oxide as the cathode material (+). The electrolyte used was potassium hydroxide, the same as in modern nickel-cadmium and alkaline batteries. The cells were extensively used in industrial and railroad applications. Nickel-cadmium batteries are still being used and have remained unchanged ever since.

In parallel with Edison's work, Jungner and Berg in Sweden, were working on the development of the nickel-cadmium cell. In place of the iron used in the Edison cell, they used cadmium, with the result that the cell operated better at low temperatures and was capable of self-discharge to a lesser degree than the Edison cell, and in addition the cell could be trickle-charged at a reduced rate. In 1949 the alkaline-manganese battery, also referred to as the alkaline battery, was developed by Lew Urry at the Eveready Battery Company laboratory in Parma, Ohio. Alkaline batteries are capable of storing higher energy within the same package size than comparable conventional dry batteries.

Zinc-mercuric oxide alkaline batteries   In 1950 Samuel Ruben invented the zinc-mercuric oxide alkaline battery (see Figure 2.11), which was licensed to the P.R. Mallory Co. The company later became Duracell, International. Mercury compounds have since been eliminated from batteries to protect the environment.

Deep-discharge batteries used in solar power backup applications in general have lower charging and discharging rate characteristics and are more efficient. A battery rated four ampere-hours (Ah) over 6 hours might be rated at 220 Ah at the 20-hour rate and 260 Ah at the 48-hour rate. The typical efficiency of a lead-acid battery is 85 to 95 percent, and that of alkaline and nickel-cadmium (NiCd) batteries is about 65 percent.

Practically all batteries used in PV systems and in all but the smallest backup systems are lead-acid type batteries. Even after over a century of use, they still offer the

Figure 2.11   Alkaline batteries.

best price-to-power ratio. Systems that use NiCd batteries are not recommended to use them in extremely cold temperatures such as at −50 F or below.

NiCd batteries are expensive to buy and very expensive to dispose off due to the hazardous nature of cadmium. We have had almost no direct experience with these (alkaline) batteries, but from what we have learned from others we do not recommend them—one major disadvantage is that there is a large voltage difference between the fully charged and discharged states. Another problem is that they are very inefficient—there is a 30 to 40 percent heat loss just during charging and discharging.

It is important to note that all batteries commonly used in deep-cycle applications are lead-acid batteries. This includes the standard flooded (wet), gelled, and absorbed glass mat (AGM) batteries. They all use the same chemistry, although the actual construction of the plates and so forth can vary considerably.

Nickel-cadmium, nickel-iron, and other types of batteries are found in some systems but are not common due to their expense and/or poor efficiency.

MAJOR BATTERY TYPES

Solar power backup batteries are divided into two categories based on what they are used for and how they are constructed. The major applications where batteries are used as solar backup include, automotive systems, marine systems, and deep-cycle discharge systems.

The major manufactured processes include flooded or wet construction, gelled, and absorbed glass mat (AGM) types. Absorbed glass mat batteries are also referred to as "starved electrolyte" or "dry" type, because instead of containing wet sulfuric acid solution, the batteries contain a fiberglass mat saturated with sulfuric acid, which has no excess liquid.

Common flooded-type batteries are usually equipped with removable caps for maintenance-free operation. Gelled-type batteries are sealed and equipped with a small vent valve that maintains a minimal positive pressure. Absorbed glass mat batteries are also equipped with a sealed regulation-type valve that controls the chamber pressure within 4 pounds per square inch (lb/in^2).

Battery types   As described earlier, common automobile batteries are built with electrodes that are grids of metallic lead containing lead oxides that change in composition during charging and discharging. The electrolyte is dilute sulfuric acid. Lead-acid batteries, even though invented nearly a century ago, are still the battery of choice for solar and backup power systems. With improvements in manufacturing, batteries can last as long as 20 years.

Nickel-cadmium, or alkaline, storage batteries, in which the positive active material is nickel oxide and the negative material contains cadmium, are generally considered very hazardous due to the cadmium. The efficiency of alkaline batteries ranges from 65 to 80 percent compared to 85 to 90 percent for lead-acid batteries. Their nonstandard voltage and charging current also make them very difficult to use.

Deep-discharge batteries used in solar power backup applications in general have lower charging and discharging rate characteristics and are more efficient.

In general, all batteries used in PV systems are lead-acid type batteries. Alkaline-type batteries are used only in exceptionally low temperature conditions of below −50°F. Alkaline batteries are expensive to buy and due to the hazardous contents are very expensive to dispose off.

BATTERY LIFE SPAN

The life span of a battery will vary considerably with how it is used, how it is maintained and charged, the temperature, and other factors. In extreme cases, it can be damaged within 10 to 12 months of use when overcharged. On the other hand if the battery is maintained properly, the life span could be extended over 25 years. Another factor that can shorten the life expectancy by a significant amount is when the batteries are stored uncharged in a hot storage area. Even dry charged batteries when sitting on a shelf have a maximum life span of about 18 months; as a result most are shipped from the factory with damp plates. As a rule, deep-cycle batteries can be used to start and run marine engines. In general, when starting, engines require a very large inrush of current for a very short time. Regular automotive starting batteries have a large number of thin plates for maximum surface area. The plates, as described earlier, are constructed from impregnated lead paste grids similar in appearance to a very fine foam sponge. This gives a very large surface area, and when deep-cycled, the grid plates quickly become consumed and fall to the bottom of the cells in the form of sediment. Automotive batteries will generally fail after 30 to 150 deep-cycles if deep-cycled, while they may last for thousands of cycles in normal starting use discharge conditions. Deep-cycle batteries are designed to be discharged down time after time and are designed with thicker plates. The major difference between a true deep-cycle battery and regular batteries is that the plates in a deep-cycle battery are made from solid lead plates and are not impregnated with lead oxide paste. Figure 2.12 shows a typical solar battery bank system.

Stored energy in batteries in general is discharged rapidly. For example, short bursts of power are needed when starting an automobile on a cold morning, which results in high amounts of current being rushed from the battery to the starter. The standard unit for energy or work is the joule (J), which is defined as 1 watt-second of mechanical work performed by a force of 1 newton (N) or 0.227 pound (lb) pushing or moving a distance of 1 meter (m). Since 1 hour has 3600 seconds, 1 watt-hour (Wh) is equal to 3600 J. The stored energy in batteries is either measured in milliampere-hours if small or ampere-hours if large. Battery ratings are converted to energy if their average voltages are known during discharge. In other words, the average voltage of the battery is maintained relatively unchanged during the discharge cycle. The value in joules can also be converted into various other energy values as follows:

Joules divided by 3,600,000 yields kilowatt-hours.

Joules divided by 1.356 yields English units of energy foot-pounds.

Joules divided by 1055 yields British thermal units.

Joules divided by 4184 yields calories.

Figure 2.12 **Deep-cycle battery.** *Courtesy of Solar Integrated Technologies.*

BATTERY POWER OUTPUT

In each instance when power is discharged from a battery, the battery's energy is drained. The total quantity of energy drained equals the amount of power multiplied by the time the power flows. Energy has units of power and time, such as kilowatt-hours or watt-seconds. The stored battery energy is consumed until the available voltage and current to levels of the battery are exhausted. Upon depletion of stored energy, batteries are recharged over and over again until they deteriorate to a level where they must be replaced by new units. High-performance batteries in general have the following notable characteristics. First, they must be capable of meeting the power demand requirements of the connected loads by supplying the required current while maintaining a constant voltage, and they must have sufficient energy storage capacity to maintain the load power demand as long as required. In addition, they must be as inexpensive and economical as possible and be readily replaced and recharged.

BATTERY INSTALLATION AND MAINTENANCE

Unlike many electrical apparatuses, standby batteries have specific characteristics that require special installation and maintenance procedures, which if not followed can impact the quality of the battery performance.

Battery types   As mentioned earlier, the majority of today's emergency power systems make use of two types of batteries, namely lead-acid and nickel-cadmium (NiCd).

Within the lead-acid family there are two distinct categories, namely flooded or vented (filled with liquid acid) and valve-regulated lead acid (VRLA, immobilized acid).

Lead-acid and NiCd batteries must be kept dry at all times and in cool locations, preferably below 70°F, and must not be stored for long in warm locations. Materials such as conduit, cable reels, and tools must be kept away from the battery cells.

Battery installation safety   What separates battery installers from the layperson is the level of awareness and respect for dc power. Energy stored in the battery cell is quite high, and sulfuric acid (lead-acid batteries) or potassium hydroxide (a base used in NiCd batteries) electrolytes could be very harmful if not handled professionally. Care should always be exercised when handling these cells. Use of chemical-resistant gloves, goggles, and a face shield, as well as protective sleeves, is highly recommended. The battery room must be equipped with an adequate shower or water sink to provide for rinsing of the hands and eyes in case of accidental contact with the electrolytes. Stored energy in a single NiCd cell of 100-Ah capacity can produce about 3000 A if short circuited between the terminal posts. Also, a fault across a lead-acid battery can send shrapnel and terminal post material flying in any direction, which can damage the cell and endanger workers.

Rack cabinet installation   Stationary batteries must be mounted on open racks or on steel or fiberglass racks or enclosures. The racks should be constructed and maintained in a level position and secured to the floor and must have a minimum of 3 feet of walking space for egress and maintenance.

Open racks are preferable to enclosures since they provide a better viewing of electrolyte levels and plate coloration, as well as easier access for maintenance. For multistep or bleacher-type racks, batteries should always be placed at the top or rear of the cabinet to avoid anyone having to reach over the cells. Always use the manufacturer-supplied connection diagram to ensure the open positive and negative terminals when charging the cells. In the event of installation schedule delays, if possible, delay delivery.

BATTERY SYSTEM CABLES

Appendix A provides code-rated dc cable tables for a variety of battery voltages and feed capacities. The tables provide American Wire Gauge (AWG) conductor gauges and voltage drops calculated for a maximum of a 2 percent drop. Whenever larger drops are permitted, the engineer must refer to NEC tables and perform specific voltage drop calculations.

CHARGE CONTROLLERS

A charge controller is essentially a current-regulating device that is placed between the solar panel array output and the batteries. These devices are designed to keep batteries charged at peak power without overcharging. Most charge controllers incorporate special electronics that automatically equalize the charging process.

dc FUSES

All fuses used as overcurrent devices, which provide a point of connection between PV arrays and collector boxes must be dc rated. Fuse ratings for dc branch circuits, depending on wire ampacities, are generally rated from 15 to 100 A. The dc-rated fuses familiar to solar power contractors are manufactured by a number of companies such as Bussman, Littlefuse, and Gould and can be purchased from electrical suppliers.

Various manufacturers identify the fuse voltage by special capital letter designations. The following are a sample of time-delay type fuse designations used by various manufacturers:

Bussman. Voltage rating up to 125 V dc and current ampacity range of 1 to 600 A–FRN-R

Bussman. Voltage rating up to 300 V dc and current ampacity range of 1 to 600 A–FRS-R

Littlefuse. Voltage rating up to 600 V dc and current ampacity range of 1 to 600 A–IDSR

Photovoltaic output as a rule must be protected with extremely fast acting fuses. The same fuses can also be utilized within solar power control equipment and collector boxes. Some of the fast-acting fuses used are manufactured by the same companies listed before:

Bussman. Midget fast-acting fuse, ampacity rating 0.1 to 30 A–ATM

JUNCTION BOXES AND EQUIPMENT ENCLOSURES

All junction boxes utilized for interconnecting raceways and conduits must be of waterproof construction and be designed for outdoor installation. All equipment boxes, such as dc collectors must either be classified as MENA 3R or NEMA 4X.

Solar Power System Wiring

This section covers solar power wiring design and is intended to familiarize engineers and system integrators with some of the most important aspects related to personnel safety and hazards associated with solar power projects.

Residential and commercial solar power systems, up until a decade ago, because of a lack of technology maturity and higher production costs, were excessively expensive and did not have sufficient power output efficiency to justify a meaningful return on investment. Significant advances in solar cell research and manufacturing technology have recently rendered solar power installation a viable means of electric power cogeneration in residential and commercial projects.

As a result of solar power rebate programs available throughout the United States, Europe, and most industrialized countries, solar power industries have flourished and expanded their production capacities in the past 10 years and are currently offering reasonably cost effective products with augmented efficiencies.

In view of constant and inevitable fossil fuel-based energy cost escalation and availability of worldwide sustainable energy rebate programs, solar power because of its inherent reliability and longevity, has become an important contender as one of the most viable power cogeneration investments afforded in commercial and industrial installations.

In view of the newness of the technology and constant emergence of new products, installation and application guidelines controlled by national building and safety organizations such as the National Fire Protection Association, which establishes the guidelines for the National Electrical Code (NEC), have not been able to follow up with a number of significant matters related to hazards and safety prevention issues.

In general, small-size solar power system wiring projects, such as residential installations commonly undertaken by licensed electricians and contractors who are trained in life safety installation procedures, do not represent a major concern. However, large installations where solar power produced by photovoltaic arrays generates several hundreds volts of dc power require exceptional design and installation measures.

An improperly designed solar power system in addition to being a fire hazard can cause very serious burns and in some instances result in fatal injury. Additionally, an improperly designed solar power system can result in a significant degradation of power production efficiency and minimize the return on investment.

Some significant issues related to inadequate design and installation include improperly sized and selected conductors, unsafe wiring methods, inadequate overcurrent protection, unrated or underrated choice of circuit breakers, disconnect switches, system grounding, and numerous other issues that relate to safety and maintenance.

At present the NEC in general covers various aspects of photovoltaic power generation systems; however, it does not cover special application and safety issues. For example, in a solar power system a deep-cycle battery backup with a nominal 24 V and 500 Ah can discharge thousands of amperes of current if short circuited. The enormous energy generated in such a situation can readily cause serious burns and fatal injuries.

Unfortunately most installers, contractors, electricians, and even inspectors who are familiar with the NEC most often do not have sufficient experience and expertise with dc power system installation; as such requirements of NEC are seldom met. Another significant point that creates safety issues is related to material and components used, which are seldom rated for dc applications.

Electrical engineers and solar power designers who undertake solar power system installation of 10 kWh or more (a nonpackaged system) are recommended to review 2005 NEC Section 690 and the suggested solar power design and installation practices report issued by Sandia National Laboratories.

To prevent the design and installation issues discussed, system engineers must ensure that all material and equipment used are approved by Underwriters Laboratories. All components such as overcurrent devices, fuses, and disconnect switches are dc rated. Upon completion of installation, the design engineer should verify, independently of the inspector, whether the appropriate safety tags are permanently installed and attached to all disconnect devices, collector boxes, and junction boxes and verify if system wiring and conduit installation comply with NEC requirements.

The recognized materials and equipment testing organizations that are generally accredited in the United States and Canada are Underwriters Laboratories (UL), Canadian Standards Association (CSA), and Testing Laboratories (ETL), all of which are registered trademarks that commonly provide equipment certification throughout the North American continent.

It should be noted that the NEC, with the exception of marine and railroad installation, covers all solar power installations, including stand-alone, grid-connected, and utility-interactive cogeneration systems. As a rule, the NEC covers all electrical system wiring and installations and in some instances has overlapping and conflicting directives that may not be suitable for solar power systems, in which case Article 690 of the code always takes precedence.

In general, solar power wiring is perhaps considered one of the most important aspects of the overall systems engineering effort; as such it should be understood and applied with due diligence. As mentioned earlier, undersized wiring or a poor choice of material application can not only diminish system performance efficiency but can also create a serious safety hazard for maintenance personnel.

WIRING DESIGN

Essentially solar power installations include a hybrid of technologies consisting of basic ac and dc electric power and electronics—a mix of technologies, each requiring specific technical expertise. Systems engineering of a solar power system requires an intimate knowledge of all hardware and equipment performance and application requirements. In general, major system components such as inverters, batteries, and emergency power generators, which are available from a wide number of manufacturers, each have a unique performance specification specially designed for specific applications.

The location of a project, installation space considerations, environmental settings, choice of specific solar power module and application requirements, and numerous other parameters usually dictate specific system design criteria that eventually form the basis for the system design and material and equipment selection.

Issues specific to solar power relate to the fact that all installations are of the outdoor type, and as a result all system components, including the PV panel, support structures, wiring, raceways, junction boxes, collector boxes, and inverters must be selected and designed to withstand harsh atmospheric conditions and must operate under extreme temperatures, humidity, and wind turbulence and gust conditions. Specifically, the electrical wiring must withstand, in addition to the preceding environmental adversities, degradation under constant exposure to ultraviolet radiation and heat. Factors to be taken into consideration when designing solar power wiring include the PV module's short-circuit current (Isc) value, which represents the maximum module output when output leads are shorted. The short-circuit current is significantly higher than the normal or nominal operating current. Because of the reflection of solar rays from snow, a nearby body of water or sandy terrain can produce unpredicted currents much in excess of the specified nominal or Isc current. To compensate for this factor, interconnecting PV module wires are assigned a multiplier of 1.25 (25 percent) above the rated Isc.

PV module wires as per the NEC requirements are allowed to carry a maximum load or an ampacity of no more than 80 percent; therefore, the value of current-carrying capacity resulting from the previous calculation is multiplied by 1.25, which results in a combined multiplier of 1.56.

The resulting current-carrying capacity of the wires if placed in a raceway must be further derated for specific temperature conditions as specified in NEC wiring tables (Article 310, tables 310.16 to 310.18).

All overcurrent devices must also be derated by 80 percent and have an appropriate temperature rating. It should be noted that the feeder cable temperature rating must be the same as that for overcurrent devices. In other words, the current rating of the devices should be 25 percent larger than the total sum of the amount of current generated from a solar array. For overcurrent device sizing NEC table 240-6 outlines the standard ampere ratings. If the calculated value of a PV array somewhat exceeds one of the standard ratings of this table, the next highest rating should be chosen.

All feeder cables rated for a specific temperature should be derated by 80 percent or the ampacity multiplied by 1.25. Cable ratings for 60, 75, and 90°C are listed in NEC tables 310.16 and 310.17. For derating purposes it is recommended that cables rated for 75°C ampacity should use 90°C column values.

Various device terminals, such as terminal block overcurrent devices must also have the same insulation rating as the cables. In other words, if the device is in a location that is exposed to a higher temperature than the rating of the feeder cable, the cable must be further derated to match the terminal connection device. The following example is used to illustrate these design parameter considerations.

A wiring design example   Assuming that the short-circuit current Isc from a PV array is determined to be 40 A, the calculation should be as follows:

1 PV array current derating = 40 × 1.25 = 50 A.
2 Overcurrent device fuse rating at 75°C = 50 × 1.25 = 62.5 A.
3 Cable derating at 75°C = 50 × 1.25 = 62.5. Using NEC table 310-16, under the 75°C columns we find a cable AWG #6 conductor that is rated for 65-A capacity. Because of ultraviolet (UV) exposure, XHHW-2 or USE-2 type cable should be chosen which has a 75-A capacity. Incidentally, the "–2" is used to designate UV exposure protection. If the conduit carrying the cable is populated or filled with four to six conductors, it is suggested, as previously, by referring to NEC table 310-15(B)(2)(a), that the conductors be further derated by 80 percent. At an ambient temperature of 40 to 45°C a derating multiplier of 0.87 is also to be applied: 75 A × 0.87 = 52.2 A. Since the AWG #6 conductor chosen with an ampacity of 60 is capable of meeting the demand, it is found to be an appropriate choice.
4 By the same criteria the closest overcurrent device, as shown in NEC table 240.6, is 60 A; however, since in step 2 the overcurrent device required is 62.5 A, the AWG #6 cable cannot meet the rating requirement. As such, a AWG #4 conductor must be used. The chosen AWG #4 conductor under the 75°C column of table 310-16 shows an ampacity of 95.

If we choose an AWG #4 conductor and apply conduit fill and temperature derating, then the resulting ampacity is $95 \times 0.8 \times 0.87 = 66$ A; therefore, the required fuse per NEC table 240-6 will be 70 A.

Conductors that are suitable for solar exposure are listed as THW-2, USE-2, and THWN-2 or XHHW-2. All outdoor installed conduits and wireways are considered to be operating in wet, damp, and UV-exposed conditions. As such, conduits should be capable of withstanding these environmental conditions and are required to be of a thick wall type such as rigid galvanized (RGS), intermediate metal conduit (IMC), thin wall electrical metallic (EMT), or schedule 40 or 80 polyvinyl chloride (PVC) nonmetallic conduits.

For interior wiring, where the cables are not subjected to physical abuse, CNM, NMB, and UF type cable are permitted. Care must be taken to avoid installation of underrated cables within interior locations such as attics where the ambient temperature can exceed the cable rating.

Conductors carrying dc current are required to use color coding recommendations as stipulated in Article 690 of the NEC. Red wire or any other color other than green and white is used for positive conductors, white for negative, green for equipment grounding, and bare copper wire for grounding. The NEC allows nonwhite grounded wires, such as USE-2 and UF-2, that are sized #6 or above to be identified with a white tape or marker.

As mentioned earlier, all PV array frames, collector panels, disconnect switches, inverters, and metallic enclosures should be connected together and grounded at a single service grounding point.

PHOTOVOLTAIC SYSTEM GROUND-FAULT PROTECTION

When a photovoltaic system is mounted on the roof of a residential dwelling, NEC requirements dictate the installation of ground-fault detection and interrupting devices (GFPD). However, ground-mounted systems are not required to have the same protection since most grid-connected system inverters incorporate the required GFPD devices.

Ground-fault detection and interruption circuitry perform ground-fault current detection, fault current isolation, and solar power load isolation by shutting down the inverter. Ground-fault isolation technology is currently going through a developmental process, and it is expected to become a mandatory requirement in future installations.

PV SYSTEM GROUNDING

Photovoltaic power systems that have an output of 50 V dc under open-circuit conditions are required to have one of the current-carrying conductors grounded. In electrical engineering, the terminologies used for grounding are somewhat convoluted and confusing. In order to differentiate various grounding appellations it would be helpful to review the following terminologies as defined in NEC Articles 100 and 250.

Grounded. Means that a conductor connects to the metallic enclosure of an electrical device housing that serves as earth.

Grounded conductor. A conductor that is intentionally grounded. In PV systems it is usually the negative of the dc output for a two-wire system or the center-tapped conductor of an earlier bipolar solar power array technology.

Equipment grounding conductor. A conductor that normally does not carry current and is generally a bare copper wire that may also have a green insulator cover. The conductor is usually connected to an equipment chassis or a metallic enclosure that provides a dc conduction path to a ground electrode when metal parts are accidentally energized.

Grounding electrode conductor. A conductor that connects the grounded conductors to a system grounding electrode, which is usually located only in a single location within the project site, and does not carry current. In the event of the accidental shorting of equipment the current is directed to the ground, which facilitates actuation of ground-fault devices.

Grounding electrode. A grounding rod, a concrete-encased ultrafiltration rate (UFR) conductor, a grounding plate, or simply a structural steel member to which a grounding electrode conductor is connected. As per the NEC all PV systems, whether grid-connected or stand-alone, in order to reduce the effects of lightning and provide a measure of personnel safety are required to be equipped with an adequate grounding system. Incidentally, grounding of PV systems substantially reduces radio-frequency noise generated by inverter equipment.

In general, grounding conductors that connect the PV module and enclosure frames to the ground electrode are required to carry full short-circuited current to the ground; as such, they should be sized adequately for this purpose. As a rule, grounding conductors larger than AWG #4 are permitted to be installed or attached without special protection measures against physical damage. However, smaller conductors are required to be installed within a protective conduit or raceway. As mentioned earlier, all ground electrode conductors are required to be connected to a single grounding electrode or a grounding bus.

EQUIPMENT GROUNDING

Metallic enclosures, junction boxes, disconnect switches, and equipment used in the entire solar power system, which could be accidentally energized are required to be grounded. NEC Articles 690, 250, and 720 describe specific grounding requirements. NEC table 25.11 provides equipment grounding conductor sizes. Equipment grounding conductors similar to regular wires are required to provide 25 percent extra ground current-carrying capacity and are sized by multiplying the calculated ground current value by 125 percent. The conductors must also be oversized for voltage drops as defined in NEC Article 250.122(B).

In some installations bare copper grounding conductors are attached along the railings that support the PV modules. In installations where PV current-carrying conductors are routed through metallic conduits, separate grounding conductors could be eliminated since the metallic conduits are considered to provide proper grounding when adequately coupled. It is, however, important to test conduit conductivity to ensure that there are no conduction path abnormalities or unacceptable resistance values.

Entrance Service Considerations for Grid-Connected Solar Power Systems

When integrating a solar power cogeneration within an existing or new switchgear, it is of the utmost importance to review NEC 690 articles related to switchgear bus capacity.

As a rule, when calculating switchgear or any other power distribution system bus ampacity, the total current-bearing capacity of the bus bars is not allowed to be loaded more than 80 percent of the manufacturer's equipment nameplate rating. In other words, a bus rated at 600 A cannot be allowed to carry a current burden of more than 480 A.

When integrating a solar power system with the main service distribution switchgear, the total bus current-bearing capacity must be augmented by the same amount as the current output capacity of the solar system. For example, if we were to add a 200-A solar power cogeneration to the switchgear, the bus rating of the switchgear must in fact be augmented by an extra 250 A. The additional 50 A represents an 80 percent safety margin for the solar power output current. Therefore, the service entrance switchgear bus must be changed from 600 to 1000 A or at a minimum to 800 A.

As suggested earlier, the design engineer must be fully familiar with the NEC 690 articles related to solar power design and ensure that solar power cogeneration system electrical design documents become an integral part of the electrical plan check submittal documents.

The integrated solar power cogeneration electrical documents must incorporate the solar power system components such as the PV array systems, solar collector distribution panels, overcurrent protection devices, inverters, isolation transformers, fused service disconnect switches, and net metering within the plans and must be considered as part of the basic electrical system design.

Electrical plans should incorporate the solar power system configuration in the electrical single-line diagrams, panel schedule, and demand load calculations. All exposed, concealed, and underground conduits must also be reflected on the plans with distinct design symbols and identification that segregate the regular and solar power system from the electrical systems.

It should be noted that the solar power cogeneration and electrical grounding should be in a single location, preferably connected to a specially designed grounding bus, which must be located within the vicinity of the main service switchgear.

Lightning Protection

In geographic locations, such as Florida, where lightning is a common occurrence, the entire PV system and outdoor-mounted equipment must be protected with appropriate lightning arrestor devices and special grounding that could provide a practical mitigation and a measure of protection from equipment damage and burnout.

LIGHTNING'S EFFECT ON OUTDOOR EQUIPMENT

Lighting surges are comprised of two elements, namely voltage and the quantity of charge delivered by lightning. The high voltage delivered by lightning surges can cause serious damage to equipment since it can break down the insulation that isolates circuit elements and the equipment chassis. The nature and the amount of damage are directly proportional to the amount of current resulting from the charge.

In order to protect equipment damage from lightning, devices know as surge protectors or arrestors are deployed. The main function of a surge arrestor is to provide a direct conduction path for lightning charges to divert them from the exposed equipment chassis to the ground. A good surge protector must be able to conduct a sufficient current charge from the stricken location and lower the surge voltage to a safe level quickly enough to prevent insulation breakdown or damage.

In most instances all circuits have a capacity to withstand certain levels of high voltages for a short time; however, the thresholds are so narrow that if charges are not removed or isolated in time, the circuits will sustain an irreparable insulation breakdown.

The main purpose of a surge arrestor device is, therefore, to conduct the maximum amount of charge and reduce the voltage in the shortest possible time. Reduction of a voltage surge is referred to as clamping, shown in Figures 2.13, 2.14 and 2.15. Voltage clamping in general depends on device characteristics such as internal resistance and the response speed of the arrestor and the point in time at which the clamping voltage is measured.

When specifying a lightning arrestor, it is necessary to take into account the clamping voltage and the amount of current to be clamped, for example, 500 V and 1000 A. Let us consider a real-life situation where the surge rises from 0 to 50,000 V in 5 nanoseconds (ns). At any time during the surge, say at 100 ns, the voltage clamping would be different from

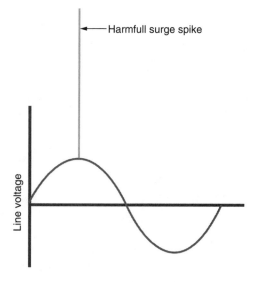

Figure 2.13 **Effect of lighting surge spike.** *Courtesy of Delta Surge Arrestor.*

Surge spike with delta surge arrestor/supressor

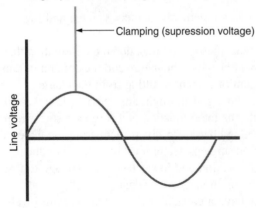

Figure 2.14 **Effect of lightning surge spike clamping.** *Courtesy of Delta Surge Arrestor.*

say the lapsed time, at 20 ns, where the voltage could have been 25,000 V; nevertheless, the voltage is clamped at the end of the specified performance period to 500 V. In general, arrestors with a high current rating have high conductivity and will remove the surge current from the circuit more rapidly and will therefore provide better equipment protection.

The following is a specification for a Delta Lightning Arrestor rated for 2300 V and designed for secondary service power equipment such as motors, electrical panels, transformers, and solar power cogeneration systems.

Model 2301–2300 series specification

Type of design: silicone oxide varistor

Maximum current capacity: 100,000 A

Maximum energy dissipations: 3000 J per pole

Maximum time of 1-mA test: 5 ns

Maximum number of surges: unlimited

Delta surge capacitor

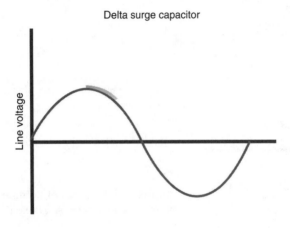

Figure 2.15 **Effect of lightning surge spike suppression.** *Courtesy of Delta Surge Arrestor.*

Figure 2.16 Lightning surge arrestor. *Courtesy of Delta Surge Arrestor.*

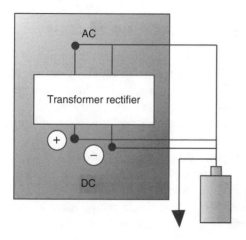

Figure 2.17 Deployment of a lightning surge arrestor in a rectifier circuit application. *Courtesy of Delta Surge Arrestor.*

Response time to clamp 10,000 A: 10 ns

Response time to clamp 25,000 A: 25 ns

Leak current at double the rated voltage: none

Case material: PVC

Central Monitoring and Logging System Requirements

In large commercial solar power cogeneration systems, power production from the PV arrays is monitored by a central monitoring system that provides a log of operation performance parameters. The central monitoring station consists of a PC-type computer that retrieves operational parameters from a group of solar power inverters by means of an RS-232 interface, a power line carrier, or wireless communication systems. Upon receipt of performance parameters, a supervisory software program processes the information and provides data in display or print format. Supervisory data obtained from the file can also be accessed from distant locations through Web networking.

Some examples of monitored data are:

- Weather-monitoring data
- Temperature
- Wind velocity and direction
- Solar power output
- Inverter output
- Total system performance and malfunction
- Direct-current power production
- Alternating-current power production
- Accumulated, daily, monthly, and yearly power production

GENERAL DESCRIPTION OF A MONITORING SYSTEM

The following central monitoring system reflects the actual configuration of the Water and Life Museum project, located in Hemet, California, and designed by the author. This state-of-the-art monitoring system provides a real-time interactive display for education and understanding of photovoltaics and the solar electric installation as well as monitoring the solar electric system for maintenance and troubleshooting purposes.

The system is made up of wireless inverter data transmitters, a weather station, a data storage computer, and a data display computer with a 26-in LCD screen. In the Water and Life Museum project configuration, the inverters, which are connected in parallel, output data to wireless transmitters located in their close proximity. Wireless transmitters throughout the site transmit data to a single central receiver located in a central data gathering and monitoring center, as shown in Figure 2.18.

Figure 2.18 **Sun tracker solar monitoring system.** *Courtesy of Schott Solar.*

The received data are stored and analyzed using the sophisticated software in computer-based supervisory systems that also serve as a data-maintenance interface for the solar power system. A weather station also transmits weather-related information to the central computer.

The stored data are analyzed and forwarded to a display computer that is used for data presentation and also stores information, such as video, sound, pictures, and text file data.

DISPLAYED INFORMATION

A standard display will usually incorporate a looping background of pictures from the site, graphical overlays of the power generation in watts and watt-hours for each building, and the environmental impact from the solar system. The display also shows current meteorological conditions.

Displayed data in general should include the following combination of items:

Project location (on globe coordinates—zoom in and out)

Current and historic weather conditions

Current positions of the sun and moon, with the date and time (local and global)

Power generation—from the total system and/or the individual buildings and inverters

Historic power generation

Solar system environmental impact

Looping background solar system photos and videos

Educational PowerPoint presentations

Installed solar electric power overview

Display of renewable energy system environmental impact statistics

The display should also be programmed to periodically show additional information related to the building's energy management or the schedule of maintenance relevant to the project:

Weather station transmitted data. Transmitted data from the weather-monitoring station should include air temperature, solar cell temperature, wind speed, wind direction, and sun intensity measured using a pyrometer.

Inverter monitoring transmitted data. Each inverter must incorporate a watt-hour transducer that will measure voltage (dc and ac), current (dc and ac), power (dc and ac), ac frequency, watt-hour accumulation, and inverter error codes and operation.

Typical central monitoring computer. The central supervising system must be configured with a CPU with a minimum of 3 GHz of processing power, 512 kilo bytes of random-access memory (RAM), and a 60-gigabyte (Gbyte) hard drive. The operating system should preferably be based on Windows XP or an equivalent system operating software platform.

Wireless transmission system specification. Data communication system hardware must be based upon a switch selectable RS-232/422/485 communication transmission protocol, have a software selectable data transmission speed of 1200 to 57,600 bits/s, and be designed to have several hop sequences share multiple frequencies. The system must also be capable of frequency hopping from 902 to 928 MHz on the FM bandwidth and be capable of providing transparent multipoint drops.

ANIMATED VIDEO AND INTERACTIVE PROGRAMMING REQUIREMENTS

A graphical program builder must be capable of animated video and interactive programming and have an interactive animation display feature for customizing the measurements listed earlier. The system must also be capable of displaying various customizable chart attributes, such as labels, trace color and thickness, axis scale, limits, and ticks. The interactive display monitor should preferably have a 30- to 42-in LCD or LED flat monitor and a 17- to 24-in touch screen display system.

Ground-Mount Photovoltaic Module Installation and Support Hardware

Ground-mount outdoor photovoltaic array installations can be configured in a wide variety of ways. The most important factor when installing solar power modules is the PV module orientation and panel incline. A ground-mount solar power installation is shown in Figure 2.19.

In general, maximum power from a PV module is obtained when the angle of solar rays impinge directly perpendicular (at a 90 degree angle) to the surface of the panels. Since solar ray angles vary seasonally throughout the year, the optimum average tilt angle for obtaining the maximum output power is approximately the local latitude minus 9 or 10 degrees (see Appendix B for typical PV support platforms and hardware and Appendix A for tilt angle installations for Los Angeles; Daggett; Santa Monica; Fresno; and San Diego, California).

In the northern hemisphere, PV modules are mounted in a north-south tilt (high end north) and in the southern hemisphere, in a south-north tilt. Appendix A also includes U.S. and world geographic location longitudes and latitudes.

To attain the required angle, solar panels are generally secured on tilted prefabricated or field-constructed frames that use rustproof railings, such as galvanized Unistrut or commercially available aluminum or stainless-steel angle channels, and fastening hardware, such as nuts, bolts, and washers. Prefabricated solar power support systems are also available from UniRac and several other manufacturers.

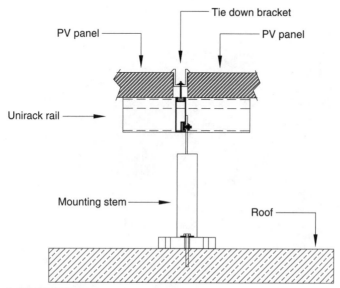

Figure 2.19 **Typical roof mount solar power installation detail.** *Courtesy of Vector Delta Design Group.*

When installing solar support pedestals, also known as stanchions, attention must be paid to structural design requirements. Solar power stanchions and pedestals must be designed by a qualified, registered professional engineer. Solar support structures must take into consideration prevailing geographic and atmospheric conditions, such as maximum wind gusts, flood conditions, and soil erosion.

A typical ground-mount solar power installation includes agricultural grounds; parks and outdoor recreational facilities; carports; sanitariums; and large commercial solar power-generating facilities, also known as solar farms (see Figure 2.20). Most solar farms are owned and operated by electric energy-generating entities such as Edison. Prior to the installation of a solar power system, structural and electrical plans must be reviewed by local electrical service authorities, such as building and safety departments. Solar power installation must be undertaken by a qualified licensed electrical contractor with special expertise in solar power installations.

A solar mounting support system profile, shown in Figure 2.21, consists of a galvanized Unistrut railing frame that is field-assembled with standard commercially available manufactured components used in the construction industry. Basic frame components in general include 2-in galvanized Unistrut channel, 90-degree and T-type connectors, spring-type channel nuts and bolts, and panel hold-down T-type or fender washers.

The main frame that supports the PV modules is welded or bolted to a set of galvanized rigid metal round pipes or square channels. The foundation support is built from

Figure 2.20 **A typical ground-mount solar power installation used in solar farms.** *Courtesy of UniRac.*

Figure 2.21   A single PV frame ground-mount solar power **support.** *Courtesy of UniRac.*

12- to 18-in-diameter reinforced concrete cast in a sauna tube. The metal support structure is secured to the concrete footing by means of expansion bolts. The depth of the footing and dimensions of channel hardware and method of PV module frame attachment are designed by a qualified structural engineer.

A typical solar power support structural design should withstand wind gusts from 80 to 120 miles per hour (mi/h). Prefabricated structures that are specifically designed for solar power applications are available from a number of manufacturers. Prefabricated solar power support structures, although somewhat more expensive, are usually designed to withstand 120-mi/h wind gusts and are manufactured from stainless steel, aluminum, or galvanized steel materials.

Roof-Mount Installations

Roof-mount solar power installations are made of either tilted or flat-type roof support structures or a combination of both. Installation hardware and methodologies also differ depending on whether the building already exists or is a new construction. Roof attachment hardware material also varies for wood-based and concrete constructions. Figure 2.22 depicts a prefabricated PV module support railing system used for roof-mount installations.

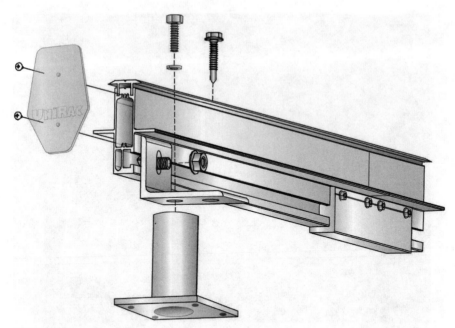

Figure 2.22 Prefabricated PV module support railing for roof-mount system. *Courtesy of UniRac.*

WOOD-CONSTRUCTED ROOFING

In new constructions, the PV module support system installation is relatively simple since locations of solar array frame pods, which are usually secured on roof rafters, can be readily identified.

Prefabricated roof-mount stands that support railings and associated hardware, such as fasteners, are commercially available from a number of manufacturers. Solar power support platforms are specifically designed to meet physical configuration requirements for various types of PV module manufacturers.

Some types of PV module installation, such as in Figure 2.23, have been designed for direct mounting on roof framing rafters without the use of specialty railing or support hardware. As mentioned earlier, when installing roof-mount solar panels, care must be taken to meet the proper directional tilt requirement. Another important factor to be considered is that solar power installations, whether ground or roof mounted, should be located in areas free of shade caused by adjacent buildings, trees, or air-conditioning equipment. In the event of unavoidable shading situations, the solar power PV module location, tilt angle, and stanchion separations should be analyzed to prevent cross shading.

LIGHTWEIGHT CONCRETE-TYPE ROOFING

Solar power installation PV module support systems for concrete roofs are configured from prefabricated support stands and railing systems similar to the ones used in

Figure 2.23 Roof-mount PV support railing system assembly detail. *Courtesy of UniRac.*

Figure 2.24 Wood roof-mount stand-off support railing system assembly detail. *Courtesy of UniRac.*

wooden roof structures. Stanchions are anchored to the roof by means of rust-resistant expansion anchors and fasteners.

In order to prevent water leakage resulting from roof penetration, both wood and concrete standoff support pipe anchors are thoroughly sealed with waterproofing compounds. Each standoff support is fitted with thermoplastic boots that are in turn thermally welded to roof cover material, such as single-ply PVC.

PHOTOVOLTAIC STANCHION AND SUPPORT STRUCTURE TILT ANGLE

As discussed earlier, in order to obtain the maximum output from the solar power systems, PV modules or arrays must have an optimum tilt angle that will ensure a perpendicular exposure to sun rays. When installing rows or solar arrays, spacing between stanchions must be such that there should not be any cross shading. In the design of a solar power system, the available roof area is divided into a template format that compartmentalizes rows or columns of PV arrays.

Electric Shock Hazard and Safety Considerations

Power arrays, when exposed to the sun, can produce several hundred volts of dc power. Any contact with an exposed or uninsulated component of the PV array can produce serious burns and fatal electric shock. As such, the electrical wiring design and installation methodology are subject to rigorous guidelines, which are outlined in NEC Article 690 (discussed later).

System components, such as overcurrent devices, breakers, disconnect switches, and enclosures, are specifically rated for the application. All equipment that is subject to maintenance and repair is marked with special caution and safety warning tags to prevent inadvertent exposure to hazards (see Appendix B for typical sign details).

SHOCK HAZARD TO FIREFIGHTERS

An important safety provision, which has been overlooked in the past, is collaborating with local fire departments when designing roof-mount solar power systems on wood structures. In the event of a fire, the possibility of a serious shock hazard to firefighters will exist in instances when roof penetration becomes necessary.

BUILDING INTEGRATED PHOTOVOLTAIC SYSTEMS (BIPV)

Figure 2.25 is a custom-designed and manufactured photovoltaic module called the Building Integrated Photovoltaic Module (BIPV). This type of solar panel is constructed by laminating individual solar cells in a desired configuration, specifically designed to achieve special visual effect and are typically deployed in solarium or

Figure 2.25 Building an integrated photovoltaic BIPV installation.
Courtesy of Atlantis Energy.

trellis type of structures. Due to the separation gap between the adjacent cells, BIPV modules when compared to standard PV modules produce less energy per square feet of area.

Under operational conditions, when solar power systems actively generate power, a line carrying current at several hundred volts could pose serious burns or bodily injury and electric shock if exposed during the roof demolition process. To prevent injury under fire hazard conditions, all roof-mount equipment that can be accessed must be clearly identified with large red-on-white labels. Additionally, the input to the inverter from the PV collector boxes must be equipped with a crowbar disconnect switch that will short the output of all solar arrays simultaneously.

Another design consideration is that, whenever economically feasible, solar array groups should incorporate shorting contacts (normally closed) that could be activated from a multicontact relay, which could by engaged during emergency conditions. Crowbar circuitry is not needed.

SAFETY INSTRUCTIONS

- Do *not* attempt to service any portion of the PV system unless you understand the electrical operation and are fully qualified to do so.
- Use modules for their intended purpose only. Follow all module manufacturers' instructions. Do *not* disassemble modules or remove any part installed by the manufacturer.

- Do *not* attempt to open the diode housing or junction box located on the back side of any factory-wired modules.
- Do *not* use modules in systems that can exceed 600 V open circuit under any circumstance or combination of solar and ambient temperature.
- Do *not* connect or disconnect a module unless the array string is open or all the modules in the series string are covered with nontransparent material.
- Do *not* install during rainy or windy days.
- Do *not* drop or allow objects to fall on the PV module.
- Do *not* stand or step on modules.
- Do *not* work on PV modules when they are wet. Keep in mind that wet modules when cracked or broken can expose maintenance personnel to very high voltages.
- Do *not* attempt to remove snow or ice from modules.
- Do *not* direct artificially concentrated sunlight on modules.
- Do *not* wear jewelry when working on modules.
- Avoid working alone while performing field inspection or repair.
- Wear suitable eye protection goggles and insulating gloves rated at 1000 V.
- Do *not* touch terminals while modules are exposed to light without wearing electrically insulated gloves.
- Always have a fire extinguisher, a first-aid kit, and a hook or cane available when performing work around energized equipment.
- Do *not* install modules where flammable gases or vapors are present.

Maintenance

In general, solar power system maintenance is minimal and PV modules often only require a rinse and mopping with mild detergent once or twice a year. They should be visually inspected for cracks, glass damage, and wire or cable damage. A periodic check of the array voltage by a voltmeter may reveal malfunctioning solar modules.

TROUBLESHOOTING

All photovoltaic modules become active and produce electricity when illuminated in the presence of natural solar or high ambient lighting. Solar power equipment should be treated with the same caution and care as regular electric power service. Unlicensed electricians or inexperienced maintenance personnel should not be allowed to work with solar power systems.

In order to determine the functional integrity of a PV module, the output of one module must be compared with that of another under the same field operating conditions.

It should be noted that the output of a PV module is a function of sunlight and prevailing temperature conditions, and as such, electrical output can fluctuate from one extreme to another.

One of the best methods to check module output functionality is to compare the voltage of one module to that of another. A difference of greater than 20 percent or more will indicate a malfunctioning module.

When measuring electric current and voltage output values of a solar power module, short-circuit current (Isc) and open-circuit voltage (Voc) values must be compared with the manufacturer's product specifications.

To obtain the Isc value a multimeter ampere meter must be placed between the positive and negative output leads shorting the module circuit. To obtain the Voc reading a multimeter voltmeter should simply be placed across the positive and negative leads of the PV module.

For larger current-carrying cables and wires, current measurements must be carried out with a clamping meter. Since current clamping meters do not require circuit opening or line disconnection, different points of the solar arrays could be measured at the same time. An excessive differential reading will be an indication of a malfunctioning array.

It should be noted that when a PV system operates at the startup and commissioning, it is seldom that occurring problems result from module malfunction or failure; rather most malfunctions result from improper connections or loose or corroded terminals.

In the event of a damaged connector or wiring, a trained or certified technician should be called upon to perform the repairs. Malfunctioned PV modules, which are usually guaranteed for an extended period of time, should be sent back to the manufacturer or installer for replacement.

Please be cautioned not to disconnect dc feed cables from the inverters unless the entire solar module is deactivated or covered with a canvas or a nontransparent material.

It is recommended that roof-mount solar power installations should have ¾-in water hose bibs installed at appropriate distances to allow periodic washing and rinsing of solar modules.

The following safety warning signs must be permanently secured to solar power system components.

WARNING SIGNS

For a solar installation system:

Electric shock hazard—Do not touch terminals—Terminals on both line and load sides may be energized in open position.

For a switchgear and metering system:

Warning—Electric shock hazard—Do not touch terminals—Terminals on both the line and load side may be energized in the open position.

For pieces of solar power equipment:

Warning—Electric shock hazard—Dangerous voltages and currents—No user-serviceable parts inside—Contact qualified service personnel for assistance.

For battery rooms and containers:

Warning—Electric shock hazard—Dangerous voltages and currents—Explosive gas—No sparks or flames—No smoking—Acid burns—Wear protective clothing when servicing typical solar power system safety warning tags.

Photovoltaic Design Guideline

When designing solar power generation systems, the designer must pay specific attention to the selection of PV modules; inverters; and installation material and labor expenses, and specifically be mindful of the financial costs of the overall project. The designer must also assume responsibility to assist the end user with a rebate procurement documentation. The following are major highlights that must be taken into consideration:

Photovoltaic module design parameters

1 Panel rated power (185, 175, 750 W, and so on)
2 Unit voltage (6, 12, 24, 48 V, and so forth)
3 Rated amps
4 Rated voltage
5 Short-circuit amperes
6 Short-circuit current
7 Open-circuit volts
8 Panel width, length, and thickness
9 Panel weight
10 Ease of cell interconnection and wiring
11 Unit protection for polarity reversal
12 Years of warrant by the manufacturer
13 Reliability of technology
14 Efficiency of the cell per unit surface
15 Degradation rate during the expected life span (warranty period) of operation
16 Longevity of the product
17 Number of installations
18 Project references and contacts
19 Product manufacturer's financial viability

Inverter and automatic transfer system

1 Unit conversion efficiency
2 Waveform harmonic distortion
3 Protective relaying features (as referenced earlier)
4 Input and output protection features
5 Service and maintenance availability and cost
6 Output waveform and percent harmonic content
7 Unit synchronization feature with utility power
8 Longevity of the product
9 Number of installations in similar types of application
10 Project references and contacts
11 Product manufacturer's financial viability

It should be noted that solar power installation PV cells and inverters that are subject to the California Energy Commission's rebate must be listed in the commission's eligible list of equipment.

Installation contractor qualification

1 Experience and technical qualifications
2 Years of experience in solar panel installation and maintenance
3 Familiarity with system components
4 Amount of experience with the particular system product
5 Labor pool and number of full-time employees
6 Troubleshooting experience
7 Financial viability
8 Shop location
9 Union affiliation
10 Performance bond and liability insurance amount
11 Previous litigation history
12 Material, labor, overhead, and profit markups
13 Payment schedule
14 Installation warrantee for labor and material

Financial Analysis

The financial analysis of solar power projects involves standard techniques applied in commercial and industrial capital equipment acquisition. At present, the state of California, due to recent energy shortages, has taken a number of measures to encourage the use of solar power systems in residential, commercial, and industrial installations. Specific financial aids available include buyback programs by municipalities and most major power utility companies, such as Southern California Edison (SCE), Santiago Gas and Electric (SGE), and Pacific Gas and Electric (PG&E). In addition, there are a number of lending institutions that under sponsorship of federal and state governments offer low-interest-rate capital equipment loans at 5 percent over a period of 5 years.

In the following chapters we will explore the details of financial analysis and econometric considerations associated with sustainable energy systems.

BUYBACK PROCEDURES AND DOCUMENTATION REQUIREMENTS

To apply for the state-, municipal-, or utility-sponsored buyback programs just referenced, the applicant must complete the following requirements.

1 The project must be located in one of the sponsoring agencies' locations.
2 The solar power system's design must be completed.
3 A licensed installation contractor must be selected.

4 Hardware installed must have a minimum of 5 years warranty.

5 Hardware must be certified by national standards (IEC 1648).

6 The applicant must acquire a municipal permit for the project.

7 A complete set of design documentation, which must include the site location, PV cell installation, power demand calculation, electrical riser diagram, and so forth, must be submitted to the sponsoring agency for evaluation.

8 The applicant should obtain the sponsoring agency's fund reservation application forms from that agency's Web site.

9 Submit the equipment purchase order from the retailer or contractor; a copy of the letter of intent, if an existing project; and a copy of one year's worth of electric bills (not applicable to new projects) to the California Energy Commission.

Upon the California Energy Commission's review, the applicant will receive a Fund Reservation Confirmation and Claim Letter. The fund reservation will be valid for a period of 18 months for systems with a rating larger than 10 kW. For power systems rated 10 kW or less, the reservation period is 9 months.

Please refer to Chapter 6 for California incentive program application forms. When starting the solar power system installation, the applicant must submit these claim forms to the California Energy Commission to obtain a rebate. The rebate is made available within 30 days. It is a common notion that the state of California sponsored rebate programs, which are intended to reduce the energy crisis, will most likely last for only a limited period of time.

3

SOLAR POWER GENERATION
PROJECT IMPLEMENTATION

Introduction

In Chapter 2 we covered the basic concepts of solar power system design, reviewed various system configurations, and outlined all major system equipment and materials required to implement a solar power design. In this chapter the reader will become acquainted with a number of solar power installations that have been implemented throughout the United States and abroad. The broad range of projects reviewed include very small stand-alone pumping stations, residential installations, solar farm installations, large pumping stations, and a few significant commercial and institutional projects. All these projects incorporate the essential design concepts reviewed in Chapter 2. Prior to reviewing the solar power projects, the design engineers must keep in mind that each solar power system design presents unique challenges, requiring special integration and implementation, that may not have been encountered before and may recur in future designs.

Designing a Typical Residential Solar Power System

A typical residential solar power system configuration consists of solar photovoltaic (PV) panels; a collector fuse box; a dc disconnect switch; some lightning protection devices; a charge controller for a battery if required; an appropriately sized inverter; the required number of PV system support structures; and miscellaneous components, such as electrical conduits or wires and grounding hardware. Additional expenses associated with the solar power system will include installation labor and associated electrical installation permits.

Prior to designing the solar power system, the designer must calculate the residential power consumption demand load. Electrical power-consuming items in a household must be calculated according to the NEC recommended procedure outlined in the following steps. The calculation is based on a 2000-ft^2 conventional single residential unit:

Step 1: Lighting load. Multiply the living space square area by 3 W: 2000 × 3 = 6000 W.

Step 2: Laundry load. Multiply 1500 W for each set of laundry appliance, which consists of a clothes washer and dryer: 1500 × 1 = 1500 W.

Step 3: Small appliance load. Multiply kitchen appliance loads rated 1500 W by 2: 1500 W × 2 = 3000 W.

Step 4: Total lighting load. Total the sum of the loads calculated in steps 1 to 3: 6000 + 1500 + 3000 = 10,500 W.

Step 5: Lighting load derating. Use the first 3000 W of the summed-up load (step 4) and add 35 percent of the balance to it: 3000 + 2,625 = 5625 W.

Step 6: Appliance loads. Assign the following load values (in watts) to kitchen appliances:

Dishwasher	1200
Microwave oven	1200
Refrigerator	1000
Kitchen hood	400
Sink garbage disposer	800
Total kitchen appliance load	4600

If the number of appliances equals five or more, the total load must be multiplied by 75 percent, which in this case is 3450 W.

Step 7: Miscellaneous loads. Loads that are not subject to power discounts include air conditioning, Jacuzzi, pool, and sauna and must be totaled as per the equipment nameplate power ratings. In this example we will assume that the residence is equipped with a single five-tone packaged air-conditioning system rated at 17,000 W.

When totaling the highlighted load, the total energy consumption is 26,075 W. 26,075 = (AC) 17000 + (APPLIANCE) 3450 + (LIGHTING POWER) 5625 = 26075. At a 240-V entrance service this represents about 100 A of load. However, considering the average power usage, the realistic mean operating energy required discounts full-time power requirements by the air conditioning, laundry equipment, and kitchen appliances; hence, the norm used for sizing the power requirement for a residential unit boils down to a fraction of the previously calculated power. As a rule of thumb, an average power demand for a residential unit is established by equating 1000 to 1500 W per 1000 ft^2 of living space. Of course this figure must be augmented by considering the geographic location of the residence, number of habitants, occupancy time of the population within the dwelling unit, and so forth. As a rule, residential dwellings in hot climates and desert locations must take the air-conditioning load into consideration.

As a side note when calculating power demand for large residential areas, major power distribution companies only estimate 1000 to 1500 W of power per household, and this is how they determine their mean bulk electric power purchase blocks.

When using a battery backup, a 30 percent derating must be applied to the overall solar power generation output efficiency, which will augment the solar power system requirement by 2500 W.

In order to size the battery bank, one must decide how many hours the overall power demand must be sustained during the absence of sun or insulation. To figure out the ampere-hour capacity of the battery storage system, the aggregate wattage worked out earlier must be divided by the voltage and then multiplied by the backup supply hours. For example, at 120 V ac, the amperes produced by the solar system, which is stored in the battery bank, will be approximately 20 A. To maintain power backup for 6 hours, the battery system must be sized at about 160 Ah.

Example of Typical Solar Power System Design and Installation Plans for a Single Residential Unit

The following project represents a complete design and estimating procedure for a small single-family residential solar power system. In order to establish the requirements of a solar power system, the design engineer must establish the residential power demand calculations based on NEC design guidelines, as shown in the following.

Project location   Palm Springs, California

Electrical engineer consultant   Vector Delta Design Group, Inc., 2325 Bonita Dr., Glendale, CA 91208

Solar power contractor   Grant Electric, 16461 Sherman Way, Van Nuys, CA 91406

Project design criteria   The residential power demand for a single-family dwelling involves specific limits of energy use allocations for area lighting, kitchen appliances, laundry, and air-conditioning systems. For example, the allowed maximum lighting power consumption is 3 W per square foot of habitable area. The laundry load allowed is 1500 W for the washer and dryer.

The first 3000 W of the total combined lighting and laundry loads are accounted at 100 percent, and the balance is applied at 35 percent. The total appliance loads, when there are more than five appliances, are also derated by 25 percent. Air conditioning and other loads such as pool, sauna, and Jacuzzi are applied at their 100 percent value.

The demand load calculation of the 1400-ft^2 residential dwelling shown below indicates a continuous demand load of about 3000 W/h. If it is assumed that the residence is fully occupied and is in use for 12 hours a day, the total daily demand load translates into 36,000 W/day.

Since the average daily insolation in southern California is about 5.5 hours, the approximate solar power system required to satisfy the daily demand load will be about 6000 W. Occupancies that are not fully inhabited throughout the day may require a somewhat smaller system.

In general, an average 8 hours of habitation time should be used for sizing the solar power system, which in this example would yield a total daily power demand of 24,000 Wh, which in turn translates into a 4000-W solar power system.

Commercial Applications

The following plans are provided for illustrative purposes only. The actual design criteria and calculations may vary depending on the geographic location of the project,

Figure 3.1 **Residential roof-mount solar PV installation, Palm Springs, California.** *Photo courtesy of Grant Electric, engineered by Vector Delta Design Group, Inc.*

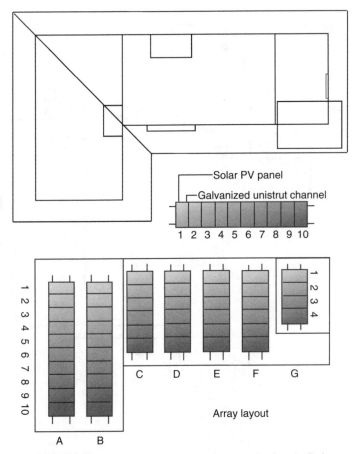

Figure 3.2 Residential solar power installation in Palm Springs, California. *Courtesy of Vector Delta Design Group, Inc.*

cost of labor, and materials, which can significantly vary from one project to another. The following projects were collaborations among the identified organizations.

TCA ARSHAG DICKRANIAN ARMENIAN SCHOOL PROJECT

Location Hollywood, California

Architect Garo Minassian and Associates, 140 Acari Dr., Los Angeles, CA 90049,

Electrical engineer consultant Vector Delta Design Group, Inc., 2325 Bonita Dr., Glendale, CA 91208,

Solar power contractor Grant Electric, 16461 Sherman Way, Van Nuys, CA 91406,

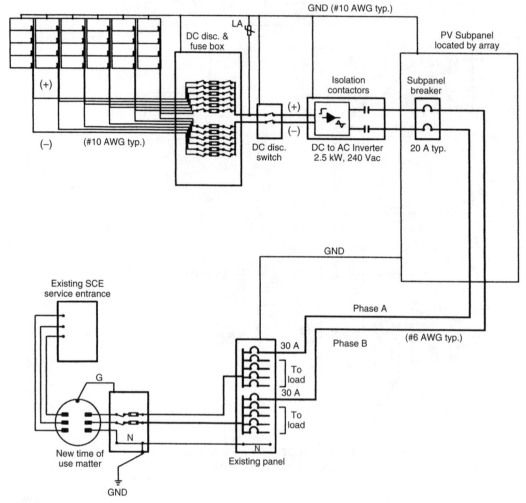

Figure 3.3 Wiring diagram for the Palm Springs residential project.

Project design criteria The project described here is a 70-kW roof-mount solar power cogeneration system, which was completed in 2004. The design and estimating procedures of this project are similar to that of the residence in Palm Springs. Diagrams and pictures of this project are shown in Figures 3.5 through 3.11.

In order to establish the requirements of a solar power system, the design engineer must determine the commercial power demand calculations based on the NEC design guidelines. The power demand estimate for this project is shown in the solar power estimate that appears in Figure 3.6. Power demand calculations for commercial systems depend on the project use, which are unique to each application. Power demand

Figure 3.4   A Xantrex inverter system from the Palm Springs residential project.

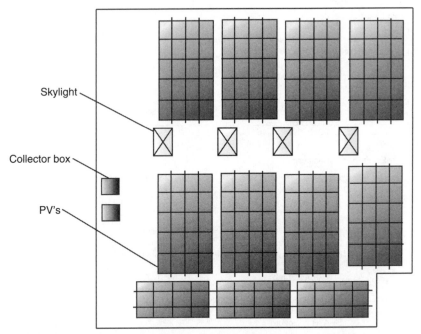

Skylight

Collector box

PV's

Figure 3.5   Roof-mount solar power systems for the TCA Dickranian Armenian School in Hollywood, California. *Courtesy of Grant Electric and Vector Delta Design Group, Inc.*

calculations for this project were calculated to be about 280 kWh. The solar power installed represents about 25 percent of the total demand load. Since the school is closed during summer, energy credited cumulated for three months is expected to augment the overall solar power cogeneration contribution to about 70 percent of the overall demand.

This project was commissioned in March 2005 and has been operating at optimum capacity, providing a substantial amount of the lighting and power requirements of the school. Fifty percent of the overall installed cost of the overall project cost was paid by CEC rebate funds.

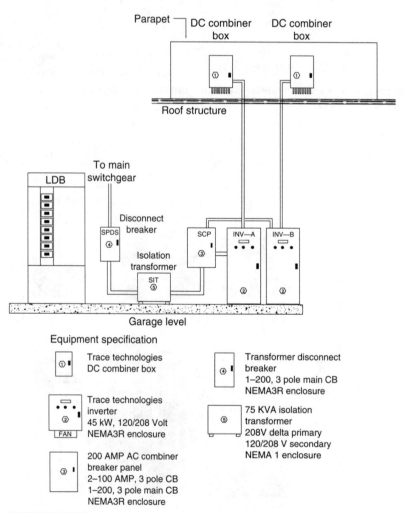

Figure 3.6 Solar power electrical system riser diagram for the TCA Dickranian Armenian School. *Courtesy of Grant Electric and Vector Delta Design Group, Inc.*

Figure 3.7 Solar power electric inverter for the TCA Dickranian Armenian School. *Courtesy of Grant Electric and Vector Delta Design Group, Inc.*

Figure 3.8 Solar power inverter and emergency generator system for the TCA Dickranian Armenian School. *Courtesy of Grant Electric and Vector Delta Design Group, Inc.*

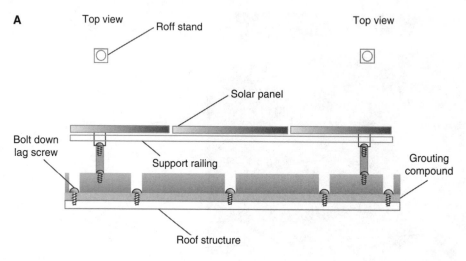

A

Top view

Roff stand

Top view

Solar panel

Bolt down
lag screw

Support railing

Grouting
compound

Roof structure

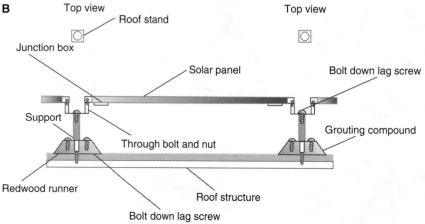

B

Top view

Roof stand

Top view

Junction box

Solar panel

Bolt down lag screw

Support

Through bolt and nut

Grouting compound

Redwood runner

Roof structure

Bolt down lag screw

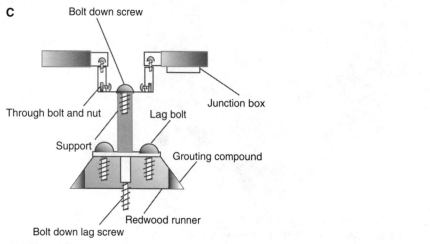

C

Bolt down screw

Through bolt and nut

Junction box

Lag bolt

Support

Grouting compound

Bolt down lag screw

Redwood runner

Figure 3.9 Solar panel support platform detail: (a) side view, (b) front
view, and (c) rail attachment detail. *Courtesy of Vector Delta Design Group, Inc.*

Figure 3.10 Roof-mount solar power standoffs at TCA Dickranian School. *Courtesy of Vector Delta Design Group, Inc.*

Figure 3.11 Roof-mount solar cogeneration system at TCA Dickranian School. *Courtesy of Vector Delta Design Group, Inc.*

BORON SOLAR POWER FARM COGENERATION SYSTEM PROJECT

Location Boron, California

Electrical engineer consultant Vector Delta Design Group, Inc., 2325 Bonita Dr., Glendale, CA 91208

Solar power contractor Grant Electric, 16461 Sherman Way, Van Nuys, CA 91406

Project design criteria The project described here is a 200-kW solar power farm cogeneration system, which was completed in 2003. The design and estimating procedures of this project are similar to the two projects already described.

In view of the vast project terrain, this project was constructed by use of relatively inexpensive, lower-efficiency film technology PV cells that have an estimated efficiency of about 8 percent. Frameless PV panels were secured on 2-in Unistrut channels, which were mounted on telephone poles that penetrated deep within the desert sand. The power produced from the solar farm is being used by the local Indian reservation. The project is shown in Figures 3.13 and 3.14.

Figure 3.12 **The Boron solar project.** *Courtesy of Grant Electric.*

Figure 3.13 **Building-integrated PV assembly.** *Courtesy of Vector Delta Design Group, Inc.*

WATER AND LIFE MUSEUM

Project location   Hemet, California

Architect   Lehere Gangi Architects, 239 East Palm Ave., Burbank, CA 91502

Electrical and solar power consultants   Vector Delta Design Group, Inc., 2325 Bonita Dr., Glendale CA 91208

Electrical contractor   Morrow Meadows, 231 Benton Court, City of Industry, CA 91789

Project description   This project is located in Hemet, California, an hour-and-a-half driving distance from downtown Los Angeles. The project consists of a 150-acre campus with a Water Education Museum, sponsored by the Metropolitan Water District and the Water Education Board; and Archaeology and Paleontology Museum, sponsored by the City of Hemet; several lecture halls; a bookstore; a cafeteria; and two auditoriums. In this installation, PV panels are assembled on specially prefabricated sled-type support structures that do not require roof penetration. Roof-mount PV arrays are strapped together with connective ties to create large island platforms that can withstand 120-mi/h winds. A group of three PV assemblies with an output power capacity of about 6 kW are connected to a dedicated inverter. Each inverter assembly on the support incorporates overcurrent protective circuitry, fusing, and power collection bussing terminals.

Figure 3.14   Water and Life Museum photovoltaic panel support assembly. *Courtesy of Vector Delta Design Group, Inc.*

The inverter chosen for this project includes all technology features, such as islanding, ac power isolation, voltage, and frequency synchronization required for grid connectivity. In addition, the inverters are also equipped with a wireless monitoring transmitter, which can relay various performance and fault-monitoring parameters to a centrally located data acquisition system.

Strategically located ac subpanels installed on rooftops accumulate the aggregated ac power outputs from inverters. Outputs of subpanels are in turn cumulated by a main ac collector panel, the output of which is connected to a central collector distribution panel located within the vicinity of the main service switchgear. Grid connection of the central ac collector panel to the main service bus is accomplished by means of a fused disconnect switch and a net meter.

The central supervisory system gathers and displays the following data:

- Project location (on globe coordinates, zoom in and out)
- Current and historic weather conditions
- Current positions of the sun and moon and the date and time (local and global)
- Power generation—total system or individual buildings and inverters
- Historic power generation
- Solar system environmental impact
- System graphic configuration data
- Educational PowerPoint presentations

TABLE 3.1   ENGINEERING COST ESTIMATE FOR THE WATER AND LIFE MUSEUM SOLAR POWER PROJECT

SOLAR POWER SYSTEM ESTIMATE						
	HRS		RATE		TOTAL	
Engineering rate		$	150.00			
Site investigation	40	$	150.00	$	6,000	
Preliminary design coordination	200	$	150.00	$	30,000	
Report preparation	40	$	150.00	$	6,000	
Travel and accommodation	4	$	250.00	$	1,000	
Sub total				$	43,000.00	$ 43,000.00
Engineering support						
	HRS		RATE		TOTAL	
Permits and rebate applications	12	$	150.00	$	1,800.00	
Project management	80	$	150.00	$	12,000.00	
Travel expenses	10	$	250.00	$	2,500.00	
Other				$	-	
Sub total				$	16,300.00	$ 16,300.00
Engineering design						
	HRS		RATE		TOTAL	
PV systems design	600	$	150.00	$	90,000.00	
Architectural design	160	$	150.00	$	24,000.00	
Structural design	80	$	150.00	$	12,000.00	
Electrical design	600	$	150.00	$	90,000.00	
Tenders and contracting	80	$	150.00	$	12,000.00	
Construction supervision	64	$	150.00	$	9,600.00	
Training manuals	80	$	150.00	$	12,000.00	
Sub total				$	249,600.00	$ 308,900.00
Renewable energy equipment						
	kW-DC		COST		TOTAL	
PV modules - per kWh-DC	542	$	3,900.00	$	2,113,800.00	
Transportation-per project	10	$	500.00	$	5,000.00	
Other		$	20,000.00	$	20,000.00	
Tax			8.25%	$	174,388.50	
Sub total				$	2,313,188.50	$ 2,313,188.50

(Continued)

TABLE 3.1   ENGINEERING COST ESTIMATE FOR THE WATER AND LIFE MUSEUM SOLAR POWER PROJECT (*Continued*)

System installation

	kW-DC	COST		TOTAL	
PV module support structure-per kWh	510	$	640.00	$	326,400.00
Transformation kWh	0			$	-
Inverter-per kWh	500	$	488.00	$	244,000.00
Electrical materials-per kW	500	$	250.00	$	125,000.00
System installation labor-per kWh	480	$	1,000.00	$	480,000.00
Transportation-per project	10	$	3,000.00	$	30,000.00
Tax			8.25%	$	57,370.50
Sub total				$	1,262,770.50	$ 1,262,770.50

Training

	HRS	RATE		TOTAL	
Training	48	$	150.00	$	7,200.00
Document production	80	$	150.00	$	12,000.00
Total Miscellaneous				$	19,200.00	$ 19,200.00

Project Cost Summary

Preliminary engineering study	$ 43,000.00
Engineering support	$ 16,300.00
Engineering design	$ 308,900.00
Renewable energy equipment	$ 2,313,188.50
System installation	$ 1,262,770.50
Training	$ 19,200.00
Total cost	**$ 3,963,359.00**
OH&P 12%	$ 475,603.1
Total Installed Cost	$ 4,438,962.08
Cost ped DC watt installed $ 8.19	
Cost ped AC watt installed (91% conv. eff.) $ 9.00	
Buyback rebate ped AC Wh -SCE $ 4.50	
Net installed AC cost per watt	

Figure 3.15   Water and Life Museum partial PV panel array.
Courtesy of Vector Delta Design Group, Inc.

Figure 3.16   Water and Life Museum PV panel module interwiring.
Courtesy of Vector Delta Design Group, Inc.

Figure 3.17 Water and Life Museum Sunny Boy inverter assembly.
Courtesy of Vector Delta Design Group, Inc.

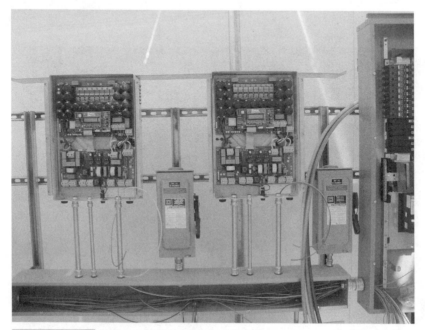

Figure 3.18 Water and Life Museum inverters. *Courtesy of Vector Delta Design Group, Inc.*

Figure 3.19 Water and Life Museum emergency generator system.
Courtesy of Vector Delta Design Group, Inc.

- Temperature
- Wind velocity and direction
- Sun intensity
- Solar power output
- Inverter output
- Total system performance and malfunction
- Direct-current power production
- Alternating-current power production
- Accumulated daily, monthly, and yearly power production

THE HEAVENLY RETREAT PROJECT

The following project is an example of excellence in solar heating and power generation design by owners Joel Goldblatt and Leslie Danzinger. The designer of the project has taken maximum advantage of applying sustainable energy systems design and has blended the technology with the natural setting of the environment.

The "Heavenly Retreat" is a unique, celestially aligned, passive and active solar home, with two built-in integrated greenhouses, located at a 9000-ft elevation in the northern New Mexico Rockies. The heavily forested land is located on the northwest face of an 11,600-ft peak facing Wheeler Peak, the highest point in New Mexico and a landmark in the southern Rockies. The property slopes toward the northwest, set amid many tall pines and fir trees. The south side of the property opens to a large garden with a solar calendar that marks the sunrises and sunsets through the year within a meditation circle and seating area. Figures 3.20 and 3.21 show the external and internal view of solar panel installation.

The designer established the passive solar design by first visiting the natural, forested area many times over a three-year period, from 1985 to 1988, to measure and observe the motions and arc of the sun and moon, as well as to establish celestial cardinal points to the north and south. The design aligns many rooms and exterior walls toward Polaris in the north, and thus establishes the solar south. Many rooms have markers and cast shadows along the floors that show when the sun is at the solar south position, thus indicating the daily time of local solar noon and creating a silent rhythm within the living spaces.

Using these studies of the sun, moon, and planetary alignments, as well as building with adobe, heavy beam, and recyclable product construction and insulation, the home is a tribute to its environment. Being embedded into the forest floor it rises with three terraced stories to a meditation room with a pyramid skylight and a geodesic dome garage. It produces its own solar electricity (3.6 kW) and solar thermal heating (180,000 Btu/h) for 10 zones of hydronic radiant floor heating. It includes three year-round producing greenhouses, and an organically cultivated summer garden. Stepping outdoors from one of three patios, trails go off into the forest with over 50 acres of alpine forest and meadows, sharing the forest with deer, elk, and many other animals. The home was completed in 2002. It is being further refined and developed to serve as a spiritual retreat for guests, offering meditation and seminars in solar design and construction.

The owner, who is also a designer of the project, spent many years contemplating about the project and undertook special design efforts and planning to honor and respect the building's surrounding environment. The project was not designed to be

Figure 3.20 **The Heavenly Retreat project.** *Photo courtesy of Mr. Joel Goldblatt, New Mexico, USA.*

just a sustainable energy design installation project but rather a holistic environmental design that could unite technology and ecology

The sustainable energy system is configured by the deployment of 11 solar thermal panels with a control system that augments water heating by use of a supplemental propane boiler. The heated water is the primary heat source, which provides space heating as well as a domestic hot water (DHW) system. At present the solar thermal system accounts for about 60 to 70 percent of the heat, thus reducing propane use by about 40 to 50 percent.

The project was constructed in two phases. The first phase was completed from 1988 to 1994, and the second phase began in 1999 and was completed in 2002. The present net power output of the photovoltaic power system is rated at 3.6 kW ac. The solar power generation consists of two systems; one is installed in the garden, and the other in the kitchen above the counter, which also serves as a solar window for the adjacent interior greenhouse. The kitchen solar power panels are building-integrated photovoltaic (BIPV) panes custom-manufactured by Atlantis Energy. They are built with quadruple-pane tempered glass panels that sandwich Shell Solar Power Max cells by a lamination process; dc-to-ac power conversion is achieved by the use of two stacked Trace SW-4048 inverters and Outback MX-60 charge controllers.

The roof-mount solar BIPV panes are secured to the building structure at an angle that matches the pitch of the rest of the roof, which is about 32 degrees up from the horizon, in close proximity to the optimal angle matching the geographic latitude for an equinox alignment that favors the winter sun angle.

Figure 3.21 **The Heavenly Retreat project.** *Photo courtesy of Mr. Joel Goldblatt, New Mexico, USA.*

The battery backup system has a 1600-A/h backup capacity. The batteries alternate between a float status and net-metering without the local utility—a rural electric cooperative.

The total system, including all underground infrastructure and the battery storage and monitoring equipment, but excluding the owner's installation time, cost about $80,000 or about $23/W, somewhat expensive compared to off-the-shelf type equipment, but it is a totally unique solar integrated solution in harmony with the indoor greenhouse environment.

Metering is dual-rate "time of use," and the utility consumption averages 1400 kWh/month. Even with recent electric rate increases, the average monthly bill is $130/month. The owner also pays a $15 monthly fee to participate in the green power program offered by the local utility. The owner estimates that the monthly saving resulting from use of the solar power system amounts to at least $200 per month. The owner's average propane bill now ranges from $80/month in the summer to about $250/month in the winter. Without solar thermal system installation, his summer bill would range from $150/month and his winter bill would be over $600/month, an average of a 60 percent energy cost savings.

At present, the state of New Mexico does not have a cash rebate program, although as of 2006, with federal and newly passed state incentives the owner will be able to claim a 30 percent average tax credit from both federal and state governments.

Elsewhere in New Mexico, Public Service of New Mexico (PNM) has instituted a 13 cent/kWh production credit feed-in tariff within its service in the Santa Fe and Albuquerque areas.

WESTERN FILTER CORPORATION, VALENCIA, CALIFORNIA

Architect   California Green Design 18025 Rancho Street, Encino, CA 91316

Description   The rooftop installation shown in Figure 3.22 consists of 1602 photovoltaic panels, each rated 167 W with a total output of 267 kW dc. The solar panel is grouped into 89 arrays of 13 panels each.

The inverters deployed were manufactured by Xantrex. The solar power panels occupy a 45,000-ft^2 area and are mounted on top of a 145,000-ft^2 building.

G/G INDUSTRIES, VALENCIA, CALIFORNIA, COMMERCIAL

This is another roof-mount solar power cogeneration system that has a designed 58-kW dc power output capacity. The system installation was completed within one month, without any disturbance to the factory operation. The photovoltaic panels installed on the roof have a low profile and are totally invisible from street level.

COHEN HOUSE, RESIDENTIAL

The solar power system in Figure 3.24 is a post-mounted installation and is installed in a luxurious house with vast land surroundings. The photovoltaic system is mounted on

Figure 3.22   **Commercial roof-mount PV installation.** *Courtesy of California Green Design.*

Figure 3.23   **A typical roof-mount solar power system installation.** *Courtesy of California Green Design.*

Figure 3.24 **A PV system installation in a natural setting.** *Courtesy of California Green Design.*

three poles, each supporting 12 panels of 165 W dc. A single SMA 2500-W inverter is mounted on a 6 ft × 6 ft × 2 ft reinforced-concrete pad. Solar support pedestals are of the sun-tracking type. A preprogrammed motor adjusts the orientation of the photovoltaic panels and maintains a perpendicular orientation to the sun's rays.

CAMPBELL HOUSE, RESIDENTIAL

This project is located in Santa Clarita Valley. The photovoltaic system is mounted on a shade canopy that, in addition to providing electric energy, serves as a solar shade in an area where summer temperatures exceed 100°F. The wooden canopy with a 5 percent tilt provides support for 30 solar panels, each rated at 167 W dc. Two SMA 2500-W inverters convert the dc power to an ac grid-connected power.

KARGODORIAN HOUSE, RESIDENTIAL

This residential project is located in Chatsworth, California, and has an 8 kW dc output. The PV system is installed in a unique setting among boulder rocks. Because of the excessive regional wind gusts, which can exceed 70 mi/h, the PV system has been placed among the rocks to provide protection from the winds. By blending the solar power system with boulder rocks, the architect has created a unique architectural monument.

Figure 3.25 Photovoltaic power in a natural setting. *Courtesy of California Green Design.*

Figure 3.26 Inverter system assembly for Figure 3.25. *Courtesy of California Green Design.*

Figure 3.27 Bottom view of a solar power canopy installation.
Courtesy of California Green Design.

Figure 3.28 Top view of solar canopy. *Courtesy of California Green Design.*

Small-Scale Solar Power Pumping

The solar submersible pump is probably the most efficient, economical, and trouble-free water pump. Figure 3.29 depicts a small submersed solar pumping system diagram. In some installations the procedure for installing a solar power pump simply involves fastening a pipe to a pump and placing the unit in a water pond, lake, well, or river. The output of the solar panel is connected to the pump, and the panel is then pointed toward the sun and up comes the water. The pumps are generally lightweight and easily moved and are capable of yielding hundreds of gallons per day at distances of over 200 ft above the source. The pumps are of a rugged design and are capable of withstanding significant abuse without damage even if they are run in dry conditions for a short time.

Pumping water with solar power is reliable and inexpensive and is a combination of a submersible pump and a solar cell panel that can be procured inexpensively.

Engineers have spent years in developing a water pumping system to meet the needs of ranchers, farmers, and homesteaders. These systems are reliable and affordable and can be set up by a person with no experience or very little mechanical or electrical know-how.

In some instances the pumping systems can be equipped with a battery bank to store energy, in which case water can be pumped at any time, morning, noon, or night and on cloudy days. The system could also be equipped with a simple float switch circuitry that will allow the pumps to operate on a demand basis. The solar pumping system described earlier requires very little maintenance.

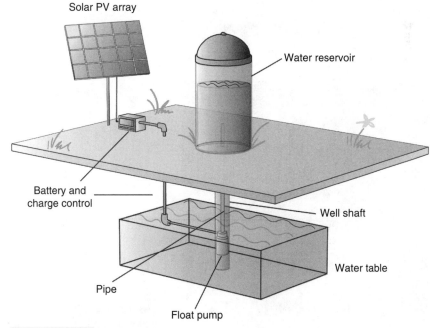

Figure 3.29 **Water well solar power pumping diagram.**

The batteries used for most systems are slightly different than ones used in cars. They are called deep-cycle batteries and are designed to be rechargeable and to provide a steady amount of power over a long period of time. Details about the design and application of battery systems have been discussed in Chapter 2 of this book. In some farming operations solar-powered water systems pump the water into large holding tanks that serve as reserve storage supply during cloudy weather or at night.

In larger installations solar modules are usually installed on special ground- or pole-mounting structures. For added output efficiency, solar panels are installed on tracker-mounting structures that follow the sun like a sunflower.

Large-Capacity Solar Power Pumping Systems

A typical large-scale solar power pumping system, presented in Figure 3.33 is manufactured and system engineered by WorldWater & Power Corporation, whose headquarters are located in New Jersey. WorldWater & Power is an international solar engineering and water management company with a unique, high-powered solar technology that provides solutions to water supply and energy problems. The company has developed patented AquaMax solar electric systems capable of operating

Figure 3.30 Submersible solar water pump in a rural Philippine village. *Courtesy of WaterWorld & Power Corp.*

Figure 3.31 Submersible solar pumping in a rural village in India.
Courtesy of Shell Solar.

pumps and motors up to 600 horsepower (hp), which are used for irrigation, refrigeration and cooling, and water utilities, making it the first company in the world to deliver mainstream solar electric pumping capacity.

The AquaMax system has the following key features.

Automatic switching technology. In general, grid power normally provides the power to the pumps. With the AquaMax system, in the event of power loss, the system automatically and instantaneously switches power fully to the solar array. In keeping with the islanding provisions of the interconnection rules, when the power to the grid is off, the pump or motor keeps operating from solar power alone without interruption. This solar pumping system is the only one of its kind; other grid-tied solar systems shut down when grid power is interrupted.

Power-blending technology. The AquaMax seamlessly blends dc power from the solar array and ac power from the grid to provide a variable-frequency ac signal to the pump or motor. This does two important things. It eliminates large power surges to the motor, and so reduces peak demand charges in the electric bill. It also increases the efficiency of the motor, so it uses less energy to operate. This power-blending technology also means that motors benefit from a "soft start" capability, which reduces wear and tear on the motor and extends its life.

Off-grid capability customers can elect to run a pump or motor off-grid on solar power alone. This may be useful if there is a time of day when, for example, running the pump or motor would incur a large demand charge that the customer wishes to avoid. The system makes operation still possible, while avoiding peak demand charges imposed by the utility.

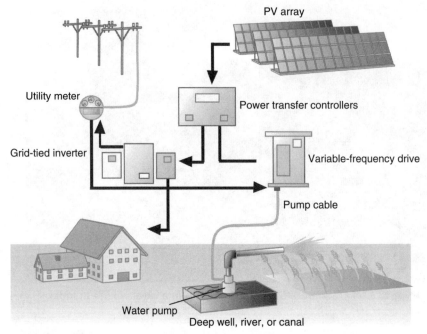

Figure 3.32 **Grid connected solar pumping system diagram.** *Courtesy of WorldWater & Power Corporation.*

Figure 3.33 **Large-scale solar pumping application in California Imperial Valley.** *Courtesy of WorldWater & Power Corporation.*

The 2003 power outages in the Northeast and Midwest highlighted a critical application of this proprietary solar technology. The systems described here are capable of driving pumps or ac motors up to 600 hp as backup for grid power or in combination with the grid or other power sources, such as diesel generators. The systems can also be assembled for mobile, emergency use, or could be used as part of a permanent power installation. In either case, they can provide invaluable power backup in emergency power outages and can operate independent of the electric grid, relying instead on the constant power of the sun.

Pump Operation Characteristics

The following discussion is presented to introduce design engineers to various issues related to pump and piping operational characteristics that affect power demand requirements. In general, the pumping and piping design should be trusted to experienced and qualified mechanical engineers.

Every cooling tower requires at least one pump to deliver water. Pump selection is based on the flow rate, total head, and ancillary issues such as type, mounting, motor enclosure, voltage, and efficiency.

The gallons per minute (gal/min) is dictated by the manufacturer of the equipment. The total head (in feet) is calculated for the unique characteristics of each project as follows:

Total head = net vertical lift + pressure drop at cooling tower exit + pressure drop in piping to pump + pressure drop from pump to destination storage compartment + pressure drop of storage + pressure drop through the distribution system + velocity pressure necessary to cause the water to attain the required velocity

The total head is usually tabulated as the height of a vertical water column. Values expressed in pounds per square inch (lb/in^2) are converted to feet by the following formula:

Head-feet = pounds per square inch $\times$ 2.31

The vertical lift is the distance the water must be lifted before it is let to fall. Typically, it is the distance between the operating level and the water inlet. Pressure drop in the piping to the pump consists of friction losses as the water passes through the pipe, fittings, and valves. Fittings and valves are converted to equivalent lengths of straight pipe (from a piping manual) and added to the actual run to get the equivalent length of suction piping.

Then the tabulated pressure drop from the piping manual for a specific length of pipe is compared to the length and the pressure drop calculated by proportion:

Pressure drop = pressure drop for specific length of pipe $\times$ equivalent pipe length/pipe

Typically, end suction pumps are selected and are of the close-coupled type (where the pump impeller fastens directly to the shaft and the pump housing bolts directly to the motor) for up to about 15 horsepower and the base-mounted type (where the separate pump and motor fasten to a base and are connected by a coupling) is used for larger sizes.

The static lift is typically the distance between the operating level in the cold water basin and the reservoir inlet near the top of the towers.

When selecting a pump, it is important to make sure the net available suction head exceeds the required net suction head. This ensures the application will not cause water to vaporize inside the pump causing a phenomenon called cavitation. Vaporization inside the pump occurs when small water particles essentially "boil" on the suction side of the pump. These "bubbles" collapse as they pass into the high-pressure side producing the classic "marbles sound" in the pump. If operated under this condition, pumps can be damaged.

Pumps are also required to operate under net positive suction head (NPSH) conditions, which means that the pump lift must be able to cope with the local barometric pressure and handle the friction losses in the suction line and vapor pressure of the water being pumped.

4

ENERGY CONSERVATION

Introduction

Energy efficiency is an issue that affects all projects. Whether you are considering a renewable energy system for supplying electricity to your home or business, or just want to save money with your current electrical service supplier, the following suggestions will help you reduce the amount of energy that you use.

If electricity is currently supplied from a local utility, increasing the energy efficiency of a project will help to conserve valuable nonrenewable resources. Investing in a renewable energy system, such as solar or wind, and increasing the energy efficiency of a project will reduce the size and cost of the solar or wind energy system needed.

There are many ways to incorporate energy efficiency into a design. Most aspects of energy consumption, within a building, have more efficient options than traditional methods.

In this chapter we will review the basic concepts of conventional electric power generation and distribution losses and provide some basic recommendations and suggestions about energy conservation measures, which could significantly increase the efficiency of energy use.

In addition to providing energy-saving suggestions, we will review automated lighting design and California Energy Commission Title 24 design compliance. In view of recent green building design measures and raised consciousness about energy conservation, we will review the United States Green Building Council's Leadership in Energy and Environmental Design (LEED)™.

General Energy-Saving Measures

The following recommendations involve simple yet very effective means of increasing energy use. By following the recommendations, energy use can be minimized noticeably without resorting to major capital investment.

LIGHTING

Providing lighting within a building can account for up to 30 percent of the energy used. There are several options for reducing this energy usage. The easiest method for reducing the energy used to provide lighting is to invest in compact fluorescent lights, as opposed to traditional incandescent lights. Compact fluorescent lights use approximately 75 percent less energy than typical incandescent lights. A 15-W compact fluorescent light will supply the same amount of light as a 60-W incandescent light, while using only 25 percent of the energy that a 60-W incandescent light would require. Compact fluorescent lights also last significantly longer than incandescent lights, with an expected lifetime of 10,000 hours on most models. Most compact fluorescent lights also come with a one-year warranty.

Another option for saving money and energy related to lighting is to use torchieres. In recent years halogen torchieres have become relatively popular. However, they create extremely high levels of heat; approximately 90 percent of the energy used by a halogen lamp is emitted as heat, not light. Some halogen lamps generate enough heat to fry an egg on the top of the lamp. These lamps create a fire hazard due to the possibility of curtains touching the lamp and igniting or a lamp falling over and igniting carpet. Great alternatives to these types of lamps are compact fluorescent torchieres. Whereas a halogen torchiere used 4 hours per day will consume approximately 438 kWh in a year, a compact fluorescent torchiere used 4 hours per day will only consume 80 kWh in a year. If you currently pay $0.11 per kilowatt-hour, this would save you over $30 per year, just by changing one lamp.

APPLIANCES

There are many appliances used in buildings that require a significant amount of energy to operate. However, most of these appliances are available in highly efficient models.

Refrigerators   Conventional refrigerators are a major consumer of energy. It is possible to make a refrigerator more effective and efficient by keeping it full. In the event a refrigerator is not fully stocked with food, one must consider keeping jugs of water in it. When a refrigerator is full, the contents will retain the cold. If a refrigerator is old, then consideration should be given to investing in a new, highly efficient, star-rated model. There are refrigerators in the market that use less than 20 kWh per month. When you compare this to the 110 kWh used per month by a conventional refrigerator, you can save over $90 per year (based on $0.11/kWh).

Clothes washers   Washing machines are a large consumer of not only electricity but water as well. By using a horizontal-axis washing machine, also known as a front loader because the door is on the front of the machine, it is possible to save money from using less electricity, water, and detergent.

Front loaders have a more efficient spin cycle than top loaders, which further increases savings due to clothes requiring less time in the dryer. These are the types of machines typically found in Laundromats. The machines are more cost effective than conventional top loaders. Another option is to use a natural gas or propane washer and dryer, which is currently more cost effective than using electric models. If you are on a solar or wind energy system, gas and propane are options that will reduce the overall electricity usage of your home.

Water heaters   Water heaters can be an overwhelming load for any renewable energy system, as well as a drain on the pocketbook for those using electricity from a local utility. The following are some suggestions to increase the efficiency of electric heaters.

- Lower the thermostat to 120 to 130°F.
- Fix any leaky faucets immediately.
- Wrap your water heater with insulation.
- Turn off the electricity to your indoor water heater if you will be out of town for three or more days.
- Use a timer to turn off the water heater during the hours of the day when no one is at home.

If you are looking for a higher-efficiency water heater, you may want to consider using a "flash" or "tankless" water heater, which heats water on demand. This method of heating water is very effective and does not require excessive electricity to keep a tank of water hot. It also saves water because you do not have to leave the water running out of the tap while you wait for it to get hot. Propane or natural gas water heaters are another option for those who want to minimize their electricity demand as much as possible.

INSULATION AND WEATHERIZATION

Inadequate insulation and air leakage are leading causes of energy waste in many homes. By providing adequate insulation in your home, walls, ceilings, and floors will be warmer in the winter and cooler in the summer. Insulation can also help act as a sound absorber or barrier, keeping noise levels low within the home. The first step to improving the insulation of a building is to know the type of existing insulation.

To check the exterior insulation, simply switch off the circuit breaker to an outlet on the inside of an exterior wall. Then remove the electric outlet cover and check to see if there is insulation within the wall. If there is not, insulation can be added by an insulation contractor who can blow cellulose into the wall through small holes, which are then plugged. The geometry of attics will also determine the ease with which additional insulation can be added. Insulating an attic will significantly increase the ability to keep heat in during winter and out during summer.

One of the easiest ways to reduce energy bills and contribute to the comfort of your home or office space is by sealing air leaks around windows and doors. Temporary or permanent weather stripping can be used around windows and doors. Use caulk to seal other gaps that are less than ¼ inch wide and expanding foam for larger gaps. Storm windows and insulating drapes or curtains will also help improve the energy performance of existing windows.

HEATING AND COOLING

Every indoor space requires an adequate climate control system to maintain a comfortable environment. Most people live or work in areas where the outdoor temperature fluctuates beyond ideal living conditions. A traditional air-conditioning or heating system can be a tremendous load on a solar or wind energy system, as well as a drain on the pocketbook for those connected to the utility grid. However, by following some of the insulation and weatherization tips previously mentioned, it is possible to significantly reduce heating losses and reduce the size of the heating system.

The following heating and cooling tips will help further reduce heating and cooling losses and help your system work as efficiently as possible. These tips are designed to increase the efficiency of the heating and/or cooling system without making drastic remodeling efforts. Table 4.1 below shows energy distribution in residential and commercial projects.

Heating   When considering the use of renewable energy systems, electric space and water heaters are not considered viable options. Electric space and water heaters require a significant quantity of electricity to operate at a time of the year when the least amount of solar radiation is available.

Forced-air heating systems also use inefficient fans to blow heated air into rooms that may not even be used during the day. They also allow for considerable leakage through poorly sealed ductwork. Ideally, an energy-independent home or office space will not require heating or cooling due to passive solar design and quality insulation. However, if the space requires a heating source, one should consider a heater that burns fuel to provide heat and does not require electricity. Some options to consider are woodstoves and gas or propane heaters.

Cooling   A conventional air-conditioning unit is an enormous electrical load on a renewable energy system and a costly appliance to use. As with heating, the ideal energy-independent home should be designed to not require an air-conditioning unit. However, since most homeowners considering renewable energy systems are not going to redesign their home or office space, an air-conditioning unit may be necessary.

If you adequately insulate your home or office space and plug any drafts or air leaks, air-conditioning units will have to run less, which thus reduces energy expenditure. Air-conditioning units must be used only when it is absolutely necessary.

Another option is to use an evaporative cooling system. Evaporative cooling is an energy-efficient alternative to traditional air-conditioning units. Evaporative cooling works by evaporating water into the airstream. An example of evaporative cooling is the chill you get when stepping out of a swimming pool and feeling a breeze. The chill you get is caused by the evaporation of the water from your body. Evaporative cooling uses this evaporation process to cool the air passing through a wetted medium.

TABLE 4.1 ENERGY DISTRIBUTION IN RESIDENTIAL AND COMMERCIAL PROJECTS	
	% OF POWER USED
Apartment Buildings	
Environmental control	70%
Lighting, receptacles	15
Water heating	3
Laundry, elevator, miscellaneous	12
Single Residential	
Environmental control	60%
Lighting, receptacles	15
Water heating	3
Laundry, pool	22
Hotel and Motels	
Space heating	60%
Air conditioning	10
Lighting	11
Refrigeration	4
Laundry, kitchen, restaurant, pool	15
Water heating, miscellaneous	
Retail Stores	
Environmental control hvac	30
Lighting	60
Elevator, security, parking	10

Early civilizations used this method by doing something as simple as hanging wet cloth in a window to cool the incoming air. Evaporative cooling is an economical and energy-efficient solution for your cooling needs. With an evaporative cooling unit there is no compressor, condenser, chiller coils, or cooling towers. Therefore, the cost of acquiring and operating an evaporative cooling unit is considerably less than for a conventional air-conditioning unit, and maintenance costs are lower due to the units requiring simpler procedures and lower-skilled maintenance workers. Also, unlike conventional air-conditioning units, evaporative cooling does not release chlorofluorocarbons (CFCs) into the atmosphere.

By following these recommendations, it is possible to turn a home or office space into an energy-efficient environment.

Power Factor Correction

The intent of the following discussion is to familiarize the reader with the basic concepts of the power factor and its effect on energy consumption efficiency. Readers interested in a further understanding of reactive power concepts should refer to electrical engineering textbooks.

In large commercial or industrial complexes where large amounts of electric power are used for fluorescent lighting or heavy machinery, the efficiency of incoming power, which is dependent upon the maintenance of the smallest possible phase angle between the current and voltage, is usually widened, thus resulting in a significant waste of energy. The cosine of the phase angle between the current and voltage, which is referred to as the power factor, is the multiplier that determines whether the electric energy is used at its maximum to deliver lighting or mechanical energy or is wasted as heat. Power (P) in electrical engineering is defined as the product of the voltage (V) and current (I) times the cosine of the phase angle or $P = V \times I \times$ cosine phase angle. When the phase angle between the current and voltage is zero, the cosine equals 1 and therefore $P = V \times I$, which represents the maximum power conversion or delivery.

The principal components of motors, transformers, and lighting ballasts are wound-copper coils referred to as inductance elements. A significant characteristic of inductors is that they have a tendency to shift the current and voltage phase angles, which results in power factors that are less than 1 and, hence, in reduced power efficiency. The performance of power usage, which is defined as the ratio of the output power to the maximum power, is therefore used as the figure of merit. The reduction of electric power efficiency resulting from reactive power is wasted energy that is lost as heat. In a reactive circuit, the phase angle between current and voltage shifts, thus, giving rise to reactive power that is manifested as unused power, which dissipates as heat.

Mitigation measures that can be used to minimize inductive power loss include the installation of phase-shifting capacitor devices that negate the phase angle created by induction coils. As a rule, the maximum power affordable for efficient use of electric power should be above 93 percent. In situations where the power factor measurements indicate a value of less than 87 percent, power losses can be minimized by the use of capacitor reactance.

A Few Words about Power Generation and Distribution Efficiency

It is interesting to observe that most of us, when using electric energy, are oblivious to the fact that the electric energy provided to our household, office, or workplace is mostly generated by extremely low efficiency conversion of fossil fuels such as coal, natural gas, and crude oils. In addition to producing substantial amounts of pollutants, electric plants when generating electric power operate with meager efficiency and deliver electricity to

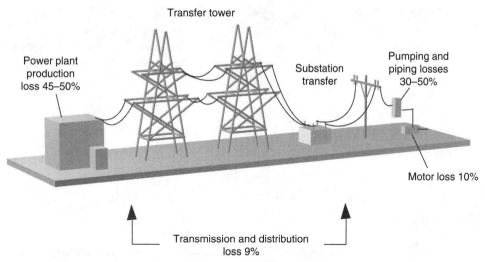

Figure 4.1 Transmission and distribution losses associated with electric fossil fuel power generation delivery losses.

the end user with great loss. To illustrate the point let us review the energy production and delivery of a typical electric-generating station that uses fossil fuel.

By setting an arbitrary unit of 100 percent for the fossil fuel energy input into the boilers, we see that due to losses resulting from power plant machinery, such as turbines, generators, high-voltage transformers, transmission lines, and substations, the efficiency of delivered electric energy at the destination is no more than 20 to 25 percent. The efficiency of energy use is further reduced when the electric energy is used by motors, pumps, and a variety of equipment and appliances that have their own specific performance losses. Table 4.2 depicts comparative losses between solar and fossil fuel power generation systems.

As evidenced by the preceding example, when comparing solar power generation with electric power generated by fossil fuel, the advantages of solar power generation in the long run become quite obvious.

A short-sighted assessment by various experts siding with conventional fossil power generation is that less burning of coal and crude oil to minimize or prevent global warming will increase the national expenditure to such a degree that governments will be prevented from meeting the society's needs for transportation, irrigation, heating, and many other energy-dependent services. On the other hand, environmentalists argue that protection of nature and the prevention of global warming warrant the required expenditure to prevent inevitable climatic deterioration.

With advances in technology, the increased output efficiency of solar PV modules and the reduction in the cost of PV modules, which would result from mass production, will within the next decade make solar power installation quite economical. National policies should take into consideration that technologies aimed at reducing global warming could indeed be a major component of the gross national income and

TABLE 4.2   SOLAR POWER AND FOSSIL FUEL POWER GENERATION COMPARISON TABLE

	SOLAR ELECTRIC POWER	FOSSIL FUEL ELECTRIC POWER
Delivery efficiency	Above 90%	Less than 30%
Maintenance	Very minimal	Considerable
Transmission lines	None required	Very extensive
Equipment life span	25–45 years	Maximum of 25 years
Investment payback	8–14 years	20–25 years
Environmental impact	No pollution	Very high pollution index
Percent of total U.S. energy	Less than 1%	Over 75%
Reliability index	Very high	Good

that savings from fossil fuel consumption could be much less than the expenditure for research and development of solar power and sustainable energy technologies.

In the recent past some industry leaders, such as DuPont, IBM, Alcan, NorskeCanada, and British Petroleum have expended substantial capital toward the reduction of carbon dioxide and greenhouse gas emissions, which has resulted in billions of dollars of savings.

For example, British Petroleum has reduced carbon dioxide emissions by 10 percent in the past 10 years and as a result has cut $650 million of expenses. DuPont by reducing 72 percent of greenhouse gases has increased its production by 30 percent, which resulted in $2 billion of savings. The United States at present uses 47 percent less energy dollars than it did 30 years ago, which results in $1 billion per day of savings.

Computerized Lighting Control

In general, conventional interior lighting control is accomplished by means of hardwired switches, dimmers, timers, lighting contactor relays, occupancy sensors, and photoelectric eyes that provide the means to turn various light fixtures on and off or to reduce luminescence by dimming.

The degree of interior lighting control in most instances is addressed by the state of California Title 24 energy regulations, which dictate specific design measures required to meet energy conservation strategies including:

■ Interior room illumination switching
■ Daylight illumination control or harvesting
■ Duration of illumination control by means of a preset timing schedule

- Illumination level control specific to each space occupancy and task environment
- Lighting zone system management
- Exterior lighting control

Figures 4.2 through 4.10 depict various wiring diagrams and lighting control equipment used to increase illumination energy consumption efficiency.

In limited spaces such as small offices, commercial retail, and industrial environments (where floor spaces do not exceed 10,000 ft^2), lighting control is undertaken by hardwiring of various switches, dimmers, occupancy sensors, and timers. However, in large environments, such as high-rise buildings and large commercial and industrial environments, lighting control is accomplished by a computerized automation system that consists of a centralized control and display system that allows for total integration of all the preceding components.

A central lighting control system embeds specific software algorithms that allow for automated light control operations to be tailored to meet specific energy and automation management requirements unique to a special environment. An automated lighting control system, in addition to reducing energy waste to an absolute minimum, allows for total operator override and control from a central location.

Because of the inherent design of a centralized lighting system, the central monitoring system offers indispensable advantages that cannot be accomplished by hard-wired systems. Some of the advantages of a centralized lighting control system are as follows:

- Remote manual or automatic on-off control of up to 2400 lighting groups within a predetermined zone.
- Remote dimming of lighting within each zone.
- Automatic sequencing control of individual groups of lights.
- Sequencing and graded dimming or step activation of any group of lights.
- Remote status monitoring of all lights within the overall complex.
- Inrush current control for incandescent lights, which substantially prolongs the life expectancy of lamps.
- Visual display of the entire system illumination throughout the complex by means of graphic interfaces.
- Free-of-charge remote programming and maintenance of the central lighting and control from the equipment supplier's or manufacturer's headquarters.
- Optional remote radio communication interfaces that allow for control of devices at remote locations without the use of conduits and cables. In some instances, radio control applications can eliminate trenching and cable installation, which can offset the entire cost of a central control system. Contrary to conventional wiring schemes where all wires from fixtures merge into switches and lighting panels, an intelligent lighting control system, such as the one described here, makes use of Type 5 cable (a bundle of four-pair twisted shielded wires), which can interconnect up to 2400 lighting control elements. A central control and monitoring unit located in an office constantly communicates with a number of remotely located intelligent control boxes that perform the lighting control measures required by Title 24 and beyond.

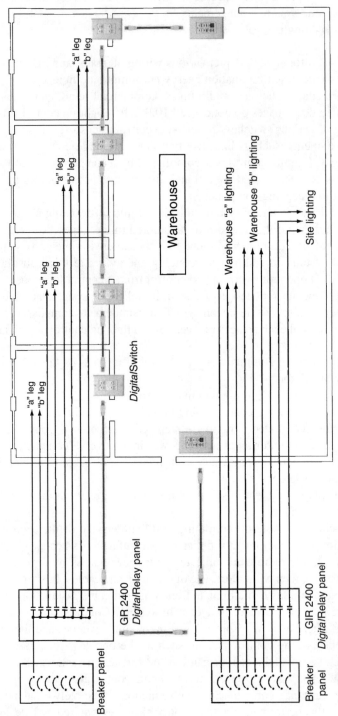

Figure 4.2 A typical centralized lighting control wiring plan. *Photo courtesy of LCD.*

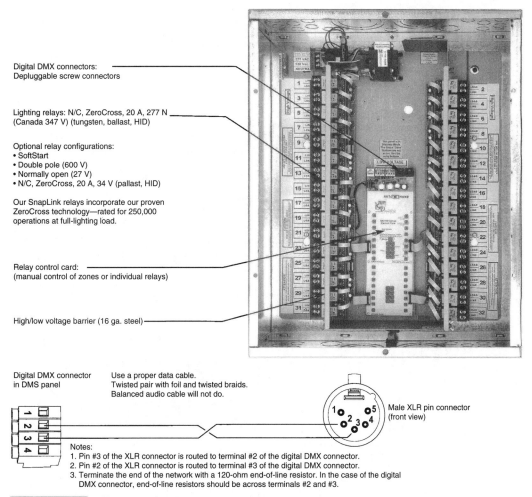

Digital DMX connectors:
Depluggable screw connectors

Lighting relays: N/C, ZeroCross, 20 A, 277 N
(Canada 347 V) (tungsten, ballast, HID)

Optional relay configurations:
• SoftStart
• Double pole (600 V)
• Normally open (27 V)
• N/C, ZeroCross, 20 A, 34 V (pallast, HID)

Our SnapLink relays incorporate our proven
ZeroCross technology—rated for 250,000
operations at full-lighting load.

Relay control card:
(manual control of zones or individual relays)

High/low voltage barrier (16 ga. steel)

Digital DMX connector   Use a proper data cable.
in DMS panel   Twisted pair with foil and twisted braids.
                             Balanced audio cable will not do.

Male XLR pin connector
(front view)

Notes:
1. Pin #3 of the XLR connector is routed to terminal #2 of the digital DMX connector.
2. Pin #2 of the XLR connector is routed to terminal #3 of the digital DMX connector.
3. Terminate the end of the network with a 120-ohm end-of-line resistor. In the case of the digital
   DMX connector, end-of-line resistors should be across terminals #2 and #3.

Figure 4.3 **DMX relay lighting control wiring plan.** *Photo courtesy of LCD.*

Since remote lighting, dimming, and occupancy sensing is actuated by means of electronically controlled relay contacts, any number of devices such as pumps, outdoor fixtures, and a various number of devices with varying voltages could be readily controlled with the same master station.

In addition to providing intelligent master control, remote station control devices and intelligent wall-mount switches specifically designed for interfacing with intelligent remote devices provide local lighting and dimming control override. Moreover, a centralized lighting control system can readily provide required interlocks between heating, ventilation, and air-conditioning (HVAC) systems by means of intelligent thermostats.

Even though central intelligent lighting control systems, such as the one described here, add an initial cost component to the conventional wiring, in the long run the extended expectancy of lamps, lower maintenance cost, added security, and

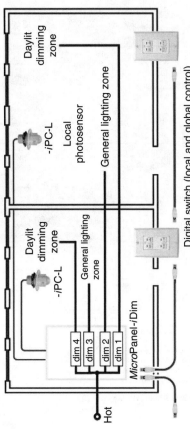

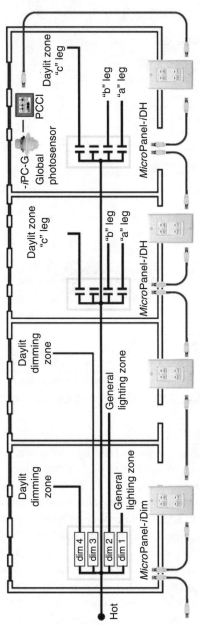

Figure 4.4 **Centralized dimming and lighting control diagram.** *Photo courtesy of LCD.*

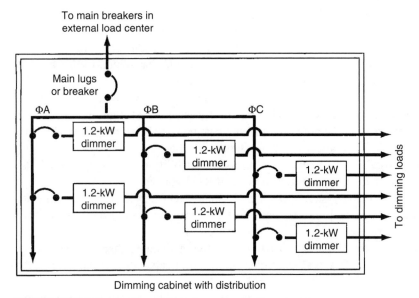

Figure 4.5 **Centralized dimming circuit diagram.** *Photo courtesy of LCD.*

considerable savings resulting from energy conservation undoubtedly justify the added initial investment.

In fact, the most valuable feature of the system is flexibility of control and ease of system expansion and reconfiguration, deployed in an application such as the Water and Life Museum project discussed in Chapter 3, should be considered as indispensable.

The major cost components of a centralized lighting control system consist of the central and remote controlled hardware and dimmable fluorescent T8 ballasts. It is a well-established fact that centralized control lighting systems pay off in a matter of a few years and provide a substantial return on investment by the sheer savings on energy consumption. Needless to say, no measure of security can be achieved without central lighting control.

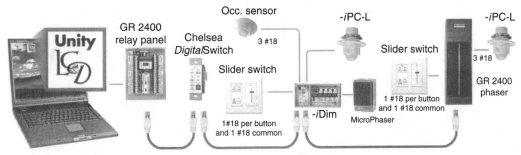

Figure 4.6 **Remote lighting control component configuration.** *Photo courtesy of LCD.*

The automated centralized lighting control system manufactured by Lighting Control Design (LCD), and shown in Figure 4.7, provides typical control components used to achieve the energy conservation measures discussed earlier. It should be noted that the lighting control components and systems presented in this chapter are also available by Lutron and several other companies.

Some of the major lighting system components available for system design and integration include centralized microprocessor-based lighting control relays that incorporate 32 to 64 addressable relay channels, 365-day programmable astronomical timers, telecommunication modems, mixed voltage output relays (120 or 277 V), manual override for each relay, and a linkup capability of more than 100 links to digital devices via category 5 patch cables and RJ45 connectors.

The preceding systems also include smart breaker panels that use solenoid-operated thermal magnetic breakers that effectively provide overcurrent protection as well as lighting control.

Overcurrent devices are usually available in single or three phases; a current rating of 15, 20, and 30 A; and an arc current interrupt capacity (AIC) of 14 kilo Amperes @ 120/208 V and 65K @ 277/480 V.

A microprocessor-based, current-limiting subbranch distribution panel provides lighting calculations for most energy regulated codes. For example, California's Title 24 energy compliance requirements dictate 45 W of linear power for track lighting, while the city of Seattle in the state of Washington requires 70 W/ft for the same track lighting system. The current limiting subpanel effectively provides a programmable circuit current-limiting capability that lowers or raises the volt-ampere (VA) rating requirement for track lighting circuits. The current-limiting capacity for a typical panel is 20 circuits, with each capable of limiting current from 1 to 15 A.

Another useful lighting control device is a programmable zone lighting control panel, which is capable of the remote control of 512 uniquely addressable lighting control relays. Groups of relays can either be controlled individually, referred to as discrete mode, or can be controlled in groups, referred to as zone mode. Lighting relays in typical systems are extremely reliable and are designed to withstand 250,000 operations at full load capacity.

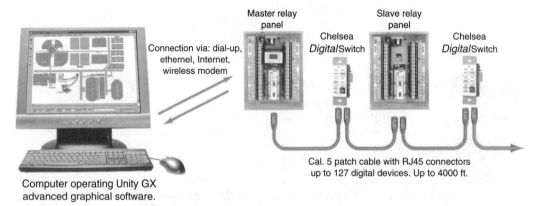

Master relay panel

Slave relay panel

Chelsea *Digital*Switch

Chelsea *Digital*Switch

Connection via: dial-up, ethernet, Internet, wireless modem

Computer operating Unity GX advanced graphical software.

Cal. 5 patch cable with RJ45 connectors up to 127 digital devices. Up to 4000 ft.

Figure 4.7   **Centralized light monitoring and control system.** *Photo courtesy of LCD.*

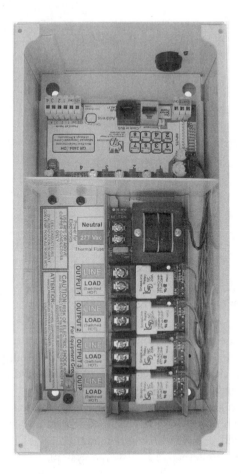

Figure 4.8 Local microprocessor-based control panel. *Photo courtesy of LCD.*

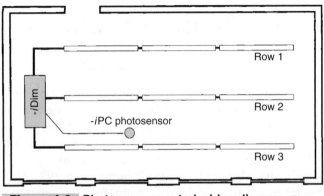

Figure 4.9 Photosensor control wiring diagram. *Photo courtesy of LCD.*

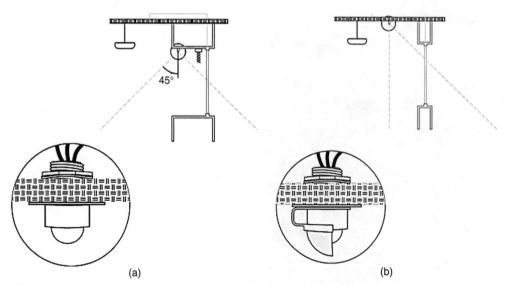

(a)   (b)

Figure 4.10 **Photo sensor control configuration scheme.** *Photo courtesy of LCD.*

For limited area lighting control a compact microprocessor-based device, referred to as a MicroControl, provides a limited capability for controlling two to four switches and dimmable outputs. All microcontrolled devices are daisy chained and communicate with a central lighting command and control system.

A desktop personal computer with a monitor located in a central location (usually the security room) communicates with all the described lighting system panels and microcontrollers via twisted shielded category 5 communication cables. Wireless modem devices are also available as an alternative hard-wired system.

Other optional equipment and devices available for lighting control include digital astronomical time clocks, prefabricated connector cables, dimmer switches, lock-type switches, indoor and outdoor photosensor devices, and modems for remote communication.

California Title 24 Electric Energy Compliance

In response to the 2000 electricity crisis, the state of California legislature mandated the California Energy Commission (CEC) to update the existing indoor lighting energy conservation standards and to develop outdoor lighting energy efficiency compliant cost-effective measures. The intent of the legislature was to develop energy conservation standards that would reduce electricity system energy consumption.

Regulations for lighting have been enforced in California since 1977. However, the measures only addressed indoor lighting through control requirements and maximum allowable lighting power.

For CEC Title 24 compliance form samples please refer to Appendix C.

SCOPE AND APPLICATION

Earlier energy regulation standards only applied to interior and outdoor lighting of buildings that were air-conditioned, heated, or cooled. The updated standards however address lighting in non-air-conditioned buildings and also covers general site illumination and outdoor lighting. The standards include control requirements as well as limits on installed lighting power. The standards also apply to internally and externally illuminated signs.

For detailed coverage of the energy control measures and regulations refer to CEC's standard publications.

Indoor Lighting Compliance

In this section we will review the requirements for indoor lighting design and installation, including controls. It is addressed primarily to lighting designers, electrical engineers, and building department personnel responsible for lighting and electrical plan checking and inspection purposes.

Indoor lighting is perhaps the single largest consumer of energy (kilowatt-hours) in a commercial building, which amounts to approximately one-third of electric energy use. The principal purpose of the standards is to mitigate excessive energy use and provide design guidelines for the effective reduction of energy use, without compromising the quality of lighting. Figure 4.11 depicts lighting energy use in a residential unit.

The primary mechanism for regulating indoor lighting energy under the standards is to limit the allowable lighting power, in watts, installed in the buildings.

Mandatory measures apply to the entire building's lighting systems and equipment consists of the use of such items as manual switching, daylight area controls, and automatic shutoff controls. The mandatory requirements are required to be met either by prescriptive or performance approaches, as will be described here.

As a rule, allowed lighting power for a building is determined by one of the following five methods:

1 *Complete building method.* This method applies to situations when the entire building's lighting system is designed and permitted at one time. This means that at least 90 percent of the building has a single primary type of use, such as retail. In the case of wholesale stores, at least 70 percent of the building area must be used for merchandise sales functions. In some instances this method may be used for an entire tenant space in a multitenant building where a single lighting power value governs the entire building.

2 *Area category method.* This method is applicable for any permit situation, including tenant improvements. Lighting power values are assigned to each of the major function areas of a building, such as offices, lobbies, and corridors.

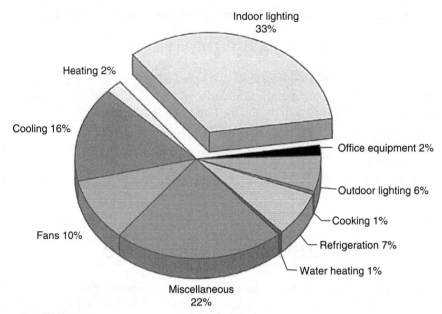

Figure 4.11 **Lighting energy use distribution.** *Courtesy of CEC.*

3 *Tailored method.* This method is applicable when additional flexibility is needed to accommodate special task lighting needs in specific task areas. Lighting power allowances are determined room by room and task by task, with the area category method used for other areas in the building.

4 *Performance approach.* This method is applicable when the designer uses a CEC-certified computer program to demonstrate that the proposed building's energy consumption, including lighting power, meets the energy budget. This approach incorporates one of the three previous methods, which sets the appropriate allowed lighting power density used in calculating the building's custom energy budget. It may only be used to model the performance of lighting systems that are covered under the building permit application.

5 *Actual adjusted lighting power method.* This method is based on the total design wattage of lighting, less adjustments for any qualifying automatic lighting controls, such as occupant-sensing devices or automatic daylighting controls. The actual adjusted lighting power must not exceed the allowed lighting power for the lighting system to comply.

LIGHTING TRADEOFFS

The intent of energy control measures is to essentially restrict the overall installed lighting power in the buildings, regardless of the compliance approach. It should be noted that there is no general restriction regarding where or how general lighting power is used, which means that installed lighting could be greater in some areas and lower in others, provided that the total lighting energy wattage does not exceed the allowed lighting power.

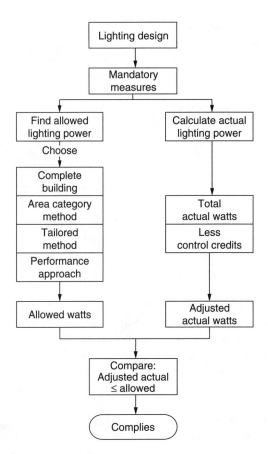

Figure 4.12 **Indoor lighting energy design guide flowchart.** *Courtesy of CEC.*

A second type of lighting tradeoff, which is also permitted under the standards, is a tradeoff of performance between the lighting system and the envelope or mechanical systems. Such a tradeoff can only be made when permit applications are sought for those systems filed under performance compliance where a building with an envelope or mechanical system has a more efficient performance than the prescriptive efficiency energy budget, in which case more lighting power may be allowed.

When a lighting power allowance is calculated using the previously referenced performance approach, the allowance is treated as if it is determined using one of the other compliance methods. It should be noted that no tradeoffs are allowed between indoor lighting and outdoor lighting or lighting located in non-air-conditioned spaces. Figure 4.12 is a flow chart presentation of steps required to achieve mandatory lighting measures.

MANDATORY MEASURES

Mandatory measures are compliance notes that must be included in the building design and on the engineering or Title 24 forms stating whether compliance is of the prescriptive or performance method building occupancy type.

The main purpose of mandatory features is to set requirements for manufacturers of building products, who must certify the performance of their products to the CEC. However, it is the designer's responsibility to specify products that meet these requirements.

LIGHTING EQUIPMENT CERTIFICATION

The mandatory requirements for lighting control devices include minimum specifications for features such as automatic time control switches, occupancy-sensing devices, automatic daylighting controls, and indoor photosensor devices. The majority of the requirements are currently part of standard design practice in California and are required for electrical plan checking and permitting.

Without exception all lighting control devices required by mandatory measures must be certified by the manufacturer before they can be installed in a building. The manufacturer must also certify the devices to the CEC. Upon certification, the device is listed in the Directory of Automatic Lighting Control Devices.

Automatic time switches   Automatic time switches, sometimes called time clocks, are programmable switches that are used to automatically shut off the lights according to preestablished schedules depending on the hours of operation of the building. The devices must be capable of storing weekday and weekend programs. In order to avoid the loss of programmed schedules, timers are required to incorporate backup power provision for at least 10 hours during power loss.

Occupancy-sensing devices   Occupancy-sensing devices provide the capability to automatically turn off all lights in an area for no more than 30 minutes after the area has been vacated. Sensor devices that use ultrasonic sensing must meet certain minimum health requirements and must have the built-in ability for sensitivity calibration to prevent false signals that may cause power to turn on and off.

Devices that use microwave detection (rarely used) principles must likewise have emission controls and a built-in sensitivity adjustment.

Automatic daylighting controls   Daylighting controls consist of photosensors that compare actual illumination levels with a reference illumination level and gradually reduce the electric lighting until the reference level has been reached. These controls are also deployed for power adjustment factor (PAF) lighting credits in the daylit areas adjacent to windows. It is also possible to reduce the general lighting power of the controlled area by separate control of multiple lamps or by step dimming.

Stepped dimming with a time delay prevents cycling of the lights, which is typically implemented by a time delay of 3 minutes or less before electric lighting is reduced or is increased.

Light control in daylight is accomplished by use of photodiode sensors. It should be noted that this requirement cannot be met with devices that use photoconductive cells. In general, stepped switching control devices are designed to indicate the status of lights in controlled zones by an indicator.

Interior photosensor device   Daylighting control systems in general use photosensor devices that measure the amount of light at a reference location. The photosensor provides light level illumination information to the controller, which in turn enables it to increase or decrease the area electric light level.

Photosensor devices must, as previously stated, be certified by the CEC. Devices having mechanical slide covers or other means that allow for adjusting or disabling of the photosensor are not permitted or certified.

Multilevel astronomical time switch controls   Areas with skylights that permit daylight into a building area are required to be calculated by the prescriptive calculation method and are required to be controlled by mandatory automatic controls that must be installed to reduce electric lighting when sufficient daylight is available. Multilevel astronomical time switch controls or automatic multilevel daylight controls specially designed for general lighting control must meet the mandatory requirements for automatic controls when the particular zone has an area greater than 2500 ft^2.

The purpose of astronomical time switch controls is to turn off lights where sufficient daylight is available. Astronomical timers accomplish this requirement by keeping track of the time since sunrise and amount of time remaining before sunset. As a basic requirement, the control program must accommodate multilevel two-step control for each zone programmed to provide independently scheduled activation and deactivation of the lights at different times.

In the event of overly cloudy or overly bright days the astronomical timers are required to have manual override capability. Usually, the override switches in a zone are configured so that lights will revert to the off position within 2 hours, unless the time switch schedule is programmed to keep the lights on.

To comply with the power consumption regulation requirements, light control is not allowed to be greater than 35 percent at time of minimum light output. Device compliance also mandates that devices be designed to display the date and time, sunrise and sunset times, and switching times for each step of control. To prevent a loss of settings due a temporary loss of power, timers are required to have a 10-hour battery backup. Astronomical timers also are capable of storing the time of day and the longitude and latitude of a zone in nonvolatile memory.

Automatic multilevel daylighting controls   Automatic multilevel daylighting controls are used to comply with the mandatory requirements for automatic daylighting controls when the daylight area under skylights is greater than 2500 ft^2. The controls must have a minimum of two control steps so that electric lighting can be uniformly reduced. One of the control steps is intended to reduce lighting power from 100 percent to 70 and 50 percent of full rated power.

Multilevel daylight control devices incorporate calibration and adjustment controls that are accessible to authorized personnel and are housed behind a switch plate cover, touch plate cover, or in an electrical box with a lock.

Under circumstances when the control is used under skylight conditions where the daylight area is greater than 2500 ft^2, the power consumption must not be greater than

35 percent of the minimum electric light output. This is achieved when the timer control automatically turns all its lights off or reduces the power by 30 percent.

Fluorescent dimming controls, even though somewhat expensive, usually meet the minimum power requirements. Controls for high-intensity discharge (HID) lamps do not meet the power requirements at minimum dimming levels; however, a multistage HID lamp switching control can.

Outdoor astronomical time switch controls   Outdoor lighting control by means of astrological time switches is permitted if the device is designed to accommodate automatic multilevel switching of outdoor lighting. Such a control strategy allows all, half, or none of the outdoor lights to be controlled during different times of the day, for different days of the week, while ensuring that the lights are turned off during the daytime.

Incidentally, this feature is quite similar to the indoor multilevel astronomical control with the exception that this control scheme offers a less stringent offset requirement from sunrise or sunset. Mandatory certification for this device requires the controller to be capable of independently offsetting on-off settings for up to 120 minutes from sunrise or sunset.

INSTALLATION REQUIREMENTS

When using automatic time switch control devices or occupant sensors for automatic daylight control, the device must be installed in accordance with the manufacturer's instructions. Devices must also be installed so that the device controls only luminaries within day-lit areas, which means that photosensors must either be mounted on the ceiling or installed in locations that are accessible only to authorized personnel, so that they must maintain adequate illumination in the area.

Certified ballasts and luminaries   All fluorescent lamp ballasts and luminaries are regulated by the Appliance Efficiency Regulations certified by the CEC and are listed in the efficiency database of these regulations.

Area controls   The best way to minimize energy waste and to increase efficiency is to turn off the lights when they are not in use. All lights must have switching or controls to allow them to be turned off when not needed.

Room switching   It is mandatory to provide lighting controls for each area enclosed by ceiling height partitions, which means that each room must have its own switches. Ganged switching of several rooms at once is not permitted. A switch may be manually or automatically operated or controlled by a central zone lighting or occupant-sensing system that meets the mandatory measure requirements.

Accessibility   It is mandatory to locate all switching devices in locations where personnel can see them when entering or leaving an area. In situations when the switching device cannot be located within view of the lights or area, the switch position and states must be annunciated or indicated on a central lighting panel.

Security or emergency Lights within areas required to be lit continuously or for emergency egress are exempt from the switching requirements. However, the lighting level is limited to a maximum of 0.5 W/ft^2 along the path of egress. Security or emergency egress lights must be controlled by switches accessible only to authorized personnel.

Public areas In public areas, such as building lobbies and concourses, switches are usually installed in areas only accessible to authorized personnel.

Outdoor Lighting and Signs

In response to the electricity crisis in 2000, the California legislature mandated the CEC to develop outdoor lighting energy efficiency standards that are technologically feasible and cost effective. The purpose of the legislature was to develop energy efficiency standards that could provide comprehensive energy conservation.

OUTDOOR ASTRONOMICAL TIME SWITCH CONTROLS

As briefly referenced earlier, outdoor lighting control by means of astrological time switches is permitted if the device is designed to accommodate automatic multilevel switching of outdoor lighting. Basically, such a control allows all, half, or none of the outdoor lights to be controlled during different times of the day, for different days of the week, while ensuring that the lights are turned off during the daytime.

Outdoor lighting and sign energy control measures are intended to conserve energy and reduce winter peak electric demand. The standards also set design directives for minimum and maximum allowable power levels when using large luminaries.

Permitted lighting power levels are based on Illuminating Engineering Society of North America (IESNA) recommendations, which are industry standard practices that have worldwide recognition. It should be noted that outdoor lighting standards do not allow tradeoffs between interior lighting, HVAC, building envelope, or water heating energy conformance requirements.

OUTDOOR LIGHTING ENERGY TRADEOFFS

Outdoor lighting tradeoffs are allowed only between the lighting applications with general site lighting illumination, which includes hardscape areas, building entrances without canopies, and outdoor sales lots.

The requirements do not permit any tradeoffs between outdoor lighting power allowances and interior lighting, HVAC, building envelope, or water heating. This includes decorative gas lighting; lighting for theatrical purposes, including stage, film, and video production; and emergency lighting powered by an emergency source as defined by the CEC.

SUMMARY OF MANDATORY MEASURES

The imposed mandatory measures on outdoor lighting include automatic controls that are designed to be turned off during daytime hours and during other times when it is not needed. The measures also require that all controls be certified by the manufacturer and listed in CEC directories. All luminaries with lamps larger than 175 W are required to have cutoff baffles so as to limit the light directed toward the ground. Luminaires with lamps larger than 60 W are also required to be high efficiency or controlled by a motion sensor.

The new CEC standards also limit the lighting power for general site illumination and for some specific outdoor lighting applications. General site illuminations specifically include parking lots, driveways, walkways, building entrances, sales lots, and other paved areas of a site. The measures also provide separate allowances for each of the previously referenced general site lighting applications and allow tradeoffs among these applications. In other words, a single aggregate outdoor lighting budget can be calculated for all the site applications together. Hardscape for automotive vehicular use, including parking lots; driveways and site roads; and pedestrian walkways, including plazas, sidewalks, and bikeways, are all considered part of general lighting.

General lighting includes building entrances and facades such as outdoor sales lots, building facades, outdoor sales frontages, service station canopies, vehicle service station hardscape, other nonsales canopies, ornamental lightings, drive-up windows, guarded facilities, outdoor dining, and temporary outdoor lighting. Site lighting is also regulated by the Federal Aviation Regulation Standards.

General lighting also covers lighting standards used in sports and athletic fields, children's playgrounds, industrial sites, automated teller machines (ATMs), public monuments, lights used around swimming pools or water features, tunnels, bridges, stairs, and ramp lighting. Tradeoffs are not permitted for specific application lighting.

Allowable lighting power for both general site illumination and specific applications are governed by four separate outdoor lighting zone requirements, as will be described later. The lighting zones in general characterize ambient lighting intensities in the surrounding areas. For example, sites that have high ambient lighting levels have a larger allowance than sites with lower ambient lighting levels. The following are title 24 CEC zone classification:

Zone – LZ1 – Government assigned area

Zone – LZ2 – Rural areas as defined by the U.S. 2000 census

Zone – LZ3 – Urban area as defined by the U.S. 2000 census

Zone – LZ4 – Currently not defined

SIGNS

Sign standards contain both prescriptive and performance approaches. Sign mandatory measures apply to both indoor and outdoor signs. Prescriptive requirements apply when the signs are illuminated with efficient lighting sources, such as electronic ballasts, while the performance requirement is applied when calculating the maximum power defined as a function of the sign surface area in watts per square foot. Figure 4.13 outlines the required steps to achieve outdoor lighting mandatory measure as defined by CEC Title 24 energy compliance.

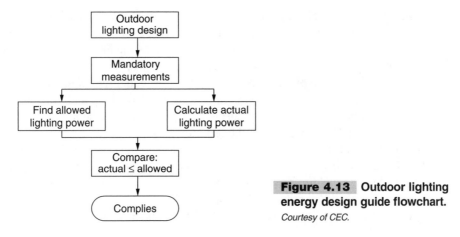

Figure 4.13 Outdoor lighting energy design guide flowchart.
Courtesy of CEC.

INSTALLED POWER

The installed power for outdoor lighting applications is determined in accordance with specific mandatory measure calculation guidelines. Luminaire power for pin-based and high-intensity discharge lighting fixtures may be used as an alternative to determine the wattage of outdoor luminaries. Luminaires with screw-base sockets and lighting systems, which allow the addition or relocation of luminaries without modification to the wiring system, must follow the required guidelines. In commercial lighting systems no power credits are offered for automatic controls; however, the use of automatic lighting controls is mandatory.

MANDATORY MEASURES

Similar to indoor lighting, mandatory features and devices must be included in all outdoor lighting project documentation, whenever applicable. The mandatory measures also require the performance of equipment to be certified by the manufacturers and that fixtures rated 100 W or larger must have high efficiency, otherwise they are required to be controlled by a motion sensor. Fixtures with lamps rated 175 W or more must incorporate directional baffles to direct the light toward the ground.

Fixture certification   Manufacturers of lighting control products are required to certify the performance of their products with the CEC. Lighting designers and engineers must assume responsibility to specify products that meet these requirements. As a rule, inspectors and code enforcement officials are also required to verify that the lighting controls specified carry CEC certification. The certification requirement applies to all lighting control equipment and devices such as photocontrols, astronomical time switches, and automatic controls.

Control devices are also required to have instructions for installation and startup calibration and must be installed in accordance with the manufacturer's directives. The control equipment and devices are required to have a visual or audio status signal that activates upon malfunction or failure.

Minimum lamp efficiency   All outdoor fixtures with lamps rated over 100 W must either have a lamp efficiency of at least 60 lumens per watt (lm/W) or must be controlled by a motion sensor. Lamp efficiencies are rated by the initial lamp lumens divided by the rated lamp power (W), without including auxiliary devices such as ballasts.

Fixtures that operate by mercury vapor principles and larger-wattage incandescent lamps do not meet these efficiency requirements. On the other hand, most linear fluorescent, metal halide, and high-pressure sodium lamps have a lamp efficiency greater than 60 lm/W and do comply with the requirements.

The minimum lamp efficiency does not apply to lighting regulated by a health or life safety statute, ordinance, or regulation, which includes, but is not limited to, emergency lighting. Also excluded are fixtures used around swimming pools; water features; searchlights or theme lighting used in theme parks, film, or live performances; temporary outdoor lighting; light-emitting diodes (LED); and neon and cold cathode lighting.

Cutoff luminaires   Outdoor luminaries that use lamps rated greater than 175 W in parking lots; hardscapes, outdoor dining, and outdoor sales areas are required to be fitted with cutoff-type baffles or filters.

Luminaries used must be specifically rated as "cutoff" in a photometric test report. A cutoff-type luminaire is defined as one where no more than 2.5 percent of the light output extends above the horizon 90 degrees or above the nadir and no more than 10 percent of the light output is at or above a vertical angle of 80 degrees above the nadir. The nadir is a point in the direction straight down, as would be indicated by a plumb line. Ninety degrees above the nadir is horizontal. Eighty degrees above the nadir is 10 degrees below horizontal.

Solar Power Facts

- Recent analysis by the U.S. Department of Energy showed that by the year 2025 half of new U.S. electricity generation could come from the sun.
- Based on current U.S. Department of Energy analysis current known worldwide crude oil reserves at the present rate of consumption will last about 50 years.
- In the United States 89 percent of the energy budget is based on fossil fuels.
- In 2000, the United States generated only 4 gigawatts (1 GW is 1000 MW) of solar power. By the year 2030, it is estimated to be 200 GW.
- A typical nuclear power plant generates about 1 GW of electric power, which is equal to 5 GW of solar power (daily power generation is limited to an average of 5 to 6 hours per day).
- Global sales of solar power systems have been growing at a rate of 35 percent in the past few years.
- It is projected that by the year 2020, the United States will be producing about 7.2 GW of solar power.
- The shipment of U.S. solar power systems has fallen by 10 percent annually but has increased by 45 percent throughout Europe.

- The annual sale growth of PV systems globally has been 35 percent.
- The present cost of solar power modules on the average is $2.33/W. By 2030 it should be about $0.38/W.
- World production of solar power is 1 GW/year.
- Germany has a $0.50/W grid feed incentive, which will be valid for the next 20 years. The incentive is to be decreased by 5 percent per year.
- In the past few years, Germany has installed 130 MW of solar power per year.
- Japan has a 50 percent subsidy for solar power installations of 3- to 4-kW systems and has about 800 MW of grid-connected solar power systems. Solar power in Japan has been in effect since 1994.
- California, in 1996, set aside $540 million for renewable energy, which has provided a $4.50 to $3.00 per watt buyback as a rebate.
- In the years 2015 through 2024, it is estimated that California could produce an estimated $40 billion of solar power sales.
- In the United States, 20 states have a solar rebate program. Nevada and Arizona have set aside a state budget for solar programs.
- Total U.S. production has been just about 18 percent of global production.
- For each megawatt of solar power we employ 32 people.
- A solar power collector in the southwest United States, sized 100×100 miles, could produce sufficient electric power to satisfy the country's yearly energy needs.
- For every kilowatt of power produced by nuclear or fossil fuel plants, ½ gallon of water is used for scrubbing, cleaning, and cooling. Solar power requires practically no water usage.
- Most sustainable energy technologies require less organizational infrastructure and less human power and therefore may result in human power resource reallocation.
- Significant impacts of solar power cogeneration are that it
  - Boosts economic development
  - Lowers the cost of peak power
  - Provides greater grid stability
  - Lowers air pollution
  - Lowers greenhouse gas emissions
  - Lowers water consumption and contamination
- A mere 6.7-mi/gal efficiency increase in cars driven in the United States could offset our share of imported Saudi oil.
- Types of solar power technology at present:
  - Crystalline
  - Polycrystalline
  - Amorphous
  - Thin- and thick-film technologies
- Types of solar power technology in the future:
  - Plastic solar cells
  - Nanostructured materials
  - Dye-synthesized cells

5

LEED™—LEADERSHIP IN ENERGY AND ENVIRONMENTAL DESIGN

Energy Use and the Environment

Ever since the creation of tools, the formation of settlements, and the advent of progressive development technologies, humankind has consistently harvested the abundance of energy that has been accessible in various forms. Up until the eighteenth-century industrial revolution, energy forms used by humans were limited to river or stream water currents, tides, solar, wind, and to very small degree geothermal energy, none of which had an adverse effect on the ecology.

Upon the discovery and harvesting of steam power and the development of steam-driven engines, humankind resorted to the use of fossil fuels, and commenced the unnatural creation of air, soil, water, and atmospheric pollutants with increasing acceleration to a degree that fears about the sustenance of life on our planet under the prevailing pollution and waste management control has come into focus.

Since global material production is made possible by the use of electric power generated from the conversion of fossil fuels, continued growth of the human population and the inevitable demand for materials within the next couple of centuries, if not mitigated, will tax the global resources and this planet's capacity to sustain life as we know it.

To appreciate the extent of energy use in human-made material production, we must simply observe that every object used in our lives from a simple nail to a supercomputer is made using pollutant energy resources. The conversion of raw materials to finished products usually involves a large number of energy-consuming processes, but products made using recycled materials such as wood, plastics, water, paper, and metals require fewer process steps and therefore less pollutant energy.

In order to mitigate energy waste and promote energy conservation, the U.S. Department of Energy, Office of Building Technology, founded the U.S. Green Building Council. The Council was authorized to develop design standards that provide for improved environmental and economic performance in commercial buildings by the use of established or advanced industry standards, principles, practices, and materials. It should be noted that the United States, with 5 percent of the world population, presently consumes 25 percent of the global energy resources.

The U.S. Green Building Council introduced the Leadership in Energy and Environmental Design (LEED™) rating system and checklist. This system establishes qualification and rating standards that categorize construction projects with certified designations, such as silver, gold, and platinum. Depending on adherence to the number of points specified in the project checklist, a project may be bestowed recognition and potentially a set amount of financial contribution by state and federal agencies.

Essentially the LEED™ guidelines discussed in this chapter, in addition to providing design guidelines for energy conservation, are intended to safeguard the ecology and reduce environmental pollution resulting from construction projects. There are many ways to analyze the benefits of LEED™ building projects. In summary, green building design is about productivity. A number of studies, most notably a study by Greg Kats of Capital-E, have validated the productivity value.

There are also a number of factors that make up this analysis. The basic concept is that if employees are happy in their workspace, such as having an outside view and daylight in their office environment, and a healthy environmental quality, they become more productive.

State of California Green Building Action Plan

The following is adapted from the detailed direction that accompanies the California governor's executive order regarding the Green Building Action Plan, also referred to as Executive Order S-20-04. The original publication, which is a public domain document can be found on the Californian Energy Commission's Web pages.

PUBLIC BUILDINGS

State buildings   All employees and all state entities under the governor's jurisdiction must immediately and expeditiously take all practical and cost-effective measures to implement the following goals specific to facilities owned, funded, or leased by the state.

Green buildings   The U.S. Green Building Council (USGBC) has developed green building rating systems that advance energy and material efficiency and sustainability, known as Leadership in Energy and Environmental Design for New Construction and Major Renovations (LEED™-NC) and LEED™ Rating System for Existing Buildings (LEED™-EB).

All new state buildings and major renovations of 10,000 ft^2 and over and subject to Title 24 must be designed, constructed, and certified by LEED™-NC Silver or higher, as described later.

Life cycle cost assessments, defined later in this section, must be used in determining cost-effective criteria. Building projects less than 10,000 ft^2 must use the same design standard, but certification is not required.

The California Sustainable Building Task Force (SBTF) in consultation with the Department of General Services (DGS), Department of Finance (DoF), and the California Energy Commission (CEC) is responsible for defining a life-cycle cost assessment methodology that must be used to evaluate the cost effectiveness of building design and construction decisions and their impact over a facility's life cycle.

Each new building or large renovation project initiated by the state is also subject to a clean on-site power generation.

All existing state buildings over 50,000 ft^2 must meet LEED^{TM}-EB standards by no later than 2015 to the maximum extent of cost effectiveness.

Energy efficiency   All state-owned buildings must reduce the volume of energy purchased from the grid by at least 20 percent by 2015 as compared to a 2003 baseline. Alternatively, buildings that have already taken significant efficiency actions must achieve a minimum efficiency benchmark established by the CEC.

Consistent with the executive order, all state buildings are directed to investigate "demand response" programs administered by utilities, the California Power Authority, to take advantage of financial incentives in return for agreeing to reduce peak electrical loads when called upon, to the maximum extent cost effective for each facility.

All occupied state-owned buildings, beginning no later than July 2005, must use the energy-efficiency guidelines established by the CEC.

All state buildings over 50,000 ft^2 must be retrocommissioned, and then recommissioned on a recurring 5-year cycle, or whenever major energy-consuming systems or controls are replaced. This is to ensure that energy and resource-consuming equipment are installed and operated at optimal efficiency. State facility leased spaces of 5000 ft^2 or more must also meet minimum U.S. EPA Energy Star standards guidelines.

Beginning in the year 2008, all electrical equipment, such as computers, printers, copiers, refrigerator units, and air-conditioning systems, that is purchased or operated by state buildings and state agencies must be Energy Star rated.

Financing and execution   The DoF, in consultation with the CEC, the State Treasurer's Office, the DGS, and financial institutions will facilitate lending mechanisms for resource efficiency projects. These mechanisms will include the use of the life cycle cost methodology and will maximize the use of outside financing, loan programs, revenue bonds, municipal leases, and other financial instruments. Incentives for cost-effective projects will include cost sharing of at least 25 percent of the net savings with the operating department or agency.

Schools

New school construction   The Division of State Architect (DSA), in consultation with the Office of Public School Construction and the CEC in California, was mandated to develop technical resources to enable schools to be built with energy-sufficient resources.

As a result of this effort, the state designated the Collaborative for High Performance Schools (CHPS) criteria as the recommended guideline. The CHPS is based on LEED™ and was developed specifically for kindergarten to grade 12 schools.

COMMERCIAL AND INSTITUTIONAL BUILDINGS

This section also includes private sector buildings, state buildings, and schools. The California Public Utilities Commission (CPUC) is mandated to determine the level of ratepayer-supported energy efficiency and clean energy generation so as to contribute toward the 20 percent efficiency goal.

LEADERSHIP

Mission of green action team   The state of California has established an interagency team know as the Green Action Team, which is composed of the director of the Department of Finance, and the secretaries of Business, Transportation, and Housing, with a mission to oversee and direct progress toward the goals of the Green Building Order.

LEED™

LEED™ project sustainable building credits and prerequisites are based on LEED™-NC2.1 New Construction. There are additional versions of LEED™ that have been adopted or are currently in development that address core or shell, commercial interiors, existing buildings, homes, and neighborhood development.

SUSTAINABLE SITES

Sustainable site prerequisite—construction activity pollution prevention
The intent of this prerequisite is to control and reduce top erosion and reduce the adverse impact on the surrounding water and air quality.

Mitigation measures involve the prevention of the loss of topsoil during construction by means of a storm water system runoff as well as the prevention of soil displacement by gust wind. It also imposes measures to prevent sedimentation of storm sewer systems by sand dust and particulate matter.

Some suggested design measures to meet these requirements include deployment of strategies, such as temporary or permanent seeding, silt fencing, sediment trapping, and sedimentation basins that could trap particulate material.

Site selection, credit no. 1   The intent of this credit is to prevent and avoid development of a site that could have an adverse environmental impact on the project location surroundings.

Sites considered unsuitable for construction include prime farmlands; lands that are lower than 5 ft above the elevation of established 100-year flood areas, as defined by the Federal Emergency Management Agency (FEMA); lands that are designated habitats for endangered species; lands within 100 ft of any wetland; or a designated public parkland.

To meet site selection requirements, it is recommended that the sustainable project buildings have a reasonably minimal footprint to avoid site disruption. Favorable design practices must involve underground parking and neighbor-shared facilities.

The point weight granted for this measure is 1.

Development density and community connectivity, credit no. 2   The intent of this requirement is to preserve and protect green fields and animal habitats by means of increasing the urban density, which may also have a direct impact on reduction of urban traffic and pollution.

A specific measure suggested includes project site selection within the vicinity of an urban area with high development density.

The point weight granted for this measure is 1.

Brownfield redevelopment, credit no. 3   The main intent of this credit is the use and development of projects on lands that have environmental contamination. To undertake development under this category, the Environmental Protection Agency (EPA) must provide a sustainable redevelopment remediation requirement permit.

Projects developed under Brownfield redevelopment are usually offered state, local, and federal tax incentives for site remediation and cleanup.

The point weight granted for each of the four measures is 1.

Alternative transportation, credit no. 4   The principal objective of this measure is to reduce traffic congestion and minimize air pollution.

Measures recommended include locating the project site within 1/2 mile of a commuter train, subway, or bus station; construction of a bicycle stand and shower facilities for 5 percent of building habitants; and installation of alternative liquid and gas fueling stations on the premises. An additional requisite calls for a preferred parking facility for car pools and vans that serve 5 percent of the building occupants, which encourages transportation sharing.

The point weight granted for this measure is 1.

Site development, credit no. 5   The intent of this measure is to conserve habitats and promote biodiversity.

Under this prerequisite, one point is provided for limiting earthwork and the destruction of vegetation beyond the project or building perimeter, 5 ft beyond walkways and roadway curbs, 25 ft beyond previously developed sites, and restoration of 50 percent of open areas by planting of native trees and shrubs.

Another point under this section is awarded for 25 percent reduction of a building footprint by what is allowed by local zoning ordinances.

Design mitigations for meeting the preceding goals involve underground parking facilities, ride-sharing among habitants, and restoring open spaces by landscape architecture planning that uses local trees and vegetation.

Storm water management, credit no. 6   The objective of this measure involves preventing the disruption of natural water flows by reducing storm water runoffs and promoting on-site water filtration that reduces contamination.

Essentially these requirements are subdivided into two categories. The first one deals with the reduction of the net rate and quantity of storm water runoff that is caused by the imperviousness of the ground, and the second relates to measures undertaken to remove up to 80 percent of the average annual suspended solids associated with the runoff.

Design mitigation measures include maintenance of natural storm water flows that include filtration to reduce sedimentation. Another technique used is construction of roof gardens that minimize surface imperviousness and allow for storage and reuse of storm water for nonpotable uses such as landscape irrigation and toilet and urinal flushing.

The point weight granted for each of the two categories discussed here is 1.

Heat island effect, credit no. 7   The intent of this requirement is to reduce the microclimatic thermal gradient difference between the project being developed and adjacent lands that have wildlife habitats.

Design measures to be undertaken include shading provisions on site surfaces such as parking lots, plazas, and walkways. It is also recommended that site or building colors a reflectance of at least 0.3 and that 50 percent of parking spaces be of the underground type.

Another design measure suggests use of Energy Star high-reflectance and high-emissivity roofing.

To meet these requirements the project site must feature extensive landscaping. In addition to minimizing building footprints, it is also suggested that building rooftops have vegetated surfaces and that gardens and paved surfaces be of light-colored materials to reduce heat absorption.

The point weight granted for each of the two categories discussed here is 1.

Light pollution reduction, credit no. 8   Essentially, this requirement is intended to eliminate light trespass from the project site, minimize the so-called night sky access, and reduce the impact on nocturnal environments. This requirement becomes mandatory for projects that are within the vicinity of observatories.

To comply with these requirements, site lighting design must adhere to Illumination Engineering Society of North America (IESNA) requirements. In California, indoor and outdoor lighting design should comply with California Energy Commission (CEC) Title 24, 2005 requirements.

Design measures to be undertaken involve the use of luminaries and lamp standards equipped with filtering baffles and low-angle spotlights that could prevent off-site horizontal and upward light spillage.

The point weight granted for this measure is 1.

WATER EFFICIENCY MEASURES

Water-efficient landscaping, credit no. 1   Basically, this measure is intended to minimize the use of potable water for landscape irrigation purposes.

One credit is awarded for the use of high-efficiency irrigation management control technology. A second credit is awarded for the construction of special reservoirs for the storage and use of rainwater for irrigation purposes.

Innovative water technologies, credit no. 2   The main purpose of this measure is to reduce the potable water demand by a minimum of 50 percent. Mitigation involves the use of gray water by construction of on-site natural or mechanical wastewater treatment systems that could be used for irrigation and toilet or urinal flushing. Consideration is also given to the use of waterless urinals and storm water usage.

The point weight granted for this measure is 1.

Water use reduction, credit no. 3   The intent of this measure is to reduce water usage within buildings and thereby minimize the burden of local municipal supply and water treatment.

This measure provides one credit for design strategies that reduce building water usage by 20 percent and a second credit for a reducing water use by 30 percent.

Design measures to meet this requirement involve the use of waterless urinals, high-efficiency toilet and bathroom fixtures, and nonpotable water for flushing toilets.

ENERGY AND ATMOSPHERE

Fundamental commissioning of building energy systems, prerequisite no. 1   This requirement is a prerequisite intended to verify intended project design goals and involves design review verification, commissioning, calibration, physical verification of installation, and functional performance tests, all of which are to be presented in a final commissioning report.

The point weight granted for this prerequisite is 1.

Minimum energy performance, prerequisite no. 2   The intent of this prerequisite is to establish a minimum energy efficiency standard for a building. Essentially, the basic building energy efficiency is principally controlled by mechanical engineering heating and air-conditioning design performance principles, which are outlined by ASHRAE/IESNA and local municipal or state codes. The engineering design procedure involves so-called building envelop calculations, which maximize energy performance. Building envelop computations are achieved by computer simulation models that quantify energy performance as compared to a baseline building.

Fundamental refrigerant management, prerequisite no. 3   The intent of this measure is the reduction of ozone-depleting refrigerants used in HVAC systems.

Mitigation involves replacement of old HVAC equipment with equipment that does not use CFC refrigerants.

Optimize energy performance, credit no. 1   The principal intent of this measure is to increase levels of energy performance above the prerequisite standard in order to reduce environmental impacts associated with excessive energy use. The various credit

TABLE 5.1   NEW AND OLD BUILDING CREDIT POINTS

% INCREASE IN ENERGY PERFORMANCE		CREDIT POINTS
NEW BUILDINGS	EXISTING BUILDINGS	
14	7	2
21	14	4
28	21	6
35	28	8
42	35	10

levels shown in Table 5.1 are intended to reduce the design energy budget for the regulated energy components described in the requirements of the ASHRAE/IESNA standard. The energy components include building envelope, hot water system, and other regulated systems defined by ASHRAE standards.

Similarly to previous design measures, computer simulation and energy performance modeling software is used to quantify the energy performance as compared to a baseline building system.

On-site renewable energy, credit no. 2   The intent of this measure is to encourage the use of sustainable or renewable energy technologies such as solar photovoltaic cogeneration, solar power heating and air-conditioning, fuel cells, wind energy, landfill gases, and geothermal and other technologies discussed in various chapters of this book. The credit award system under this measure is based on a percentage of the total energy demand of the building. See Table 5.2.

Additional commissioning, credit no. 3   This is an enforcement measure to verify whether the designed building is constructed and performs within the expected or intended parameters. The credit verification stages include preliminary design documentation review, construction documentation review when construction is completed,

TABLE 5.2   ENERGY SAVING CREDIT POINTS

% TOTAL ENERGY SAVINGS	CREDIT POINTS
5	1
10	2
20	3

selective submittal document review, establishment of commissioning documentation, and finally the post-occupancy review.

It should be noted that all these reviews must be conducted by an independent commissioning agency.

Point weight credit awarded for each of the four categories is 1.

Enhanced refrigerant management, credit no. 4   This measure involves installation of HVAC, refrigeration, and fire suppression equipment that do not use hydrochlorofluorocarbon (HCFC) agents.

Point weight credit awarded for this category is 1.

Measurement and verification, credit no. 5   This requirement is intended to optimize building energy consumption and provide a measure of accountability. The design measures implemented include the following:

- Lighting system control, which may consist of occupancy sensors, photocells for control of daylight harvesting, and a wide variety of computerized systems that minimize the energy waste related to building illumination. A typical discussion of lighting control is covered under California Title 24 energy conservation measures, with which all building lighting designs must comply within the state.
- Compliance of constant and variable loads, which must comply with motor design efficiency regulations.
- Motor size regulation that enforces the use of variable-speed drives (VFD).
- Chiller efficiency regulation measures that meet variable-load situations.
- Cooling load regulations.
- Air and water economizer and heat recovery and recycling.
- Air circulation, volume distribution, and static pressure in HVAC applications.
- Boiler efficiency.
- Building energy efficiency management by means of centralized management and control equipment installation.
- Indoor and outdoor water consumption management.

Point weight credit awarded for each of the four categories is 1.

Green power, credit no. 6   This measure is intended to encourage the use and purchase of grid-connected renewable energy cogenerated energy derived from sustainable energy such as solar, wind, geothermal, and other technologies described throughout this book. A purchase-and-use agreement of this so-called green power is usually limited to a minimum of a 2-year contract. The cost of green energy use is considerably higher than that of regular energy. Purchasers of green energy, who participate in the program, are awarded a Green-e products certification.

MATERIAL AND RESOURCES

Storage and collection of recyclables, prerequisite no. 1   This prerequisite is a measure to promote construction material sorting and segregation for recycling and landfill deposition. Simply put, construction or demolition materials such as glass, iron, concrete, paper, aluminum, plastics, cardboard, and organic waste must be separated and stored in a dedicated location within the project for further recycling.

Building reuse, credit no. 1   The intent of this measure is to encourage the maximum use of structural components of an existing building that will serve to preserve and conserve a cultural identity, minimize waste, and reduce the environmental impact. It should be noted that another significant objective of this measure is to reduce the use of newly manufactured material and associated transportation which ultimately results into energy use and environmental pollution.

Credit is given for implementation of the following measures:

■ One credit for maintenance and reuse of 75 percent of the existing building.
■ Two credits for maintenance of 100 percent of the existing building structure shell and the exterior skin (windows excluded).
■ Three credits for 100 percent maintenance of the building shell and 50 percent of walls, floors, and the ceiling.

This simply means that the only replacements will be of electrical, mechanical, plumbing, and door and window systems, which essentially boils down to a remodeling project.

Construction waste management, credit no. 2   The principal purpose of this measure is to recycle a significant portion of the demolition and land-clearing materials, which calls for implementation of an on-site waste management plan. An interesting component of this measure is that the donation of materials to a charitable organization also constitutes waste management.

The two credits awarded under this measure include one point for recycling or salvaging a minimum of 50 percent by weight of demolition material and two points for salvage of 75 percent by weight of the construction and demolition debris and materials.

Material reuse, credit no. 3   This measure is intended to promote the use of recycled materials, thus, reducing the adverse environmental impact caused by manufacturing and transporting new products. For using recycled materials in a construction, one credit is given to the first 5 percent and a second point for a 10 percent total use. Recycled materials used could range from wall paneling, cabinetry, bricks, construction wood, and even furniture.

Recycled content, credit no. 4   The intent of this measure is to encourage the use of products that have been constructed from recycled material. One credit is given if 25 percent of the building material contains some sort of recycled material or 40 percent of minimum by weight use of so-called postindustrial material content. A second point is awarded for an additional 25 percent recycled material use.

Regional materials, credit no. 5   The intent of this measure is to maximize the use of locally manufactured products, which minimizes transportation and thereby reduces environmental pollution.

One point is awarded if 20 percent of the material is manufactured within 500 miles of the project and another point is given if the total recycled material use reaches 50 percent. Materials used in addition to manufactured goods also include those that are harvested, such as rock and marble from quarries.

Rapidly renewable materials, credit no. 6   This is an interesting measure that encourages the use of rapidly renewable natural and manufactured building materials. Examples of natural materials include strawboards, woolen carpets, bamboo flooring, cotton-based insulation, and poplar wood. Manufactured products may consist of linoleum flooring, recycled glass, and concrete as an aggregate.

Point weight awarded for this measure is 1.

Certified wood, credit no. 7   The intent of this measure is to encourage the use of wood-based materials. One point is credited for the use of wood-based materials such as structural beams and framing and flooring materials that are certified by Forest Council Guidelines (FSC).

INDOOR ENVIRONMENTAL QUALITY

Minimum indoor air quality (iaq) performance, prerequisite no. 1   This prerequisite is established to ensure indoor air quality performance to maintain the health and wellness of the occupants. One credit is awarded for adherence to ASHRAE building ventilation guidelines such as placement of HVAC intakes away from contaminated air pollutant sources such as chimneys, smoke stacks, and exhaust vents.

Point weight awarded for this measure is 1.

Environmental tobacco smoke (ETS) control, prerequisite no. 2   This is a prerequisite that mandates the provision of dedicated smoking areas within buildings, which can effectively capture and remove tobacco and cigarette smoke from the building. To comply with this requirement, designated smoking rooms must be enclosed and designed with impermeable walls and have a negative pressure (air being sucked in rather than being pushed out) compared to the surrounding quarters. Upon completion of construction, designated smoking rooms are tested by the use of a tracer gas method defined by ASHRAE standards, which impose a maximum of 1 percent tracer gas escape from the ETS area.

This measure is readily achieved by installing a separate ventilation system that creates a slight negative room pressure.

Point weight awarded for this measure is 1.

Outdoor air delivery monitoring, credit no. 1   As the title implies, the intent of this measure is to provide an alarm monitoring and notification system for indoor and outdoor spaces. The maximum permitted carbon dioxide level is 530 parts per million.

To comply with the measure HVAC systems are required to be equipped with a carbon dioxide monitoring and annunciation system, which is usually a component of building automation systems.

Point weight awarded for this measure is 1.

Increased ventilation, credit no. 2   This measure is intended for HVAC designs to promote outdoor fresh air circulation for building occupants' health and comfort. A credit of one point is awarded for adherence to the ASHRAE guideline for naturally ventilated spaces where air distribution is achieved in a laminar flow pattern. Some HVAC design strategies used include displacement and low-velocity ventilation, plug flow or under-floor air delivery, and operable windows that allow natural air circulation.

Construction (IAQ) air quality management plan, credit no. 3   This measure applies to air quality management during renovation processes to ensure that occupants are prevented from exposure to moisture and air contaminants. One credit is awarded for installation of absorptive materials that prevent moisture damage and filtration media to prevent space contamination by particulates and airborne materials.

A second point is awarded for a minimum of flushing out of the entire space, by displacement, with outside air for a period of 2 weeks prior to occupancy. At the end of the filtration period a series of test are performed to measure the air contaminants.

Low-emitting materials, credit no. 4   This measure is intended to reduce indoor air contaminants resulting from airborne particulates such as paints and sealants. Four specific areas of concern include the following: (1) adhesives, fillers, and sealants; (2) primers and paints; (3) carpet; and (4) composite wood and agrifiber products that contain urea-formaldehyde resins.

Each of these product applications are controlled by various agencies such as the California Air Quality Management District, Green Seal Council, and Green Label Indoor Air Quality Test Program.

Point weight awarded for each of the four measures is 1.

Indoor chemical and pollutant source control, credit no. 5   This is a measure to prevent air and water contamination by pollutants. Mitigation involves installation of air and water filtration systems that absorb chemical particulates entering a building. Rooms and areas such as document reproduction rooms, copy rooms, and blueprint quarters, which generate trace air pollutants, are equipped with dedicated air exhaust and ventilation systems that create negative pressure. Likewise, water circulation, plumbing, and liquid waste disposal are collected in an isolated container for special disposal. This measure is credited a single point.

Controllability of systems, credit no. 6   The essence of this measure is to provide localized distributed control for ventilation, air conditioning, and lighting. One point is awarded for autonomous control of lighting and control for each zone covering 200 ft^2 of area with a dedicated operable window within 15 ft of the perimeter wall. A second point is given for providing air and temperature control for 50 percent of the nonperimeter occupied area. Both of these measures are accomplished by centralized or local area lighting control and HVAC building control systems. The above measures are intended to control lighting and air circulation. Each of the above two measures is awarded one point.

Thermal comfort, credit no. 7   The intent of this measure is to provide environmental comfort for building occupants. One credit is awarded for thermal and humidity control for specified climate zones and another for the installation of a permanent central temperature and humidity monitoring and control system.

Daylight and views, credit no. 8   Simply stated, this measure promotes architectural space design that allows for maximum outdoor views and interior sunlight exposure. One credit is awarded for spaces that harvest indirect daylight for 75 percent of spaces occupied for critical tasks. A second point is awarded for direct sight of vision glazing from 90 percent of normally occupied work spaces. It should be noted that copy rooms, storage rooms, mechanical equipment rooms, and low-occupancy rooms do not fall into these categories. In other words 90 percent of the work space is required to have direct sight of a glazing window.

Some architectural design measures taken to meet these requirements include building orientation, widening of building perimeter, deployment of high-performance glazing windows, and use of solar tubes.

INNOVATION AND DESIGN PROCESS

Innovation in design, credit no. 1   This measure is in fact a merit award for an innovative design that is not covered by LEED™ measures and in fact exceeds the required energy efficiency and environmental pollution performance milestone guidelines. The four credits awarded for innovation in design are: (1) identification of the design intent, (2) meeting requirements for compliance, (3) proposed document submittals that demonstrate compliance, and (4) a description of the design approach used to meet the objective.

LEED™-accredited professional, credit no. 2   One point is credited to the project for a design team that has a member who has successfully completed the LEED™ accreditation examination.

CREDIT SUMMARY

Sustainable sites	10 points
Water efficiency	3 points
Energy and atmosphere	8 points
Material and resources	9 points
Indoor environmental quality	10 points
Innovation in design	2 points

The grand total is 42 points.

OPTIMIZED ENERGY PERFORMANCE SCORING POINTS

Additional LEED™ points are awarded for building efficiency levels, as shown in Table 5.3.

TABLE 5.3 LEED™ BUILDING EFFICIENCY POINTS		
% INCREASE IN ENERGY PERFORMANCE		
NEW BUILDINGS	EXISTING BUILDINGS	POINTS
15	5	1
20	10	2
25	15	3
30	20	4
35	25	5
40	30	6
45	35	7
50	40	8
55	45	9
60	50	10

Project certification is based on the cumulated points, as shown in Table 5.4.

Los Angeles Audubon Nature Center— A LEED™-Certified Platinum Project

The following project is the highest-ranked LEED™-certified building by the U.S. Green Building Council within the United States. The pilot project, known as Debs Park Audubon Center, is a 282-acre urban wilderness that supports 138 species of birds. It is located in the center of the city of Los Angeles, California, and was commissioned on January 13, 2004. Based on the Building Rating System TM2.1, the project received a platinum rating, the highest possible. Figure 5.1 shows a view of the Los Angeles Audubon Center's roof mount solar PV system.

TABLE 5.4 LEED™ CERTIFICATION CATEGORIES AND ASSOCIATED POINTS	
LEED™ certified	26–32 points
Silver level	33–38 points
Gold level	39–51 points
Platinum level	52–69 points

Figure 5.1 Los Angeles Audubon Center. *Photograph courtesy of LA Audubon Society.*

The key to the success of the project lies in the design considerations given to all aspects of the LEED™ ranking criteria, which include sustainable building design parameters such as the use of renewable energy sources, water conservation, recycled building materials, and maintenance of native landscaping. The main office building of the project is entirely powered by an on-site solar power system that functions "off the grid." The building water purification system is designed such that it uses considerably less water for irrigation and bathrooms. To achieve the platinum rating the building design met 52 out of the total available 69 LEED™ energy conservation points outlined in Table 5.3.

The entire building, from the concrete foundation and rebars to the roof materials, was manufactured from recycled materials. For example, concrete-reinforcement rebars were manufactured from melted scrap metal and confiscated handguns. All wood material used in the construction of the building and cabinetry were manufactured from wheat board, sunflower board, and Mexican agave plant fibers.

A 26-kW roof-mount photovoltaic system provides 100 percent of the center's electric power needs. A 10-ton solar thermal cooling system installed by SUN Utility Network Inc., provides a solar air-conditioning system believed to be the first of its kind in southern California. The HVAC system provides the total air-conditioning needs of the office building. The combination of the solar power and the solar thermal air-conditioning system renders the project completely self-sustainable requiring no power from the power grid. Figure 5.2 shows a roof mount solar photovoltaic panel installation.

The cost of this pilot project upon completion was estimated to be about $15.5 million. At present, the project houses a natural bird habitat, exhibits, an amphitheater, and a hummingbird garden. The park also has a network of many hiking trails enjoyed by local residents.

The thermal solar air-conditioning system, which is used only in few countries, such as Germany, China, and Japan, utilizes an 800-ft² array of 408 Chinese-manufactured

Figure 5.2 **Los Angeles Audubon Center grid-independent solar power generation, system.** *Photograph courtesy of LA Audubon Society.*

Sunda vacuum tube solar collectors. Each tube measures 78-in long and has a 4-in diameter, and each encloses a copper heat pipe and aluminum nitride plates that absorb solar radiation. Energy trapped from the sun's rays heats the low-pressure water that circulates within and is converted into a vapor that flows to a condenser section. A heat exchanger compartment heats up an incoming circulating water pipe through the manifold which allows for the transfer of thermal energy from the solar collector to a 1200-gal insulated high-temperature hot-water storage tank. When the water temperature reaches 180°F, the water is pumped to a 10-ton Yamazaki single-effect absorption chiller. A lithium bromide salt solution in the chiller boils and produces water vapor that is used as a refrigerant, which is subsequently condensed; its evaporation at a low pressure produces the cooling effect in the chiller. Figure 5.3 depicts the solar thermal heating and air-conditioning system diagram.

This system also provides space heating in winter and hot water throughout the year. Small circulating pumps used in the chiller are completely energized by the solar photovoltaic system.

It is estimated that the solar thermal air-conditioning and heating system relieves the electric energy burden by as much as 15 kW. The cost of energy production at the Audubon Center is estimated to be $0.04 per kilowatt, which is substantially lower than the rates charged by the city of Los Angeles Department of Water and Power. It should be noted that the only expense in solar energy cost is the minimal maintenance and investment cost, which will be paid off within a few years. The following are architectural and LEED™ design measures applied in the Los Angeles Audubon Center.

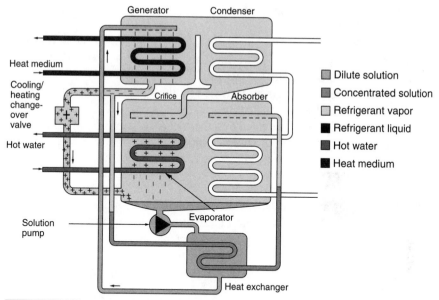

Figure 5.3 Los Angeles Audubon Center, solar thermal heating and air-conditioning system. *Graph courtesy of SUN Utility Network, Inc.*

Architectural green design measures

■ Exterior walls are ground-faced concrete blocks, exposed on the inside, insulated and stuccoed on the outside.

■ Steel rebars have 97 percent recycled content.

■ 25 percent fly ash is used in cast-in-place concrete and 15 percent in grout for concrete blocks.

■ More than 97 percent of construction debris is recycled.

■ Aluminum-framed windows use 1-in-thick clear float glass with a low-emittance (low-E) coating.

■ Plywood, redwood, and Douglas fir members for pergolas are certified by the Forestry Stewardship Council.

■ Linoleum countertops are made from linseed oil and wood flour and feature natural jute tackable panels that are made of 100 percent recycled paper.

■ Burlap-covered tackable panels are manufactured from 100 percent recycled paper.

■ Batt insulation is formaldehyde-free mineral fiber with recycled content.

■ Cabinets and wainscot are made of organic wheat boards and urea-formaldehyde-free medium-density fiberboard.

■ Engineered structural members are urea-formaldehyde-free.

■ Synthetic gypsum boards have 95 percent recycled content.

■ Ceramic tiles have recycled content.

■ Carpet is made of sisal fiber extracted from the leaves of the Mexican agave plant.

Green energy operating system

- 100 percent of the electric power for lighting is provided by an off-grid polycrystalline photovoltaic solar power system. The system also includes a 3- to 5-day battery-backed power storage system.
- To balance the electric power provided by the sun, all lighting loads are connected or disconnected by a load-shedding control system.
- Heating and cooling is provided by a thermal absorption cooling and heating system.
- Windows open to allow for natural ventilation.

Green water system

- All the wastewater is treated on-site without a connection to the public sewer system.
- Storm water is kept on-site and diverted to a water-quality treatment basin before being released to help recharge groundwater.
- Two-stage, low-flow toilets are installed throughout the center.
- The building only uses 35 percent of the city water typically consumed by comparable structures.

TriCom Office Building

This project is a commercial and industrial use type building, which has 23,300 ft² of building space. The TriCom Building incorporates three commercial uses of space, namely, executive office, showroom, and warehouse space.

The project was completed in 2003 at a cost of about $3.3 million. It was designed by Caldwell Architects and was constructed by Pozzo Construction. The solar power was designed and installed by Sun Utility, Solar Company.

The TriCom Building has received LEED™ certification by the U.S. Green Building Council, and it is the first certified building in Pasadena. It is a prototype for Pasadena and was undertaken with the partnerships of Pasadena Water and Power, a local landscape design school, and other entities.

The project site is in the expanded enterprise zone of the city of Pasadena, which was part of the redevelopment project. To promote alternative transportation, the building is located ¼ mile from nine bus lines and ¾ mile from the light rail. A bicycle rack and an electric vehicle charger is on site to encourage additional alternative modes of transportation.

The landscape is composed of drought-tolerant plants eliminating the need for a permanent irrigation system and thus conserving local and regional potable water resources. Significant water savings are realized by faucet aerators and dual-flush and low-gallons per flush (GPF) toilets.

A 31-kW roof-mount photovoltaic solar power cogeneration system, installed by Solar Webb Inc., provides over 50 percent of the building's demand for electric energy. The building construction includes efficient lighting controls, increased insulation, dual-glazed windows, and Energy Star–rated appliances that reduce energy consumption. Figure 5.4 shows the roof-mount solar PV system.

Figure 5.4 TriCom roof-mounted solar power cogeneration system.
Photograph courtesy of Solar Integrated Technologies, Los Angeles, CA.

Approximately 80 percent of the building material, such as concrete blocks, the rebar, and the plants were manufactured or harvested locally, thereby minimizing the environmental pollution impacts that result from transportation.

In various areas of the building flooring, the cover is made from raw, renewable materials such as linseed oil and jute. The ceiling tiles are also made from renewable materials such as cork. Reused material consisted of marble and doors from hotels, used tiles from showcase houses, and lighting fixtures that augment the architectural character.

The carpet used is made of 50 percent recycled content, the ceiling tiles are made from 75 percent recycled content, and the aluminum building signage is made from 94 percent recycled content. Figure 5.5 shows the inverter system assembly.

To enhance environmental air quality all adhesives, sealants, paints, and carpet systems contain little or no volatile organic compounds (VOCs), which makes the project comply with the most rigorous requirements for indoor air quality. Operable windows also provide the effective delivery and mixing of fresh air and support the health, safety, and comfort of the building occupants.

Warehouse, Rochester, New York

This LEED™ gold-rated facility is equipped with a lighting system that utilizes dc fluorescent ballasts, roof-integrated solar panels, occupancy sensors, and daylight sensors for the highest possible efficiency. The building, including the innovative

Figure 5.5 **TriCom inverter system assembly.** *Photograph courtesy of Solar Integrated Technologies, Los Angeles, CA.*

lighting design, was designed by William McDonough and Partners of Charlottesville, Virginia.

The facility has 6600 ft² of office space and 33,000 ft² of warehouse. The warehouse roof is equipped with skylights and 21 kW of solar panels bonded to the roof material (SR2001 amorphous panels by Solar Integrated Technologies). A canopy in the office area is equipped with 2.1 kW of Sharp panels.

The power from the solar panels is distributed in three ways:

- 2.2 kW is dedicated to the dc lighting in the office.
- 11.5 kW powers the dc lights in the warehouse.
- 11.5 kW is not needed by the lighting system, so it is inverted to alternating current and used elsewhere in the building or sold back to the utility.

The entire system consists of 35 NPS-1000 Power Gateways from Nextek Power Systems in Hauppauge, NY. These devices take all the available power from the solar panels and send it directly to the lighting use without significant losses. Additional power, when needed at night or on cloudy days, is added from the grid.

In the office, six NPS-1000 Power Gateway modules power 198 T-8, four-foot fluorescent lamps, illuminating most areas at 1.1 W/ft². Each of the fixtures is equipped with a single high-efficiency dc ballast for every two lamps. Most of the fixtures are controlled by a combination of manual switches, daylight sensors, and occupancy sensors in 13 zones.

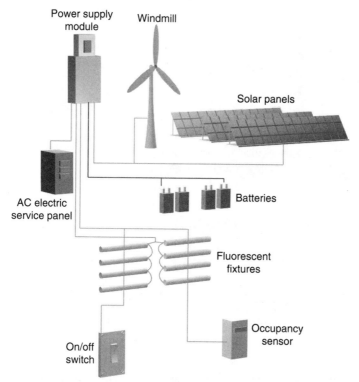

Figure 5.6 Nextek Power Systems solar power integration diagram.

In the warehouse area 29 NPS-1000 Power Gateway power 158, six-lamp T-8 fixtures. These fixtures have a low, medium, and high settings for two-, four-, and six lamp-type fixtures, which are dimmed by daylight, and occupancy sensors are located throughout the area. The goal of the control architecture is to maintain a lighting level of 0.74 W/ft^2, using daylight when available, whenever the area is occupied.

Figures 5.6 and 5.7 depict NEXTEK dc power system lighting controls.

The logic of the lighting system is designed for optimum efficiency. Sources of light and power are prioritized such that:

■ First, daylight from the skylights is used.
■ Second, if daylighting is not sufficient and the area is occupied, then power from the solar panels is added.
■ Third, if daylight and solar power is not enough, then the additional power required is taken from the grid.

A number of factors contribute to the value of this system:

■ Using the electricity generated by the solar panels to power the lighting eliminates inverter losses and improves efficiency by as much as 20 percent.

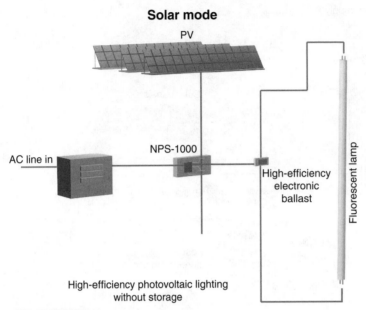

Figure 5.7 **Direct solar lighting.** *Courtesy of Nextek Power Systems.*

- The low-voltage control capability of the dc ballasts eliminate rectification losses and enables the innovative control system to be installed easily.
- Roof-integrated solar panels reduce installation costs and allow the cost of the roof to be recovered using a 5-year accelerated depreciation formula.

WEB-BASED MONITORING

The Nextek data collection and monitoring system provides a Web-based display of power generated, power used, and weather. Additionally, the system also displays performance data and identifies anomalies in the system, such as burned-out lamps and sensors that are not operating properly.

VALUE

The Green DC, or Green Distribution Center, has been a successful effort on this distribution center's part to bring more sustainable business practices into its facilities. The payback on its investment at the Rochester location, after rebates and accelerated depreciation, is approximately 12.6 years. The remaining system output will produce energy for 7.5 years, producing $60,000 in value at today's rates in Rochester. It is important to note that in areas where the avoided cost of peak power is higher than $0.10/kWh, this return of investment can drop to under 6 years, meaning that in those areas the facility would enjoy free peak power from the solar PV array for at least 14 years after the investment is returned. This equates to a $112,000 benefit at today's rates.

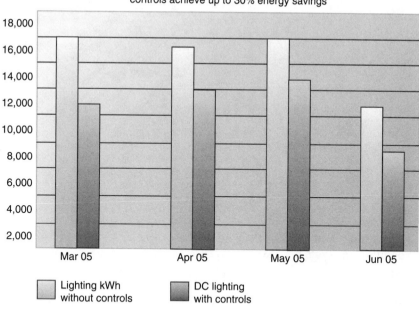

Results to date show that occupancy sensors and dimming controls achieve up to 30% energy savings

Lighting kWh without controls

DC lighting with controls

	kWh saved	$ Saved with controls	% kWh or $ saved
March 05	4,970	497.02	31%
April 05	3,458	345.80	22%
May 05	3,429	342.85	22%
June 05	3,248	324.75	30% Thru 6/19
	15,104	1,510.43	

- Adjustments made in June will enjure closer to 30% savings going forward.

- In other parts of the U.S. the cost of energy is over $0.20

Figure 5.8 **Occupancy sensors efficiency chart.** *Courtesy of Nextek Power Systems.*

Water and Life Museum, Hemet, California

Architect: Gangi Lehere Architects, 231 e. Palm Ave., Burbank, CA 91502
Electrical and Solar Power Systems Engineers: Vector Delta Design Group, Inc., Glendale, CA 91208

The Water and Life Museum design was a collaboration between architects Michael B. Lehrer FAIA and Mark Gangi AIA, partnering with Gangi Development's construction

Figure 5.9 **Water and Life Museum solar power cogeneration system.** *Photo courtesy of Lehrer & Gangi Architects and Vector Delta Design Group, Inc.*

division, an entity called Lehrer & Gangi Design Build, which is a sustainable think tank of complex architecture.

The project was designed as a sustainable campus and is located within a recreational park at the entrance to Metropolitan's Diamond Valley Lake. It is an example of a LEED™-rated sustainable project.

The project consists of a large composite plan of six buildings, including the Center for Water Education Museum, the Western Center of Archaeology and Paleontology Museum, a museum store, a museum café, and two conference room buildings, which are sited to produce a series of outdoor spaces from grand to intimate. A grand piazza, a campus way, and various courts strategically placed between buildings create a civic sense to the campus. Figure 5.9 shows a panoramic view of the solar power arrays of the Water and Life Museum, Hemet, California.

The project was conceived as a LEED™-designed silver category campus. If this is achieved, it will be the highest level attainable given the project's parameters.

LEED™ DESIGN CONSIDERATIONS

Sustainable sites   Many of the points listed in sustainable sites were not applicable to the Water and Life Museum project, such as urban redevelopment or brown fields. The site selected was excess property from the dam construction. The project is located on a recreation ground covering 1200 acres, which in the near future will become a recreation park that will include a golf course, a recreational lake, a swim and sports complex, and a series of bike trails, horse trails, and campgrounds. The museum complex is the gateway to the recreational grounds and is intended to become the civic center of the area.

The building is designed to accommodate bicycle storage for 5 percent of building occupants as well as to provide shower and locker facilities. The project design encourages alternative transportation to the site, which is intended to reduce negative environmental impacts from automobile use.

Because of the expected volume of pedestrian and bicycle visitors, vehicle parking spaces were reduced to provide adequate space for bicycle stands, which meet local zoning requirements.

The architecture of the grounds blends magnificent building shapes, open spaces, and interpretive gardens throughout the campus. The small footprint of the building occupying the open space of the land is a significant attribute that qualifies the project for LEED™ points since it resulted in reduced site disturbance.

Braided streams weaving throughout the site provide a thematic storey of water in southern California. The streams are also designed to mitigate storm water management for the site. The braided streams contain pervious surfaces that convey rainwater to the water table.

The parking grove of the project consists of shading trees and dual-colored asphalt. The remainder of the paving is light-colored, acid-washed concrete, and lithocrete of a light color. The roofs of the buildings are covered with single-ply white membrane and are shaded by solar panels. These light-shaded surfaces reduce the heat islanding effect.

The Water and Life Museum is located within the radius of the Palomar Observatory, which has mandatory light pollution restrictions. As a result, all lamppost fixtures are equipped with full cutoff fixtures, and shutoff timing circuits.

Water efficiency   The mission of the Center for Water Education is to transform its visitors into stewards of water. To this end, the campus is a showcase of water efficiency. The campus landscaping consists of California native plants. The irrigation systems deployed are state-of-the-art drip systems that use reclaimed water. Interpretive exhibits throughout the museum demonstrate irrigation technology from Native Americans to that which is satellite controlled. Each building is equipped with waterless urinals and dual-flush toilets.

Energy and atmosphere   Energy savings begins with the design of an efficient envelope and then the employment of sophisticated mechanical systems. The Water and Life Museum is located in a climate that has a design load of 105°F in the summer time. The project's structures provide shading of the building envelope. High-performance glass and a variety of insulation types create the most efficient building envelope possible. The building's exterior skin is constructed from three layers of perforated metal strips. The rooftop of all buildings within the campus is covered with high-efficiency photovoltaic solar power panels. The east elevation building has eight curtain walls, which bridge the 10 towers. Each curtain wall is composed of 900 ft^2 of high-performance argon-filled glass. To compensate for heat radiation, a number of translucent megabanners are suspended in front of each curtain wall. The banners are located above the finished grade, which preserves the beautiful views of the San Jacinto mountain range.

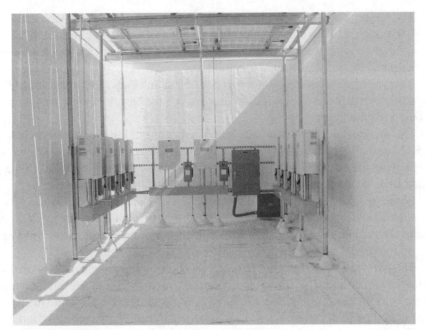

Figure 5.10 **Water and Life Museum inverter system assembly.** *Photo courtesy of Lehrer & Gangi Architects and Vector Delta Design Group Inc.*

Mechanical system The mechanical system uses a combination of radiant flooring, which is used for both heating and cooling, and forced-air units which run from the same chiller and boiler. The combination of the efficient envelope and the sophisticated mechanical system provides a project that is 38 percent more efficient than dictated by Title 24. This not only gains the project many LEED™ points in this category, but also provides a significant cost savings in the operation of these facilities.

Solar power cogeneration The Water and Life Museum boasts one of the largest institutional rooftop photovoltaic installations of its kind in the United States (549 kW of dc power), which provides approximately 70 percent of the electric energy needs of the Water Education Museum. The energy provided by the solar power system exceeds the maximum LEED™ point qualification requirements. The solar generation complex is also intended for use as a teaching demonstration system. The campus loggias are populated with building-integrated photovoltaics. These rooftop solar power panels provide a brilliant display of shading during daylight. A special exhibit system within the museum space is devoted to displaying the solar power cogeneration, which consists of an interactive display that shows climatology and solar power energy generation statistics.

The solar power and the electrical system were designed by Vector Delta Design Inc., which also incorporated numerous energy-saving design concepts and devices for controlling interior and exterior lighting.

Materials and resources   Lehrer & Gangi Design Build specified local and recycled materials wherever possible and kept a careful watch during the construction phase on how construction wastes were handled. Materials selected were that are low emitting volatile organic compounds (VOC). All contractors had restrictions regarding the types of materials that they were allowed to use. The mechanical engineering design deployed a three-dimensional model for the project, which tested the system's thermal comfort to comply with ASHRAE 55 requirements. The architectural building design ensured that 90 percent of spaces had views of natural light.

Innovation and design process   One of the significant points applied to the project design includes use of the buildings as learning centers for teaching sustainability to visitors. Within the museum, exhibit spaces are uniquely devoted to solar cogeneration, which is presented by means of interactive displays where visitors can observe real-time solar and meteorological statistical data on the display.

Hearst Tower

Hearst Tower is the first building to receive a gold LEED™ certification from the United States Green Building Council. This towering architectural monument is located in New York around 57th Street and Eighth Avenue. The site was originally built in the 1920s and was a six-storey office structure that served as the Hearst Corporation's headquarters. Construction began in 1927 and was completed in 1928 at a cost of $2 million.

The original architecture consisted of a four-storey building, set back above a two-storey base. Its design consisted of columns and allegorical figures representing music, art, commerce, and industry. As it was an important heritage monument, in 1988 the building was designated as a landmark site by the New York Landmarks Preservation Commission.

ARCHITECTURE

In 2001 the Hearst organization commissioned Foster and Partners, Architects and Cantor Seinuk, Structural Engineers to design a new headquarters at the site of the existing building. The new headquarters is a 46-storey, 600-ft-tall office tower with 856,000 ft² area. One of the unique features of the architectural design was the requirement to preserve the six-storey historical landmark façade. The old building had a 200- by 200-ft horseshoe-shaped footprint, which was totally excavated while keeping the landmark façade.

The new architectural design concept called for a tower with a 160- by 120-ft footprint. To maintain the historical façade, the design also called for a seven-storey-high interior atrium.

The structural design utilizes a composite steel and concrete floor with a 40-ft interior column-free span for open office planning. The tower has been designed with two

48	1	1	2	16	**Total Project Score**

Certified 26 to 32 points   Silver 33 to 38 points   Gold 39 to 51 points

9	0	0	1	4	**Sustainable Sites**	Possible Points **14**

Y	$	$$	?	N	Credit	Description	
Y					Prereq 1	Erosion & Sedimentation Control - KPFF	0
1					Credit 1	Site Selection - AMK with Wendy Peach	1
				1	Credit 2	Urban Redevelopment	1
				1	Credit 3	Brownfield Redevelopment	1
				1	Credit 4.1	Alternative Transportation, Public Transportation Access	1
1					Credit 4.2	Alt. Transportation, Bicycle Storage & Changing Rooms - JZ	1
			1		Credit 4.3	Alt. Transportation, Alternative Fuel Refueling Stations - JG	1
1					Credit 4.4	Alt. Transportation, Parking Capacity- JG	1
				1	Credit 5.1	Reduced Site Disturbance, Protect or Restore Open Space	1
1					Credit 5.2	Reduced Site Disturbance, Development Footprint JG	1
1					Credit 6.1	Stormwater Management, Rate and Quantity- KPFF	1
1					Credit 6.2	Stormwater Management, Treatment- KPFF	1
1					Credit 7.1	Landscape & Ext Design to Reduce Heat Islands Non-Roof - ML+A, JG MATERIAL SELECTION	1
1					Credit 7.2	Landscape & Exterior Design to Reduce Heat Islands Roof -JG-	1
1					Credit 8	Light Pollution Reduction- Peter Gevorkian	1

4	0	1	0	0	**Water Efficiency**	Possible Points **5**

Y	$	$$	?	N	Credit	Description	
1					Credit 1.1	Water Efficient Landscaping Reduce by 50% - ML+A	1
1					Credit 1.2	Water Efficient Landscaping, No Potable Use or No Irrigation - ML+A and KPFF (JZ confirmed that off-site reclaimed water will satisfy this point: 3/22/04 e-mail)	1
1					Credit 2	Innovative Wastewater Technologies - IBE	1
1					Credit 3.1	Water Use Reduction, 20% Reduction - IBE	1
		1			Credit 3.2	Water Use Reduction, 30% Reduction - IBE	1

14	1	0	0	2	**Energy & Atmosphere**	Possible Points **17**

Y	$	$$	?	N	Credit	Description	
Y	Y				Prereq 1	Fundamental Building Systems Commissioning -IBE	0
Y					Prereq 2	Minimum Energy Performance - IBE	0
Y					Prereq 3	CFC Reduction in HVAC&R Equipment -IBE	0
2					Credit 1.1	Optimize Energy Performance, 2.50-12.50% -IBE	2
2					Credit 1.2	Optimize Energy Performance, 12.51% - 22.50% -IBE	2
2					Credit 1.3	Optimize Energy Performance, 22.51% - 32.50% -IBE	2
2					Credit 1.4	Optimize Energy Performance, 32.51 - 42.50% -IBE	2
2					Credit 1.5	Optimize Energy Performance, > 47.51% -IBE	2
1					Credit 2.1	Renewable Energy, 5% - Peter Gevorkian	1
1					Credit 2.2	Renewable Energy, 10% - Peter Gevorkian	1
1					Credit 2.3	Renewable Energy, 20% - Peter Gevorkian	1
				1	Credit 3	Additional Commissioning	1
1					Credit 4	Ozone Depletion - IBE(FIRE EXTINGUSHERS PASS)	1
				1	Credit 5	Measurement & Verification	1
	1				Credit 6	Green Power - AMK (if we are bordering silver, than yes)	1

Figure 5.11 LEED™ score sheet for Museum of Water and Life. *Courtesy of Gangi Lehrer Architects.*

Platinum 52 or more points

Y	$	$$	?	N			
7	**0**	**0**	**0**	**6**	**Materials & Resources**	Possible Points	**13**
Y	/////	/////	/////	/////	Prereq 1	**Storage & Collection of Recyclables-JG**	0
				1	Credit 1.1	**Building Reuse**, Maintain 75% of Existing Shell	1
				1	Credit 1.2	**Building Reuse**, Maintain 100% of Existing Shell	1
				1	Credit 1.3	**Building Reuse**, Maintain 100% Shell & 50% Non-Shell	1
1					Credit 2.1	**Construction Waste Management**, Divert 50% - **JG**	1
1					Credit 2.2	**Construction Waste Management**, Divert 75% - **JG**	1
				1	Credit 3.1	**Resource Reuse**, Specify 5%	1
				1	Credit 3.2	**Resource Reuse**, Specify 10%	1
1					Credit 4.1	**5% Recycled Content -AMK to prvoide materials and recyclable content**	1
1					Credit 4.2	**10% Recycled Content-AMK**	1
1					Credit 5.1	**Local/Regional Materials**, 20% Manufactured Locally - **JG**	1
1					Credit 5.2	**Local/Regional Materials**, of 20% Above, 50% Harvested Locally-**JG**	1
				1	Credit 6	**Rapidly Renewable Materials**	1
1					Credit 7	**Certified Wood - AMK to provide spec to CSI--JG-get JBH to sign templete and do calculation**	1

Y	$	$$	?	N			
11	**0**	**0**	**1**	**2**	**Indoor Environmental Quality**	Possible Points	**15**
Y	/////	/////	/////	/////	Prereq 1	**Minimum IAQ Performance - IBE**	0
Y	/////	/////	/////	/////	Prereq 2	**Environmental Tobacco Smoke** (ETS) **Control: AMK**	0
1					Credit 1	**Carbon Dioxide** (CO_2) **Monitoring - IBE**	1
			1		Credit 2	**Increase Ventilation Effectiveness- IBE**	1
1					Credit 3.1	**Construction IAQ Management Plan**, During Construction - **AMK** (DURING CONSTRUCTION PHOTOS)	1
1					Credit 3.2	**Construction IAQ Management Plan**, Before Occupancy- **AMK** (DURING CONSTRUCTION PHOTOS)	1
1					Credit 4.1	**Low-Emitting Materials**, Adhesives & Sealants- **AMK**	1
1					Credit 4.2	**Low-Emitting Materials**, Paints- **AMK**	1
1					Credit 4.3	**Low-Emitting Materials**, Carpet- **AMK**	1
1					Credit 4.4	**Low-Emitting Materials**, Composite Wood- **AMK**	1
1					Credit 5	**Indoor Chemical & Pollutant Source Control- L+G, IBE**	1
				1	Credit 6.1	**Controllability of Systems**, Perimeter	1
				1	Credit 6.2	**Controllability of Systems**, Non-Perimeter	1
1					Credit 7.1	**Thermal Comfort**, Comply with ASHRAE 55-1992 - **IBE**	1
1					Credit 7.2	**Thermal Comfort**, Permanent Monitoring System - **IBE**	1
1					Credit 8.1	**Daylight & Views**, Daylight 75% of Spaces - **AMK**	1
			1		Credit 8.2	**Daylight & Views**, Views for 90% of Spaces- **AMK**	1

Y	$	$$	?	N			
3	**0**	**0**	**0**	**2**	**Innovation & Design Process**	Possible Points	**5**
						L+G to look at Q & A	
1					Credit 1.1	**Innovation in Design**: Construction Waste Management @ 95% -JG	1
1					Credit 1.2	**Innovation in Design**: Education program - Design Craftsmen SEND CIR	1
				1	Credit 1.3	**Innovation in Design**: Renewable Resources	1
				1	Credit 1.4	**Innovation in Design**:	1
1					Credit 2	**LEED™ Accredited Professional**	1
1						**PV Renewable Energy**, 70% - Peter Gevorkain	

Figure 5.11 (*Continued*)

Figure 5.12 **Old Hearst Building.** *Photo courtesy of Hearst Corporation.*

zones. The office zone starts 100-ft above street level at the 10th floor rising to the 44th-floor level. Below the 10th floor, the building houses the entrance and the lobby at the street level, the cafeteria, and an auditorium on the 3rd floor with an approximately 80-ft-high interior open space. At the 7th-floor elevation, the tower is connected to the existing landmark façade by a horizontal skylight system spanning approximately 40 ft from the tower columns to the existing façade.

STRUCTURAL DESIGN

The structural design is based upon a network of triangulated trusses of diagrid networks connecting all four faces of the tower, which has resulted in a highly efficient tube structure. The diagonal nodes are formed by the intersection of the diagonal and

Figure 5.13 **New Hearst Tower.** *Photo courtesy of Hearst Corporation.*

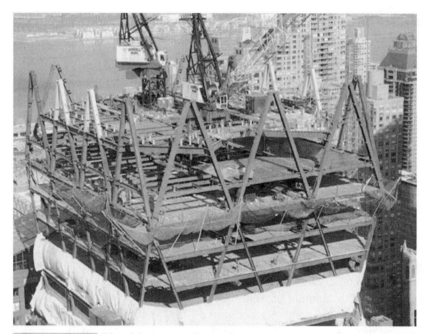

Figure 5.14 **Diagrid structural members.** *Photo courtesy of Hearst Corporation.*

horizontal structural elements. Structurally the nodes act as hubs for redirecting the member forces.

The inherent lateral stiffness and strength of the diagrids provide a significant advantage for general structural stability under gravity, wind, and seismic loading conditions. As a result of the efficient structural system, the construction consumed 20 percent less steel in comparison to conventional moment frame structures.

Below the 10th-floor level, the structure is designed to respond to the large unbraced height by using a megacolumn system around the perimeter of the tower footprint, supporting the tower perimeter structure. The megacolumn is constructed of steel tube sections that are strategically filled with concrete.

FACTS AND FIGURES

Gross area	856,000 ft^2
Typical floor	20,000 ft^2
Building height	597 ft
Number of stories	46
Client	Hearst Corporation
Architect	Foster and Partners
Associate architect	Adamson
Structural engineers	Seinuk Group
Lighting designer	George Sexton
Development manager	Tishman Speyer Properties

GREEN FACTS

For New York City, the benefits include a significant reduction in pollution and increased conservation of the city's vital resources, including water and electricity.

During construction the owners and the builder went to great lengths to collect and separate recyclable materials. As a result, about 85 percent of the original structure was recycled for future use.

The architectural design uses an innovative diagrid system that has created a sense of a four-storey triangle on the façade. The tower is the first North American high-rise structure that does not use horizontal steel beams. As a result of the structural design, the building has used 2000 tons less steel, a 20 percent savings over typical office buildings.

The computerized building automated lighting control system minimizes electric energy consumption based on the amount of natural light available at any given time by means of sensors and timers.

The air-conditioning, and space heating system uses high-efficiency equipment that utilizes outside air for cooling and ventilation for 75 percent of the time throughout the year.

For water conservation, the building roof has been designed to collect rainwater, which reduces the amount of water dumped into the city's sewer system during rainfall. The harvested rainwater is stored in a 14,000-gal reclamation tank located in the basement of the tower. It is used to replace the water lost through evaporation in the office air-conditioning system. It is also used to irrigate indoor and outdoor plants and trees, thus reducing water usage by 50 percent.

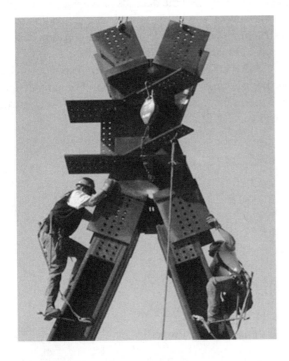

Figure 5.15 Diagrid structural **node.** *Photo courtesy of Hearst Corporation.*

The interior paint used is of the low-VOC type, and furniture used is totally formaldehyde free. All concrete surfaces are treated and sealed with low-toxicity sealants.

Conclusion

In summary, the main objective of LEED™ is a combination of energy-saving and environmental protection measures that are intended to minimize the adverse effects of construction and development. Some of the measures discussed above represent a significant cost impact, the merits of which must be weighed and analyzed carefully.

6

CALIFORNIA ENERGY COMMISSION
REBATE INCENTIVE PLANS

Introduction

This section covers partial highlights of the California Energy Commission (CEC) rebate incentive plan requirements. Complete coverage of the CEC incentive plans can be obtained by visiting the CEC Web site www.energy.ca.gov. The intent of this chapter is to familiarize the reader with the procedures of applying for and procuring California's renewable energy rebate. For additional details the readers should refer to the CEC's publication entitled *Emerging Renewables Program Handbook* (CEC-300-2005-001ED4F). The Emerging Renewables Program (ERP) was created by the CEC to help develop a self-sustaining market for renewable energy systems that supply on-site electricity needs across California. Through this program, the CEC provides funding to partially offset the cost of purchasing and installing various renewable energy system technologies. The goal of the ERP is to reduce the net cost of renewable energy systems for consumers and to stimulate sales of such systems. Essentially, the ERP provides electricity consumers with a financial incentive to install renewable energy systems on their property. The financial incentives are based on system size, technology, and type of installation, and are only paid when the system installation is complete and operational.

To qualify for the incentive program the consumer and the renewable energy system component must satisfy a number of CEC-established requirements.

One of the most important criteria is that the consumer must receive electricity distribution service at the site of installation from an existing electrical service provider that contributes funds to support the program. In California, electrical and gas service providers that take part in the rebate program are Pacific Gas & Electric Company and (PG&E), Southern California Edison Company (SCE), and San Diego Gas & Electric Company (SDG&E).

The renewable energy system installed must utilize renewable technologies, such as solar photovoltaic modules, solar thermal-electric devices, fuel cells, or small wind turbines. These systems must be interconnected to the utility distribution grid and use components that are tested and certified by the CEC. Renewable energy system equipment must be new and have a minimum of a 5-year warranty.

To take part in the rebate program, the applicants submit a reservation request form (CEC-1038 R1) and supporting documentation to reserve a fixed amount of reserved CEC rebate funds. Upon reviewing the reservation request forms the CEC forwards the applicant a payment claim form (CEC-1038 R2) that specifies the amount of funds reserved and the date when the reservation expires. Figures 6.3 and 6.4 show a sample of completed CEC reservation form.

Upon completion of the system installation and operational tests, the applicant submits the payment claim form and supporting documentation to the CEC. Upon verification of system eligibility and program requirements, the CEC reviews the amount reserved and makes payment of this amount.

Eligibility Requirements

At present the following technologies are eligible for ERP funding:

- Photovoltaic systems that convert sunlight to electric energy.
- Solar thermal electric technologies that convert sunlight to heat, a medium used to power a generator to produce electricity.
- Fuel cell technologies that convert sewer gases, landfill gases, or other renewable sources of hydrogen or hydrogen-rich gas into electricity.
- Wind turbine technologies that produce electricity and have a power output range of 50 kW or less.
- Eligibility for renewable energy systems is restricted to private use only, and the system may not be owned by an electrical corporation, as defined in the California Public Utilities Code.

GRID INTERCONNECTION

Eligible renewable energy systems must be permanently interconnected to an existing electrical distribution grid utility that serves the customer's electric load; these companies include PG&E, SCE, SDG&E, and BVE. The system interconnection must comply with applicable electric codes and utility interconnection requirements.

NEW EQUIPMENT

All system hardware components must be new and must not have been used previously. Equipment purchased or installed more than 18 months prior to applying for a reservation is not eligible.

CEC-1038 R1. (1-2005)

| **R1** | **RESERVATION APPLICATION FORM** *EMERGING RENEWABLES PROGRAM* | ☐ Modify Existing Record # _____
 ☐ Affordable Housing Project
 ☐ New Construction |

1. Physical Site of System Installation

Street Address:

City: | State: | Zip

2. Purchaser Name and Mailing Address

Phone: (   ) | Fax: (   )

3. Equipment Seller (Must be registered)

Company:

City: | CEC ID (if known)

Phone: (   ) | Fax: (   )

4. System Installation (Write "Owner" if not hiring contractor)

Company:

City: | License No.:

Phone: | Fax:

5. Electric Utility (Attach all pages of monthly statement)

☐ PG&E ☐ SCE ☐ SDG&E ☐ BVE | Service ID:

Billing Period: | KWh Used:

Note: If new construction attach building permit. Permit No. _____

Submit complete application to:

California Energy Commission
Emerging Renewables Program (MS-45)
1516 Ninth Street
Sacramento, CA 95814-5512

6. Equipment (PV modules, turbines, inverters, meters)

	Quantity	Manufacturer, Model (see CEC lists)
Generating Equipment		
Inverters, Meters		

Estimated annual energy production _____ kWh/Year

7. Rebate and Other Incentives

System Rated Output _____ watts

Total System Cost: $ _____

Expected Rebate: $ _____

Pay Rebate to: ☐ Purchaser ☐ Seller

Reassign payment? ☐ Yes ☐ No
If yes, submit form 1038 R5 with payment request.

Other Incentives: $ _____
Source/Record No.: _____

8. Declaration

The undersigned parties declare under penalty of perjury that the information in this form and the supporting documentation submitted herewith is true and correct to the best of their knowledge and that the following is true:

1. All system equipment is new and unused and has been purchased within the last 18 months.
2. The generating system is intended primarily to offset Purchaser's electrical needs at the site of installation
3. The Purchaser's intent is to operate the system at the above site of installation for its useful life or the duration of the lease agreement and
4. The generating system will be interconnected with the distribution system of the electric utility identified above.

The undersigned parties further acknowledge that they are aware of the requirements and conditions of receiving funding under the Emerging Renewables Program (ERP) and agree to comply with all such requirements and conditions as provided in the Energy Commission's ERP Guidebook and Overall Program Guidebook as a condition to receiving funding under the ERP. The undersigned Purchaser authorizes the Energy Commission during the term of the ERP to exchange information on this form with the Purchaser's electric utility to verify compliance with the requirements of the ERP.

Purchaser Signature	**Equipment Seller Signature**
Print Name: _____	Print Name: _____
Signature: _____ Date: _____	Signature: _____ Date: _____

Necessary Supporting Documentation.

1. All pages of a monthly electric utility bill.
2. Agreements to purchase and install equipment.
3. Payee Data Record (Form STD-204) if payee identified has not previously been paid by the Energy Commission.
4. If not a standard rebate application, attach other required documentation as specified in the ERP Guidebook.

Figure 6.1 Reservation Application Form. *Southern California Water Company [doing business as Bear Valley Electric Service (BVE)].*

CEC- 1038 R2 (1-2005)

R2	**REBATE PAYMENT CLAIM FORM**
	EMERGING RENEWABLES PROGRAM

RENEWABLE ENERGY PROGRAM
CALIFORNIA ENERGY COMMISSION

Mail complete payment claim to:
California Energy Commission
ERP, Payment Claim
1516 Ninth Street (MS-45)
Sacramento, CA 95814-5512

Record Number _____

Payee Number _____

[CEC use only]

[CEC use only]

Tot.Elig.Cost: $_____ Date CFA: _____

SRO watts: _____ Rebate @ _____ = $_____

1. Confirmation of Reservation Amount

_____ has been granted a reservation of $ _____ for a _____ kW renewable energy generating system. The reservation will expire on _____. The system is being installed at _____ and is expected to produce _____ (kWh per year). The payment will be made to the _____.

The generation system must be completed and the claim submitted with the appropriate documentation by the deadline. Claims must be postmarked by the expiration date or the reservation will expire. This reservation is non-transferable. System must be installed at the installation address and sold to the above.

2. System Equipment (Modules, Wind Turbines, Inverters, kWh Meters)

Number	Manufacturer	Model	
_____	_____	_____	Total System Price $ _____
_____	_____	_____	Amount paid by purchaser to date: $_____
_____	_____	_____	Orientation: *(Circle One)* W, SW, S, SE, E, Other
_____	_____	_____	Tilt: *(Circle one)* None, 1-15, 15-30, >30 Degrees
			Tracking system type: _____

3. Modifications

Has any of the information in section 1 or 2 above changed? ☐ Yes ☐ No
If yes note the changes before claiming payment.

_____
_____
_____

The undersigned parties declare under penalty of perjury that the information in this form and the supporting documentation submitted herewith is true and correct to the best of their knowledge. The parties further declare under penalty of perjury that the following statements are true and correct to the best of their knowledge:

(1) The electrical generating system described above and in any attached documents meets the terms and conditions of the Energy Commission's Emerging Renewables Program and has been installed and is operating satisfactorily as of the date stated below.
(2) The electrical generating system described above and in any attached documents is property interconnected to the utility distribution grid and has or will be issued utility approval to operate the system as interconnected to the distribution grid.
(3) The rated electrical output of the generating system, the physical location of the system, and the equipment identified were installed as stated above.
(4) Except as noted above, there were no changes in the information regarding the seller, installer, purchaser, generating system specifications, installation location, or price from that information provided in the Reservation Request Form originally submitted by the undersigned.

The undersigned parties further acknowledge that they are aware of the requirements and conditions of receiving funding under the Emerging Renewables Program (ERP) and agree to comply with all such requirements and conditions as provided in the Energy Commission's ERP Guidebook and Overall Program Guidebook as a condition to receiving funding under the ERP. As specified in the ERP Guidebook, the undersigned Purchaser authorizes the Energy Commission during the term of the ERP to exchange purchaser information on this form with the Purchaser's electric utility in order to verify compliance with the ERP requirements. If a copy of the utility "letter of authorization to operate" the system is not submitted with this payment claim form, the undersigned Purchaser understands that he/she is obligated to submit a copy of this letter to the Energy Commission once it is received.

Purchaser	*Seller*	*Is payment assigned to*
Print Name: _____	Print Name: _____	*another party?* ☐ Yes ☐ No
Signature: _____	Signature: _____	If yes, attach the payment
Date: _____	Date: _____	assignment form (CEC-1038 R5) with original signatures.

IMPORTANT - Necessary Supporting Documentation

1. Final building permit and final inspection signoff; 2. Final invoice(s) confirming the total amount paid for the system equipment and installation; 3. Five-year warranty (CEC-1038 R3 form); 4. Utility letter of authorization to interconnect the system or the Purchaser's authorization form to access Purchaser's utility data; 5. Utility bill or other proof of electrical service and consumption at the site of installation if not previously provided; 6. Payee Data Record (STD-204)

Figure 6.2 CEC Rebate Payment Claim Form.

SELF-GENERATION INCENTIVE PROGRAM
Reservation Request Form

Program Administrator
555 West Fifth Street, GT-22H4
Los Angeles,CA 90013-1011
Fax: 213-244-8222

A Sempra Energy utility

Reservation Number: _____ (leave blank)

1. Applicant Information

Company Name: The Center for Water Education

Tax Payer ID: 26-0009717

Contact Name: Gilbert Ivey

Address: 700 N. Alameda Street

Los Angeles, CA 90012

Phone: (213) 217-6622

E-mail: givey@mwdh2o.com

2. Host Customer Data

Company Name: The Center for Water Education

Tax Payer ID: 26-0009717

Contact Name: Gilbert Ivey

Address: 700 N. Alameda Street

Los Angeles, CA 90012

Facility Address 2325 Searl Parkway

(If different) Hemet,   CA   92546

Phone: (213)217-6622

E-mail: givey@mwdh2o.com

Electric Utility: Edison

Electric Utility Account Number: _____

Gas Utility: _____

Gas Utility Account Number: _____

Attach a copy of a recent electric billing statement. Host Customer must be the customer of record whose name appears on the electric bill.

3. Host Customer Demand

Host Customer highest peak demand over the past 12 months: ____0____ kW

Existing Demand Reduction Obligation - (___0___) kW
(from other interruptible/curtailmeable/demand reduction programs or utility tariffs)

Estimated Future Added Demand +431.8 kW
(documentation supporting added demand estimates must be provided)

Net Host Customer Peak Demand = 431.8 kW
Demand Reduction Program/Tariff:

4. Generating System Information

Level	
1	☐ Fuel Cell (Renewable Fuel) ☒ PV ☐ Wind
2	☐ Fuel Cell (Non-Renewable Fuel)
3-R	(Renewable Fuel) ☐ IC Engines ☐ Small/Micro-turbines
3-N	(Non - Renewable Fuel) ☐ IC Engines ☐ Small/Micro-turbines
	Fuel Type: _____
	☐ Hybrid*

* Please check the Hybrid box if more than one type of generator technology is proposed and include a separate reservation request form for each generation technology.

Manufacturer: Sharp Electronics

Model Number: NT-S5EIU

Power Rating** (kW/Unit) PTC 0.1633 kw/unit

Number of Units: 2955

Total System Output (kW): 483 kw
(Number of Units x Power Rating**)

PV & Wind systems: provide following information for the Inverter:

Inverter Manufacturer/Model No: XANTREX PV - 20208

Inverter Efficiency: 96 %

** Power Rating for PV = PTC x Inverter Efficiency; for Fuel Cells, Micro-turbines, IC Engines = At ISO Conditions; for Wind = Max. Point on Manufacturers Power Curve ≤ 30 MPH.

5. Other Incentives

Have you applied or plan to apply for other rebates, financial incentives or equipment incentives? ☐ Yes ☒ No

If yes provide amount(s): $ _____

Provide source, type, and reservation #: _____

6. Incentive Calculation

Total Eligible Project Costs:	$ 4,173,120
Less other incentive(s):	$ - (0)
Total eligible project costs (other incentives subtracted):	$ = _4,173,120_
Self-Generation Program incentive amount: Calculated based on the lesser of: ☐ Per watt or ☒ % of eligible project cost	$ _4,173,120_
Less CEC Emerging Renewable Buydown Incentive(s)***:	$ - (2,086,560)
Requested Self-Generation Program incentive amount	$ = **2,086,560**

*** CEC Buydown Incentive is subtracted from Self-Generation Incentive Amount. All other incentives are subtracted from Total Eligible Project Costs.

Note: See instructions in the Self-Generation Incentive Program Handbook for completing this form. This form must be typed or hand written neatly. If all required attachments are not included at the time of Put with signatures submission the application may be returned to the applicant.

Rev 3 January 1, 2003

Figure 6.3 A sample of CEC self-generation Incentive Program Reservation Request Form, page 1.

SELF-GENERATION INCENTIVE PROGRAM

Reservation Request Form

APPLICANT & HOST CUSTOMER AGREEMENTS

The undersigned agree that -

A. Applicant and Host Customer agree to release the Program Administrator, its affiliates, subsidiaries, current and future parent company, officers, managers, directors, agents, and employees from all claims, demands, losses, damages, costs, expenses, and liability (legal, contractual, or otherwise), which arise from or are in any way connected with any: (1) injury to or death of persons, including but not limited to employees of the Program Administrator, Host Customer, or Applicant; (2) injury to property or other interests of the Program Administrator, Host Customer, Applicant, or any third party; (3) violation of local, state, or federal common law, statute, or regulation, including but not limited to environmental laws or regulations, natural resources; (4) generation system performance shortfall; so long as such injury, violation, or shortfall (as set forth in (1) - (4) above) arises from or is in any way connected with the Project, including Applicant's performance of or failure to perform the Project, however caused, regardless of any strict liability or negligence of the Program Administrator, its officers, managers, or employees.

B. Host Customer understands that the Program Administrator has made no warranty or representation regarding the qualifications of the Applicant, and that the Host Customer is solely responsible for the selection of the Applicant to implement the Project. Host Customer understands that the Applicant may be an independent contractor and is not authorized to make any representations on behalf of the Program Administrator.

C. The Host Customer agrees that the Program Administrator will have no role in resolving any disputes between it and the Applicant.

D. The Host Customer has the authority to contract, on behalf of the legal owners of the Project Site, for installation of the generating system, or has obtained the permission of the legal owner of the Project Site to install the generating system under contract with the Applicant.

E. Host Customer understands that the Program requires inspections and measurements of the performance of the proposed generating system. Host Customer shall permit Program Administrator, its employees, contractors, and agents, during normal business hours, to: (a) install all necessary performance measurement equipment on the System in order to enable Program Administrator to accomplish performance evaluations; and (b) demonstrate, inspect, monitor, and photograph the System. These data and field measurement documentation are not for purposes of enforcement and shall not be released to outside parties, except as may be required by the California Public Utilities Commission (CPUC).

F. Host Customer shall use its best efforts to accommodate the scheduling requirements of Program Administrator and its agents for all field measurements. Host Customer shall agree to allow all information provided as part of the reservation claim process to be entered into a statewide database that will permit tracking of application for this and other incentive programs. Access to this database will be limited to Program Administrators and the CEC.

G. The undersigned declare under penalty of perjury under the laws of the State of California that 1) the information provided in this form is true and correct to the best of my knowledge, 2) the above described generating system is new and intended to offset part or all of the Host Customer's electrical needs at the site of installation, 3) the site of installation is located within the utility's service territory, 4) the self generating equipment is not intended to be used as a backup generator, and 5) the Host Customer has received a copy of this completed form.

Total System Output: _____463.68_____ **kW**

Requested Incentive Amount: $_____2,086,560_____

Payment should be made to: ☒ **Applicant**   ☐ **Host Customer**

Applicant	**Host Customer (if different than Applicant)**
Print Name The Center for Water Education	Print Name
Title Gilbert Ivey, Executive Vice President	Title
Signature _____ Date _____	Signature _____ Date _____

There are other submittal requirements; if necessary, add a second form. Attach a copy of the proof of utility service and system sizing calculation. Reservation will not be processed without the required attachments.

Rev 3 January 1, 2003

Figure 6.4 A sample of CEC self-generation Incentive Program Reservation Request Form, page 2.

RENEWABLE ENERGY SYSTEM SIZE

Eligible systems must be sized so that the amount of electricity produced by the system primarily offsets part or all of the customer's electrical needs. Electricity produced by the system may not be more than twice the expected electric demand load needs of the electricity consumed by the project. This criterion does not apply to systems of less than 10 kW.

SYSTEM INSTALLATION

System installation and integration must be executed under a written contract by a licensed California contractor who holds an active A, B, C-10, or C-46 license for photovoltaic systems.

Systems may also be self-installed by the owner; however, if the owner does not have the licensing requirements, the system is subject to a lesser rebate. All installation must be in conformance with the manufacturer's specifications and meet all applicable National Electrical Code (NEC) and Uniform Building Code (UBC) standards.

WARRANTY REQUIREMENTS

All system installations are required to have a minimum of a 5-year warranty to protect the purchaser against system or component breakdown. The warranty is required to cover and provide for no-cost repair or replacement of the system or system components, which also includes labor for 5 years. The warranty provided must be a combination of support by the manufacturer and installer.

Self-installed systems must also have a minimum of a 5-year warranty to protect the purchaser against breakdown of electric components; however, the warranty need not cover the labor costs associated with removing or replacing major components.

SYSTEM PERFORMANCE METER

All systems installed require a performance meter so that the customer can determine the amount of energy produced by the system. These meters are listed with the CEC, which measures the total energy produced by the system in kilowatt-hours.

The meters are designed to retain the kilowatt-hour production data in the event of a power outage and have a display window to allow the customer to view the power output production.

EQUIPMENT DISTRIBUTORS

Eligible ERP manufacturers and companies who sell system equipment must provide the CEC with the following information on the equipment seller information form (CEC-1038 R4). For all CEC rebate forms see Appendix C.

■ Business name, address, phone, fax, and e-mail address
■ Owner or principal contact
■ Business license number

- Contractor license number (if applicable)
- Proof of good standing in the records of the California secretary of state, as required for corporate and limited liability entities
- Reseller's license number

AUDITS AND INSPECTIONS

During the course of project execution the CEC conducts audits of the applications it receives and verifies that the information provided in the applications are true and correct. The CEC also conducts field inspections to verify that the systems installed are operating properly and within the specified limits of the reservation requested.

In the event of a payment request that appears to be questionable, the CEC may stop review of the application containing the questionable information to investigate further or request additional documentation from the contractor, equipment seller, and/or purchaser to verify the accuracy of the questionable information.

Incentives Program

This program offers two types of incentives. The first type of incentive is a rebate, which is based on the generating capacity of a system and is paid in a lump sum. The second type is a performance-based incentive, which is based on the amount of electricity generated by a system and is paid over a 3-year period. The latter is offered through a pilot program.

REBATES OFFERED

The rebates offered by the CEC vary by system size, technology, and type of installation. The incentive is intended to reduce the purchase cost of the eligible system and reduce electric energy usage from service providers. Table 6.1 lists the rebate levels available by the CEC as of January 1, 2005. These rebate levels will be reduced over time.

TABLE 6.1   REBATES AVAILABLE FOR EMERGING TECHNOLOGIES

TECHNOLOGY*	REBATE
Photovoltaic systems	$2.80/W
Solar thermal-electric systems	$3.20/W
Fuel cells using renewable fuel	$3.20/W

*Size under 30 kW.

ADDITIONAL INCENTIVES

Rebate incentives that may be received from sources other than the ERP system may affect the rebate amount received from the CEC. Fifty percent of incentives received must be subtracted from the rebate amounts listed in the rebate table if the incentives are from other utility programs, such as a state of California–sponsored incentive program or a federal government–sponsored incentive program, other than tax credits. Applicants will not receive reservations or participate in the California Public Utilities Commission–approved Self-Generation Incentive Program, Rebuild San Diego Program, or any other rebate program using ratepayer funds.

At present, approximately $118,125,000 in funding has been allocated to the ERP for 2002 through 2006. Any funding added to the ERP will be allocated to systems of less than 30 kW unless otherwise specified. The rebate levels for all technology types are reduced by 20 cents/W every 6 months.

It should be noted that the reservation period or expiration date of an approved reservation submitted after January 19, 2005, might not be extended under any circumstance.

REBATE RESERVATION

To apply for a rebate reservation, applicants must submit the following:

■ A completed reservation request form (CEC-1038 R1)
■ A copy of the agreement(s) to purchase and install a system
■ Evidence that the site electricity load is supplied by an eligible utility
■ A payee data record (Form STD-204) for the rebate recipient

Reservation request form   The reservation request form (CEC-1038 R1) identifies information about the proposed system and specifies what information must be submitted with the application. The purchaser of the system must always sign the reservation request form. If the equipment seller is designated as the payee, the seller (retailer or wholesaler) must also sign. It should be noted that eligible equipment sellers must have filed business information with the CEC to be eligible to participate in the program.

Evidence of purchase agreement   Application forms must accompany evidence of an agreement to purchase and install a system. The document must demonstrate whether the system is to be installed by the owner or a contractor. Additional information required to complete the application form include the following:

The quantity, make, and model number (as shown on the CEC list of eligible equipment) for the photovoltaic modules, wind turbines, or other generating equipment and for the inverters and system performance meters

The total purchase price of the system before applying the rebate

Language indicating the purchaser's commitment to buy the system

Printed names and signatures of the purchaser and equipment seller's authorized representative

Installation contracts must also comply with the Contractors State License Board (CSLB) requirements that include the following:

The name, address, and contractor's license number of the company performing the system installation

The site address for the system installation

A description of the work to be performed

The total agreed-upon price to install the system

The payment terms (payment dates and dollar amounts)

Printed names and signatures of the purchaser and the company's authorized representative

As mentioned earlier, qualifying contractors must have a valid A, B, C-10, or C-46 contractor's license.

Owner- or self-installed system   For owner-installed systems, the applicant must provide an equipment purchase agreement, as previously described.

In cases where there is no signed agreement to purchase equipment, the purchaser may provide invoices or receipts showing that at least 10 percent of the system equipment purchase price has been paid to the seller.

Electrical service provider eligibility   For proof of electrical service the purchaser must either contact the utility or request a log of power demand consumption for the past 12 months or must submit a recent copy of the utility bill showing the service address of the installation site, the name of the applicant, electric energy usage, and the utility name. The utility bill should not be older than 6 months from the date of application.

The completed reservation request application must be delivered by mail to:

ERP Reservation Request California Energy Commission
1516 9th St MS-45
Sacramento, CA
95814-5512

CLAIMING A REBATE PAYMENT

To receive a rebate payment, the owner must comply with all program requirements and make a complete claim for payment before the expiration of the reservation.

Payment claim form   Upon completion of the required documents, the CEC sends a copy of the payment claim form (CEC-1038 R2) to the purchaser and designated payee to confirm the amount of funding reserved on the purchaser's behalf. Upon review of the form by the purchaser, it is returned to the CEC by registered mail. Original signatures are required to process a payment.

Evidence of final payment for system installation   The owners or applicant must, upon completion of the system installation, submit a final system cost documentation, which identifies the final amount paid for the installation of the system. The documentation must include proof of the final amount paid by the applicant to the equipment seller or contractor.

The final amount paid for the system must match the cost information identified in the payment claim form, which must accompany final invoices and a copy of the final agreement. Both, the amount paid by the purchaser to the contractor and the extent to which the CEC's rebate lowered the cost of the system must be clearly indicated.

The Energy Commission will conduct spot checks to verify that payments were made as identified in the final invoices or agreements provided by equipment sellers or installers. As part of these spot checks, the Energy Commission will require applicants to submit copies of cancelled checks, credit card statements, or equivalent documentation to substantiate payments made to the equipment seller or installer. When submitting this documentation, applicants are encouraged to redact their personal account numbers or other sensitive information identified in the documentation. Applicants must explain the difference if the final amount paid by the applicant is different from the amount of the purchase or installation shown in any agreement or invoice or in the previously submitted reservation request form.

Building permit and final inspection   Upon completion of installation, the owner must submit a copy of the building permit and the final inspection sign-off for the system installation.

Warranty   A full 5-year warranty form (CEC-1038 R3) must be completed and signed by the installer (see Figure 6.5). In the event the applicant is unable to obtain warranty coverage for labor, the application will be treated as an owner-installed system and be given a 15 percent lower rebate.

Evidence of utility service eligibility   On new construction, when an electrical or gas utility is not available at the time the reservation request, the applicant must obtain account numbers from gas and electric service providers and submit a copy to the utility company that provides the rebate funds.

System interconnection with utility grid   The solar power contractor must demonstrate that the renewable energy system installed has been interconnected to the utility distribution grid and that the utility has approved this interconnection for the system. The owner must demonstrate this by submitting a letter of authorization to

CEC- 1038 R3 (1-2005)

R3	**MINIMUM WARRANTY FORM** *EMERGING RENEWABLES PROGRAM*

System Information

This warranty applies to the following _____ kW renewable energy electric generating system
Description: _____
Located at: _____

What is Covered

This five year warranty is subject to the terms below (check one of the boxes):

☐ **All** components of the generating system **AND** the system's installation. Said warrantor shall bear the full cost of diagnosis, repair and replacement of any system or system component, at no cost to the customer. This warranty also covers the generating equipment against breakdown or degradation in electrical output of more than ten percent from the originally rated output (PTC rating for modules, manufacturers rating for wind turbines); or

☐ System's installation <u>only.</u> Said warrantor shall bear the full cost of diagnosis, repair and replacement of any system or system component, exclusive of the manufacturer's coverage. (Copies of five-year warranty certificates for the major system components (i.e., solar modules, wind turbines, etc. and inverter- <u>MUST</u> be provided with this form).

General Terms

This warranty extends to the original purchaser and to any subsequent purchasers or owners at the same location during the warranty period. For the purpose of this warranty, the terms "purchaser," "subsequent owner," and "purchase" include a lessee, assignee of a lease, and a lease transaction. This warranty is effective from _____ (date of completion of the system installation).

Exclusions

This warranty does not apply to:
- Damage, malfunction, or degradation of electrical output caused by failure to properly operate or maintain the system in accordance with the printed instructions provided with the system.
- Damage, malfunction, or degradation of electrical output caused by any repair or replacement using a part or service not provided or authorized in writing by the warrantor.
- Damage malfunction, or degradation of electrical output resulting from purchaser or third party abuse, accident, alteration, improper use, negligence or vandalism, or from earthquake, fire, flood, or other acts of God.

Obtaining Warranty Service

Contact the following warrantor for service or instructions:

Name: _____   Phone: (   )
Company: _____   Fax: (   )
Address: _____

Signature: _____   Date: _____

Figure 6.5 CEC Minimum Warranty Form.

interconnect the system from the utility. For new constructions, the utility provider must provide a written confirmation that a net meter has been set at the site. The owner must also authorize the CEC to exchange applicant information with the applicant's utility company to verify compliance with these interconnection and program requirements.

Submitting a payment claim   Once a system has been installed and grid-connected and is fully operational, the following documents must be submitted to the CEC to claim a rebate payment:

Rebate payment claim form (CEC-1038 R2)

Documentation confirming what equipment and labor was purchased and fully paid

Building permit and final inspection sign-off

Five-year warranty for the system, labor, and materials

It should be noted that if the payment request application were incomplete, the CEC would request clarification or provision of all missing or unclear information that must be submitted within 60 days. The request for payment may be denied if all the requested information is not received within the time period.

Assignment of rebate payment   The rebate payment may be assigned by the owner to a third party by completing the Reservation Payment Assignment Form (CEC-1038 R5) and submitting it with the Rebate Payment Claim Form (CEC-1038 R2).

Changes that affect the rebate amount   Modifications to an approved reservation may be made prior to a payment claim or when the payment claim is submitted. When a modification includes parameters that affect incentive amounts, a new incentive amount will be calculated based on the program parameters at the time of the modification request, when the supporting documentation is deemed complete. Parameters affecting the incentive include the installation type, system size, and technology. If any change results in the installed system differing in its rated electric output or other parameters from the system originally specified in the reservation request form (CEC-1038 R1), a new rebate payment amount will be calculated.

If any change occurs that would have decreased the original rebate calculation, the amount reserved will also be decreased by the same factor. For example, if the installed system is smaller in output than originally specified in the reservation request form, the new rebate amount will be determined by prorating the amount reserved downward (using the same rebate level that was used to calculate the original rebate amount). Similarly, if the installation type changes from a professional install to an owner install, the incentive is reduced by 15 percent.

REPORTING SYSTEM PERFORMANCE

Renewable energy system performance, which is measured in kilowatt-hours, could be collected and reported to the CEC, if so desired by the owner and electrical utility,

CEC- 1038 R5 (07-2004)

R5	**RESERVATION PAYMENT ASSIGNMENT FORM**
	EMERGING RENEWABLES PROGRAM

RENEWABLE
ENERGY
PROGRAM
CALIFORNIA ENERGY COMMISSION

Record Number _____
Payee ID Number _____

Reservation Information
  Payee Name: _____
  Payee Address: _____
  _____
  Payee Contact: _____
  Payee Phone #: _____

Assignment Request

I, _____, the designated payee or authorized representative of
the payee, hereby assign the right to receive payment for the above noted reservation under the
Emerging Renewables Program to the following individual or entity:

  Name: _____
  Address: _____
  _____
  Phone #: _____

I request that payment be forwarded to this individual or entity at the address noted. Upon request
proof of payment will be forwarded to me.

Acknowledgement

As the designated payee or authorized representative, I understand that I remain responsible for
complying with the requirements of the Emerging Renewables Program and will remain liable for any
tax consequences associated with the reservation payment, despite the payment's assignment. I
further understand that I may revoke this payment assignment at any time prior to the Energy
Commission's processing of the payment by providing written notice to the Energy Commission's
Technology Market Development Office. Such notice shall be provided to: Emerging Renewables
Program, California Energy Commission, 1516 9th Street, MS-45, Sacramento, CA 95814-5512.

Executed on: _____

Signature: _____
Name: _____
Title: _____

**This completed form may be submitted with either the Reservation Request Form (CEC-1038 R1) or the
Payment Claim Form (CEC-1038 R2) for standard rebates. This form may not be submitted by telefax,
as original signatures are needed to process assignment requests.**

Figure 6.6 CEC Reservation Payment Assignment Form.

by a Web-based monitoring system administered by a private party. Owners wishing
to use this option should enter into a special agreement with their utility provider to
ensure that data are recorded on a monthly basis and reported to the CEC each quarter.
In using this option, owners assume full responsibility for all costs associated with
their utility's collection and reporting.

RESERVATION PERIOD

The reservation period is established in two parts: a 12-month preliminary reservation, during which applicants must purchase and install the proposed system, and a 3-year final reservation period, in which applicants should provide quarterly invoices to the CEC to claim incentive payments for their renewable energy cogeneration. Both of these options are fixed and cannot be extended under any circumstances.

Applicants that cannot meet these deadlines after issuance of a reservation will be denied rebate funds. In such an event, the applicants must reapply for a new rebate.

To claim incentive payments, quarterly invoices must be submitted to the CEC using a special invoice form (CEC-1038 R10). Any addition or deletion of equipment from the system during the reservation period must be reported to the CEC.

Special Funding for Affordable Housing Projects

California Assembly Bill 58 mandates the CEC to establish an additional rebate for systems installed on affordable housing projects. These projects are entitled to qualify for an extra 25 percent rebate above the standard rebate level, provided that the total amount rebated does not exceed 75 percent of the system cost. The eligibility criteria for qualifying are as follows:

The affordable housing project must adhere to California health and safety codes.

The property must expressly limit residency to extremely low, very low, lower, or moderate income persons and must be regulated by the California Department of Housing and Community Development.

Each residential unit (apartments, multifamily homes, etc.) must have individual electric utility meters.

The housing project must be at least 10 percent more energy efficient than the current standards specified.

Special Funding for Public and Charter Schools

A special amendment to the CEC mandate, enacted in February 4, 2004, established a Solar Schools Program to provide a higher level of funding for public and charter schools to encourage the installation of photovoltaic generating systems at more

Solar Schools Reservation Request Form

RENEWABLE
ENERGY
PROGRAM

CALIFORNIA ENERGY COMMISSION

The ERP guidebook is available at
www.consumerenergycenter.org/erprebate]
or by calling the Energy Commission's
Call Center at 1-800-555-7794:

Completed forms must be mailed to:
ERP Schools
California Energy Commission
1516 9th Street, MS-45
Sacramento, CA 95814-5512

Payee Designation:
Reserve rebate for: ☒ School District

1. School District

District Name:
Contact Person:

Mailing Address:

Business: (   )   Fax: (   )
Email:

2. Participating Schools

School Name:

Address:

Business: (   )   Fax: (   )
Email:

Estimated size of photovoltaic system: _____ watts.

School Name:

Address:

Business: (   )   Fax: (   )
Email:

Estimated size of photovoltaic system: _____ watts.

School Name:

Address:

Business: (   )   Fax: (   )

Email:

Estimated size of photovoltaic system: _____ watts.

3. Utility Provider

Electric Utility Provider: ☐ PG&E   ☐ SCE
                              ☐ SDG&E

4. Program Information

How did you hear about our program?
Have you previously applied for funding rebates from the ERP?
          ☐ yes   ☐ no   If yes, reservation # _____

5. Statement of energy efficient measures

For each of the participating schools identified on this form check the appropriate box:

☐   At least 80% of the classrooms use high efficiency fluorescent lighting (T8 lamps and electronic ballasts).

☐   Other energy efficient measures with equivalent or greater energy savings to Item 1 (as determined by the Commission) are in use (documentation attached).

6. Incentive Requested

Sum of Total watts _____

Total Incentive Requested $_____

☐ Full reservation request (CEC-1038 R1 and supporting documentation for each site attached) or:

☐ Preliminary 6 month reservation request

NOTE: Sum of all new systems installed in school district may not exceed 30 kW.

7. Information for Additional Schools

☐ See attached for information on other schools submitted with this application (name, address, phone, estimated size of system, signature of school official).

Declaration: On behalf of the school district, the undersigned declares under penalty of perjury that 1) the information provided in this form and the supporting documentation submitted herewith is true and correct to the best of his/her knowledge, 2) the above described generating system is intended primarily to offset part or all of the school's electrical needs at the site of installation, 3) the site of installation is located within service territory of an eligible electric utility. The undersigned acknowledges that he/she is are aware of the requirements and conditions of receiving funding under the Emerging Renewables Program (ERP) and agrees, on behalf of the school district, to comply with all such requirements and conditions as provided in the Energy Commission's ERP Guidebook and Overall Program Guidebook as a condition to receiving funding under the ERP. As specified in the ERP Guidebook, the undersigned, on behalf of the school district, authorizes the Energy Commission during the term of the ERP to exchange school district information on this form with the school district's electric utility in order to verify compliance with the ERP requirements.

School District representative	CEC Review
Print Name _____	
Signature _____	
Date _____	

IMPORTANT: Attach the following minimum documentation: (1) a signed copy of the board resolution(s) indicating support for the solar project and intent to purchase a photovoltaic system; (2) a copy of a monthly electricity statement; (3) evidence that energy efficient measures equivalent to the use of high efficiency fluorescent lighting in 80 percent of the school's classrooms. This form will not be processed without the required attachments.

Figure 6.7   CEC Solar Schools Reservation Request Form.

school sites (see the Solar Schools Reservation Request Form shown in Figure 6.7). At present the California Department of Finance has allocated a total of $2.25 million for this purpose. To qualify for the additional funds, the schools must meet the following criteria.

Public or charter schools must provide instruction for kindergarten or any of the grades 1 through 12.

The schools must have installed high-efficiency fluorescent lighting in at least 80 percent of classrooms.

The schools must agree to establish a curriculum tie-in plan to educate students about the benefits of solar energy and energy conservation.

Principal Types of Municipal Leases

There are two types of municipal bonds. One type is referred to as a "tax-exempt municipal lease," which has been available for many years and is used primarily for the purchase of equipment and machinery that has a life expectancy of 7 years or less. The second type is generally known as an "energy efficiency lease" or a "power purchase agreement" and is used most often on equipment being installed for energy efficiency purposes and is used where the equipment has a life expectancy of greater than 7 years. Most often this type of lease applies to equipment classified for use as a renewable energy cogeneration, such as solar PV and solar thermal systems. The other common type of application that can take advantage of municipal lease plans includes energy efficiency improvement of devices such as lighting fixtures, insulation, variable-frequency motors, central plants, emergency backup systems, energy management systems, and structural building retrofits.

The leases can carry a purchase option at the end of the lease period for an amount ranging from $1.00 to fair market value and frequently have options to renew the lease at the end of the lease term for a lesser payment over the original payment.

TAX-EXEMPT MUNICIPAL LEASE

A tax-exempt municipal lease is a special kind of financial instrument that essentially allows government entities to acquire new equipment under extremely attractive terms with streamlined documentation. The lease term is usually for less than 7 years. Some of the most notable benefits are:

- Lower rates than conventional loans or commercial leases
- Lease-to-own. There is no residual and no buyout
- Easier application, such as same-day approvals
- No "opinion of counsel" required for amounts under $100,000
- No underwriting costs associated with the lease

ENTITIES THAT QUALIFY FOR A MUNICIPAL LEASE

Virtually any state, county, or city municipal government and their agencies, such as law enforcement, public safety, fire, rescue, emergency medical services, water port authorities, school districts, community colleges, state universities, hospitals, and 501 organizations qualify for municipal leases. Equipment that can be leased under a municipal lease include essential use equipment and remediation equipment such as vehicles, land, or buildings. Some specific examples are listed here:

- Renewable energy systems
- Cogeneration systems
- Emergency backup systems
- Microcomputers and mainframe computers
- Police vehicles
- Networks and communication equipment
- Fire trucks
- Emergency management service equipment
- Rescue construction equipment such as aircraft helicopters
- Training simulators
- Asphalt paving equipment
- Jail and court computer-aided design (CAD) software
- All-terrain vehicles
- Energy management and solid waste disposal equipment
- Turf management and golf course maintenance equipment
- School buses and paratransit
- Water treatment systems
- Modular classrooms, portable building systems, and school furniture such as copiers, fax machines, closed circuit television surveillance equipment
- Snow and ice removal equipment
- Sewer maintenance

The transaction must be statutorily permissible under local, state, and federal laws and must involve something essential to the operation of the project.

DIFFERENCE BETWEEN A TAX-EXEMPT MUNICIPAL LEASE AND A COMMERCIAL LEASE

Municipal leases are special financial vehicles that provide the benefit of exempting banks and investors from federal income tax, allowing for interest rates that are generally far below conventional bank financing or commercial lease rates. Most commercial leases are structured as rental agreements with either nominal or fair-market-value purchase options.

Borrowing money or using state bonds is strictly prohibited in all states, since county and municipal governments are not allowed to incur new debts that will obligate payments that extend over multiyear budget periods. As a rule, state and municipal government budgets are formally voted into law; as such there is no legal authority to bind the government entities to make future payments.

As a result, most governmental entities are not allowed to sign municipal lease agreements without the inclusion of nonappropriation language. Most governments, when using municipal lease instruments, consider obligations as current expenses and do not characterize them as long-term debt obligations.

The only exceptions are bond issues or general obligations, which are the primary vehicles used to bind government entities to a stream of future payments. General obligation bonds are contractual commitments to make repayments. The government bond issuer guarantees to make funds available for repayment, including raising taxes if necessary. In the event, when adequate sums are not available in the general fund, "revenue" bond repayments are tied directly to specific streams of tax revenue. Bond issues are very complicated legal documents that are expensive and time consuming and in general have a direct impact on the taxpayers and require voter approval. Hence, bonds are exclusively used for very large building projects such as creating infrastructure like sewers and roads.

Municipal leases automatically include a nonappropriation clause; as such they are readily approved without counsel. Nonappropriation language effectively relieves the government entity of its obligation in the event funds are not appropriated in any subsequent period, for any legal reason.

Municipal leases can be prepaid at any time without a prepayment penalty. In general, a lease amortization table included with a lease contract shows the interest principal and payoff amount for each period of the lease. There is no contractual penalty, and a payoff schedule can be prepared in advance. It should also be noted that equipment and installation can be leased.

Lease payments are structured to provide a permanent reduction in utility costs when used for the acquisition of renewable energy or cogeneration systems. A flexible leasing structure allows the municipal borrower to level out capital expenditures from year to year. Competitive leasing rates of up to 100 percent financing are available with structured payments to meet revenues that could allow the municipality to acquire the equipment without having current fund appropriation.

The advantages of a municipal lease program include

- Enhanced cash flow financing allows municipalities or districts to spread the cost of an acquisition over several fiscal periods leaving more cash on hand.
- A lease program is a hedge against inflation since the cost of purchased equipment is figured at the time of the lease and the equipment can be acquired at current prices.
- Flexible lease terms structured over the useful life span of the equipment can allow financing of as much as 100 percent of the acquisition.
- Low-rate interest on a municipal lease contract are exempt from federal taxation, have no fees, and have rates often comparable to bond rates.
- Full ownership at the end of the lease most often includes an optional purchase clause of $1.00 for complete ownership.

Because of budgetary shortfalls, leasing is becoming a standard way for cities, counties, states, schools, and other municipal entities to get the equipment they need today without spending their entire annual budget to acquire it.

Municipal leases are different from standard commercial leases because of the mandatory nonappropriation clause, which states that the entity is only committing to funds through the end of the current fiscal year, even if they are signing a multiyear contract.

7

ECONOMICS OF SOLAR

POWER SYSTEMS

Introduction

Perhaps the most important task of a solar power engineer is to conduct preliminary engineering and financial feasibility studies, which are necessary for establishing an actual project design. The essence of the feasibility study is to evaluate and estimate the power generation and cost of installation for the life span of the project. The feasibility study is conducted as a first step in determining the limitations of the solar project's power production and return on investment, without expending a substantial amount of engineering and labor effort. The steps needed to conduct the preliminary engineering and financial study are presented in this chapter.

Preliminary Engineering Design

Conduct a field survey of the existing roof or mounting area. For new projects, review the available roof-mount area and mounting landscape. Care must be taken to ensure that there are no mechanical, construction, or natural structures that could cast a shadow on the solar panels. Shades from trees and sap drops could create an unwanted loss of energy production. One of the solar PV modules in a chain, when shaded, could act as a resistive element that will alter the current and voltage output of the whole array.

Always consult with the architect to ensure that installation of solar panels will not interfere with the roof-mount solar window, vents, and air-conditioning unit ductwork. The architect must also take into consideration roof penetrations, installed weight, anchoring, and seismic requirements.

Upon establishment of solar power area clearances, the solar power designer must prepare a set of electronic templates representing standard array configuration assemblies.

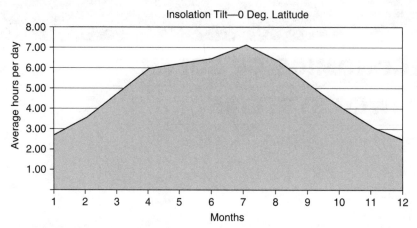

Figure 7.1 Insolation graph for Los Angeles PV panels mounted at a 0-degree tile angle.

Solar array templates then could be used to establish a desirable output of direct current (dc) power. It should be noted that, when laying blocks of PV arrays, consideration must be given to the desirable tilt inclination to avoid cross shadowing. In some instances, the designer must also consider trading solar power output efficiency to maximize the power output production. As mentioned in Chapter 2, the most desirable mounting position for a PV module to realize maximum solar insolation is about the latitude minus 10 degrees. For example, the optimum tilt angle in New York will be 39 degrees; whereas, in Los Angeles, it will be about 25 to 27 degrees. The sun exposures caused by various insolation tilts over the course of the year in Los Angeles are shown in Figures 7.1 through 7.5. To avoid cross shading, the adjacent profiles of two solar rows of arrays could be determined by simple trigonometry that could determine the geometry of the tilt by the angle

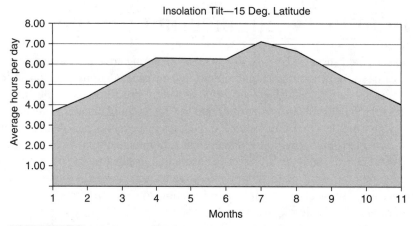

Figure 7.2 Insolation graph for Los Angeles PV panels mounted at 15-degree tile angle.

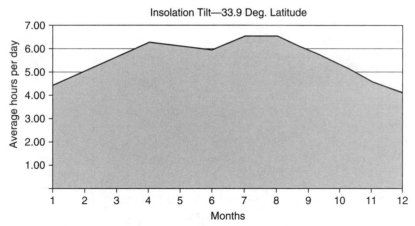

Figure 7.3 Insolation graph for Los Angeles PV panels mounted at 33.9-degree tile angle.

of the associated sine (shading height) and cosine (tandem array separation space) of the support structure incline. It should be noted that flatly laid solar PV arrays may incur about a 9 to 11 percent power loss, but the number of installed panels could exceed 30 to 40 percent on the same mounting space.

An important design criterion when laying out solar arrays is grouping the proper number of PV modules that would provide the adequate series-connected voltages and current required by inverter specifications. Most inverters allow certain margins for dc inputs that are specific to the make and model of a manufactured unit. Inverter power capacities may vary from a few hundred to many thousands of watts. When designing a solar power system, the designer should decide about the use of specific PV and inverter makes and models in advance, thereby establishing the basis of the overall configuration.

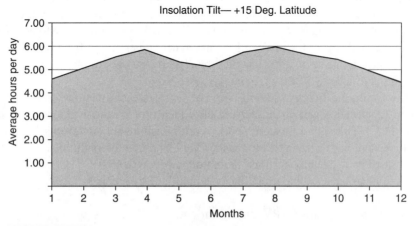

Figure 7.4 Insolation graph for Los Angeles PV panels mounted at 15-degree tile angle.

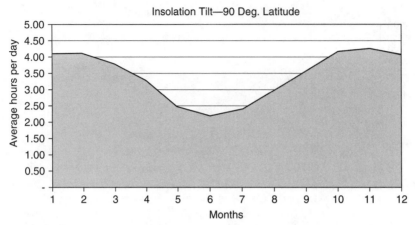

Figure 7.5 Insolation graph for Los Angeles PV panels mounted at ~90-degree tile angle.

It is not uncommon to have different sizes of solar power arrays and matching invert-ers on the same installation. In fact, in some instances, the designer may, for unavoidable occurrences of shading, decide to minimize the size of the array as much as possible, thus limiting the number of PV units in the array, which may require a small-size power capacity inverter. The most essential factor that must be taken into consideration is to ensure that all inverters used in the solar power system are completely compatible.

When laying out the PV arrays, care should be taken to allow sufficient access to array clusters for maintenance and cleaning purposes. In order to avoid deterioration of power output, solar arrays must be washed and rinsed periodically. Adequately spaced hose bibs should be installed on rooftops to facilitate flushing of the PV units in the evening time only, when the power output is below the margin of shock hazard.

Upon completing the PV layout, the designer should count the total number of solar power system components, and by using a rule of thumb must arrive at a unity cost estimate such as dollars per watt of power. That will make it possible to better approx-imate the total cost of the project. In general, net power output from power PV arrays, when converted to ac power, must be subjected to a number of factors that can degrade the output efficiency of the system.

The California Energy Commission (CEC) rates each manufacturer-approved PV unit by a special power output performance factor referred to as the power test condi-tion (PTC). This figure of merit is derived for each manufacturer and PV unit model by extensive performance testing under various climatic conditions. These tests are performed in a specially certified laboratory environment.

Design parameters that affect the system efficiency are as follows:

- Geographic location (PV units work more efficiently under sunny but cool temperatures)
- Latitude and longitude

- Associated yearly average insolation
- Temperature variations
- Building orientation (north, south, and the like)
- Roof or support structure tilt
- Inverter efficiency
- Isolation transformer efficiency
- DC and ac wiring losses resulting from density of wires in conduits
- Solar power exposure
- Long wire and cable runs
- Poor, loose, or corroded wire connections
- AC power transmission losses to the isolation transformers
- Poor maintenance and dust and grime collection on the PV modules

Meteorological Data

When the design is planned for floor-mount solar power systems, designers must investigate natural calamities such as extreme wind gusts, periodic or seasonal flooding, and snow precipitation. For meteorological data contact the NASA Surface Meteorology and Solar Energy Data Set Web site at eosweb.larc.nasa.gov/sse/RETSreen. To search for meteorological information on this Web site, the inquirer must provide the latitude and longitude for each geographic location. For example, to obtain data for Los Angeles, California, at latitude 34.09 and longitude 118.4, the statistical data provided will include the following recorded information for each month of the year for the past 10 years:

- Average daily radiation on horizontal surface [kWh/(m^2 × day)]
- Average temperature (°C)
- Average wind speed (m/s)

To obtain longitude and latitude information for a geographic area refer to the Web site www.census.gov/cgi-bin/gazetter. For complete listings of latitude and longitude data please refer to Appendix A. The following are a few examples for North American metropolitan area locations:

Los Angeles, California	34.09 N/118.40 W
Toronto, Canada	43.67 N/–79.38 W
Palm Springs, California	33.7 N/116.52 W
San Diego, California	32.82 N/117.10 W

To obtain ground surface site insolation measurements refer to the Web site http://eosweb.lac.nasa.gov/sse.

A certified registered structural engineer must design all solar power installation platforms and footings. Upon completing and integrating the preliminary design parameters previously discussed, the design engineer must conduct a feasibility analysis of the solar power cogeneration project. Some of the essential cost components of a solar power system required for final analysis are:

- Solar PV module (dollars per dc watts)
- Support structure hardware
- Electric devices, such as inverters, isolation transformers, and lightning protection devices; and hardware such as electric conduits, cables, and grounding wire

Additional costs may include:

- Material transport and storage
- Possible federal taxes and state sales taxes
- Labor wages (prevailing or nonprevailing) and site supervision (project management)
- Engineering design, which includes electrical, architectural, and structural disciplines
- Construction drawings and reproduction
- Permit fees
- Maintenance training manuals and instructor time
- Maintenance, casualty insurance, and warranties
- Spare parts and components
- Testing and commissioning
- Overhead and profit
- Construction bond and liability insurance
- Mobilization cost, site office, and utility expenses
- Liquidated damages

Energy Cost Factor

Upon completion of the preliminary engineering study and solar power generation potential, the designer must evaluate the present costs and project the future costs of the electric energy for the entire life span of the solar power system.

To determine the present value of the electric energy cost for an existing building, the designer must evaluate the actual electric bills for the past 2 years. It should be noted that the general cost per kilowatt-hour of energy provided by service distributors consists of an average of numerous charges such as commissioning, decommissioning, bulk purchase, and other miscellaneous cost items that generally appear on electric bills (that vary seasonally) but go unnoticed by consumers.

The most significant of the charges, which is in fact a penalty, is classified as *peak hour energy*. This charge occurs when the consumer's power demand exceeds the established boundaries of energy consumption as stipulated in tariff agreements. In order to maintain a stable power supply and cost for a unit of energy (a kilowatt-hour), service distributors, such as Southern California Electric (SCE) and other power-generating entities, generally negotiate a long-term agreement whereby the providers guarantee distributors a set bulk of energy for a fixed sum. Since energy providers have a limited power generation capacity, limits are set as to the amount of power that is to be distributed for the duration of the contract. A service provider such as SCE uses statistics and demographics of the territories served to project power consumption demands, which then form the baseline for the energy purchase agreement. When energy consumption exceeds the projected demand, it becomes subject to much costlier tariffs, which are generally referred to as the *peak bulk energy rate*.

Project Cost Analysis

As indicated in the preliminary solar power cogeneration study, the average installed cost per watt of electric energy is approximately $9 as shown in Figures 7.6a through 7.6b. The unit cost encompasses all turnkey cost components, such as engineering design documentation, solar power components, PV support structures, electric hardware, inverters, integration labor, and labor force training. Structures in that cost include roof-mount support frames and simple carport canopies, only. Special architectural monuments if required may necessitate some incremental cost adjustment. As per the CEC, all solar power cogeneration program rebate applications applied for before December of 2002 were subject to a 50 percent subsidy. At present, rebate allotments are strictly dependent on the amount of funding available at the time of application and are granted on a first come, first serve basis.

SYSTEM MAINTENANCE AND OPERATIONAL COSTS

As mentioned earlier, solar power systems have a near-zero maintenance requirement. This is due to solid-state technology, lamination techniques, and the total absence of mechanical or moving parts. However to prevent marginal degradation in output performance from dust accumulation, solar arrays require a biyearly rinsing with a water hose.

Figure 7.6 is a detailed estimate, designed by the author, for a solar power project for the Water and Life Museum located in Hemet, California. As discussed in Chapter 3 the project consists of two museum campuses with a total of seven buildings, each constructed with roof-mount solar power PV systems.

The costing estimate reflected in Figure 7.6 represents one of the main buildings referred to as the Water Education Museum and is a project funded by the Los Angeles Metropolitan Water District (MWD). The solar power generation of the entire campus is 540 W dc. Net ac power output including losses is estimated to be approximately 480 kW. At present the entire solar power generation system is paid for by the MWD; as a result the entire power generated by the system will be used by the Water and Life Museum, which represents approximately 70 to 75 percent of the overall electric demand load.

Economic Analysis

INITIAL COSTS & CREDITS

Engineering Rate

	Hours	Rate	Total	%
		$150.00 per hour		
Site Investigation	40	$ 150.00	$ 6,000.00	46.88%
Preliminary Design Coordination	24	150.00	3,600.00	28.13%
Report Preparation	8	150.00	1,200.00	9.38%
Travel & Accommodations	1	2,000.00	2,000.00	15.63%
Other			-	0.00%
Sub Total			$12,800.00	100.00%

DEVELOPMENT

	Hours	Rate	Total	%
Permits & Rebate Applications	8	$ 150.00	$ 1,200.00	5.41%
Project Management	120	150.00	18,000.00	81.08%
Travel Expenses	1	2,000.00	2,000.00	9.01%
Other	1	1,000.00	1,000.00	4.50%
Sub Total			$22,200.00	100.00%

ENGINEERING

	Hours	Rate	Total	%
PV Systems Design	90	$150.00	$ 13,500.00	10%
Architectural Design	90	150.00	13,500.00	10%
Structural Design	90	150.00	13,500.00	10%
Electrical Design	420	150.00	63,000.00	48%
Tenders & Contrating	48	150.00	7,200.00	5%
Construction Supervision	94	150.00	14,100.00	11%
Training Manuals	48	150.00	7,200.00	5%
Sub Total			$132,000.00	100%

RENEWABLE ENERGY EQUIPMENT

	Hours	Rate	Total	%
PV Modules (per kWh dc)	255	$3,900.00	$994,500.00	92%
Transportation	1	5,000.00	5,000.00	0%
Other			-	0%
Tax (Equipment Only)	8.25%		82,046.25	8%
Sub Total			$081,546.25	100%

INSTALLATION EQUIPMENT

	Hours	Rate	Total	%
PV Module Support Structure (per kWh)	255	$ 500.00	$127,500.00	18%
Inverter (per kWh)	320	488.00	156,160.00	22%
Electrical Materials (per kW)	320	250.00	80,000.00	11%
System Installation Labor (per kWh)	320	1,000.00	320,000.00	45%
Transportation	1	3,000.00	3,000.00	0%
Other	0		-	0%
Tax (Equipment Only)	8.25%		30,001.95	4%
Sub Total			$716,661.95	100%

Figure 7.6 Material and equipment estimate. Water Education Museum.

MISCELLANEOUS					
Training	48	$150.00	$	7,200.00	100%
Contingencies (if applicable)	0			-	0%
Sub Total			$	7,200.00	100%

OVERHEAD & PROFIT	18%	$ 355,033.48	100%

TOTAL PROJECT COST ESTIMATE:	$2,327,441.68	100%

Miscellaneous

	Hrs	Rate	Total
Training	48	$150.00	$7,200.00
Contingencies			
Total Miscellaneous			**$7,200.00**

Energy demand analysis

Lighting (kWh)	413
HVAC (kWh)	296

Avg. hourly energy (kWh)	709	
Avg. cost per kWh, current cost		$ 0.19
Avg. monthly energy expenditure		48,495.60
Avg. yearly energy expenditure projection (kWh)		581,947.20

Projected energy cost

Projected cost escalation

		5 YEARS	10 YEARS
Energy cost escalation rate	6%	$ 34,916.83	$37,011.84
Inflation rate	3.0%	17,458.42	17,982.17
Total yearly cost escalation		52,375.25	54,994.01

Solar power energy generation

Energy generation capacity (kWh Ac)		212	212
Average daily solar irradiance hr		5.5	5.5
Yearly energy generation (kWh)		425037	425037
5-Year inflation rate per kWh	15.0%	$ 0.19	$ 0.22
Yearly cost savings by PV modules		80,757.01	92,870.56
Yearly aggregate savings		133,132	147,865

Figure 7.6 (*Continued*)

Projected cost escalation			15 YEARS	20 YEARS	25 YEARS
Energy cost escalation rate			$ 39,232.55	$ 41,586.51	$ 44,081.70
Inflation rate			18,521.63	19,077.28	19,649.60
Total yearly cost escalation			57,754.19	60,663.79	63,731.30
Solar power energy generation					
Energy generation capacity (kWh Ac)			212	212	212
Average daily solar irradiance hr			5.5	5.5	5.5
Yearly energy generation kWh			425037	425037	425037
5-Year Inflation Rate per (kWh)			$ 0.25	$ 0.29	$ 0.33
Yearly cost savings by PV modules			106,801.15	122,821.32	141,244.52
Yearly aggregate savings			164,555	183,485	204,976

Project Cost Summary			COST	%
Engineering	DC kW	$/WATT	$ 132,000	5.78%
Renewable energy equipment	255	$ 3,900.00	994,850	43.54%
Renewable energy equipment tax	8.25%		82,075	3.59%
Trial transportation/crates	7	$ 500.00	3,500.00	0.15%
Installation equipment			716,711	31.37%
Miscellaneous			7,200	0.32%
Total			1,936,336	
OH&P		18.0%	348,540	15.25%
Total Initial Cost			2,284,876	100.00%
Additional allowable expenses				
Total allowable cost				
Buyback rebate (Wh SCE)			$ 4.50	Discount/W
Total Wh dc subject to rebate			255,090	dc watts
Maximum rebate				
Total rebate amount		50%	$1,142,438	
Net cost initial cost			1,142,438	
Cost per dc watts			8.96	

Figure 7.6 (*Continued*)

Feasibility Study Report

As mentioned in Chapters 3 and 4, the key to designing a viable solar power system begins with preparation of a feasibility report. A feasibility report is essentially a preliminary engineering design report that is intended to inform the end user about significant aspects of a project. The document therefore must include a thorough definition of the entire project material and financial perspective.

A well-prepared report must inform and educate the client and provide a realistic projection of all engineering and financial projections to enable the user to weigh all aspects of a project from start to finish. The report must include a comprehensive technical and financial analysis of all aspects of the project, including particulars of local climatic conditions, solar power system installation alternatives, grid-integration requirements, electric power demand, and economic cost projection analysis. The report must also incorporate photographs, charts, and statistical graphs to illustrate and inform the client about the benefits of the solar power or sustainable energy system proposed.

The following is a feasibility report prepared by the author for a public recreational facility which included a community swimming pool, baseball and football fields, and several tennis courts.

Valley Wide Recreation and Park District

901 West Esplanade Ave.
San Jacinto, CA 92581
Subject: Solar power preliminary study for Diamond Valley Lake Aquatic Facility, Community Building, Hemet, California
December 14, 2004

Dear Sirs,
The following solar power feasibility study reflects analysis of alternate approaches to grid-connected solar power cogeneration for the Diamond Valley Lake Aquatic Facility. The solar photovoltaic systems proposed are intended to provide a comprehensive electric energy cost saving solution for the duration of the guaranteed life of the equipment, which theoretically should span over 25 years and beyond.

Considering the 25,000-ft^2 extent of the recreation facility, which includes the community center, tennis courts, and baseball and soccer fields, we have proposed an alternate study that would allow the solar power cogeneration systems to expand in a modular fashion as the need arises.

In view of the fact that the present scope of the project includes only the pool, the bathrooms, and a small office building, the projected electric power demand is limited to that needed for pool filtration and the building's internal and external lights. Since the objective of a solar power cogeneration system is to save energy expenditure on the whole campus, we have based our analysis on the best possible power demand projection that would allow for tailoring of a solar power system that may meet the demand needs of the entire scope of the recreation facility.

The following are the power demand projections, which are based on the existing civil engineering site plan:

- Present office, bathroom, and pool—185 kWh (as reflected in the electrical plans)
- Community center (future), 25,000 ft^2, 20 W/ft^2—500 kWh (projected)

- Eight baseball fields, lighting load at 5000 W per field—40 kWh
- Eight soccer fields, lighting load at 5000 W per field—40 kWh
- Seven volleyball courts, lighting load at 1000 W per field—7 kWh
- Two basketball courts, lighting load at 3000 W per field—6 kWh
- Six tennis courts, lighting load at 3000 W per field—18 kWh
- Concessions building demand load—10 kWh
- Pathway lighting—10 kWh
- Projected total demand load—806 kWh
- Current demand load at 480 V, three-phase—2000-A service

As you may be aware, Southern California Edison (SCE) only provides a single service for each client, and in order to accommodate a 2000-A projected demand, the electrical service switchgear and equipment room must be designed in a fashion so as to provide sufficient expansion space and required underground conduits to meet the overall infrastructure needs. Likewise, integration of a solar power cogeneration will mandate special provisions for incorporating grid-connection solar power transformers, inverters, and isolation transformers that should be housed within the main electrical room.

In general, the solar power cogeneration contribution, when intended to curb electric energy consumption, should be tailored in proportion to the overall power demand of the complex. The effective size of solar power cogenerators in a typical installation should be about 20 to 30 percent of the overall demand load.

Solar power generation could also be designed to provide complete self-sufficiency for the entire complex during daytime operation, but also under a net metering agreement the system could be readily sized to produce surplus power that could be fed into the power grid if so desired.

A grid-connected solar power cogeneration system under a special service agreement with the primary electric power service provider could generate energy credit that will provide adequate compensation for all energy use during absence of full insolation, such as during cloudy or rainy days and during nights.

The power production capacity of a solar cogenerator system is directly proportional to the number of solar power panels installed, the efficiency of conversion devices, and the daily insolation, which is the effective daily solar aperture time for a geographic location. With reference to the local atmospheric conditions, the insolation time for Hemet is 5.5 to 6 hours of sunshine per day. The average insolation time (when sun rays impact the photovoltaic panel perpendicularly) is based on the mean value of sunny days throughout four seasons.

PROJECT SITE CLIMATIC CONDITIONS AND SURFACE METEOROLOGY

- The approximate latitude and longitude of the site are 33.50 N/116.20 W.
- The average annual temperature variation for the past 10 years has been recorded to be about 16.4°C.
- The average annual recorded wind speed is 3.58 m/s.

■ The average daily solar radiation on horizontal surfaces [kWh/(m² × day)] is about 6 hours. On the average, recorded sunshine hours in summer exceed 8 hours and under worst-case conditions in winter, insolation will be about 3.2 hours.

With reference to the preceding site climatic data, yearly average solar temperature conditions, and daily solar irradiance, the site can be considered an ideal location for solar power cogeneration. Additionally, because of the specific characteristics of solar photovoltaic systems (both single and polycrystalline) it would be possible to harvest the optimum amount of electric energy.

PROPOSED SOLAR POWER COGENERATION CONFIGURATIONS

With the specific nature of the Diamond Valley Lake Aquatic Facility project in mind, solar power cogeneration would provide not only a significant amount of electric energy, but due to the aesthetic appearance of the polycrystalline PV panels, in addition to roof-mount arrays and parking lot carports, they could also be used in verandas, picnic areas, or park bench solar shades. Solar-powered lampposts with integrated battery packs could also be used for common area and parking lot illumination. In fact, with the availability of multicolor PV panels, it would be possible to integrate the landscape architecture and solar power to create a unique blend of technology and natural beauty.

Because of the availability of the vast surface area within the campus, it is also quite possible to set up a solar power park that could generate a significant amount of electric energy beyond the present and future needs of the project.

In view of the limited roof area and the direction of the incline, the net available area that would be suitable for solar power panel installation will be limited to about 1500 to 2000 ft², which would effectively yield about 15 to 20 kWh of energy. At an average insolation time of 5.5 hours (conservative), energy contribution for the system will be as follows:

■ Hourly electric power energy produced—15 to 20 kWh
■ Daily total energy produced—82 to 110 kWh
■ Total demand as per electric power calculations—182 kWh
■ Total daily demand over 8 hours daily operation—1400 kWh
■ Percent solar power contribution—6 to 8 percent
■ Rebated energy—about 20,000 W (net ac power produced)
■ Total installed cost—$160,000

Option I. Floor-mount solar power for existing aquatic facility

■ Estimated CEC rebate amount—$72,000 (at $3.5/W) to $83,000 (at $4.0/W). (The rebate amount is subject to availability of funds at SCE or Southern California Gas.)
■ Federal tax credit based on the balance of the net installed cost being 10%—$8600 to $9500.
■ Net installed expenditure—$80,000 (at $3.5/W rebate) to $71,000 (at $4.00/W rebate).
■ Saving over the next 25 years—about $600,000.
■ Installed cost per ac watts—$7.6 to $8.0.
■ Payback on investment—10 to 15 years.

Figure 7.7 Typical solar power farm installation. *Graphics courtesy of UNIRA.*

Figure 7.8 A typical tilted roof solar power system. *Photo courtesy of Atlantis Energy.*

Option II. Roof-mount solar power for community center building   In view of the limited roof area and the direction of the incline, the net available area that would be suitable for solar power panel installation will be limited to about 12,500 ft^2, or 50 percent of the roof. If the roof design is made flat, then the effective area could be as high as 20,000 ft^2, which could yield substantially more power. At an average insolation time of 5.5 hours (conservative), energy contribution for the system will be as follows:

■ Hourly electric power energy produced—135 kWh.
■ Daily total energy production—720 kWh.
■ Total demand as per previous electric power projection—500 kWh.
■ Total daily demand over 8 hours daily operation—4000 kWh.
■ Solar power contribution—18 percent.
■ Rebated energy—about 135,000 W (net ac power produced).
■ Total installed cost—$1,087,000.
■ Estimated CEC rebate amount—$478,000 (at $3.5/W) to $546,000 ($4.0/W). (The rebate amount is subject to availability of funds at SCE or Southern California Gas.)
■ Federal tax credit based on the balance of the net installed cost being 10 percent—$54,000 to $610,000.
■ Net installed expenditure—$519,000 (at $3.5/W rebate) to $460,000 (at $4.00/W rebate).
■ Saving over the next 25 years—about $2,900,000.00.
■ Installed cost per ac watts—$0.6 to $8.0.
■ Payback on investment—8 to 12 years.

Option III. Parking stall solar power canopies   Another viable solar power installation option which is commonly used throughout the United States, Europe, and Australia is parking canopies. Solar power canopies are modular prefabricated units manufactured by a number of commercial entities that can be tailored to meet specific architectural aesthetics in a specific setting. Canopies when installed side by side could provide a any desired amount of solar power.

In general, the cost per wattage of fabricated solar power canopies is estimated to be about $1.50 to $2.00 depending on the structural design requirements dictated by the architecture. A typical parking stall when fully covered represents 100 ft^2 of usable area, which translates into about 1000 to 1500 W/h of solar power. As per the preceding estimate a group of 10 parking stalls can produce 15 kWh of energy and could provide as much power as the economics permit.

Depending on the cost of the canopy structure, the installed cost of a 1-kWh system could range from $9000 to $9500. The rebate amount would be about 40 percent, which would translate into $3000 to $3800 of CEC rebate.

POLLUTION ABATEMENT CONSIDERATION

According to a 1999 study report by the U.S. Department of Energy, 1 kW of energy produced by a coal-fired electric power–generating plant requires about 5 lb of coal.

Figure 7.9 A typical roof-installed solar power installation. *Photo courtesy of PowerLight Corporation.*

Likewise, generation of 1.5 kWh of electric energy per year requires about 7400 lb of coal that in turn produces 10,000 lb of carbon dioxide (CO_2).

Roughly speaking the calculated projection of the power demand for the project totals about 2500 to 3000 kWh. This will require between 12,000,000 and 15,000,000 lb of coal, thereby producing about 16,000,000 to 200,000,000 lb of carbon dioxide and contributing toward air pollution and global warming from greenhouse gases.

Solar power in turn if implemented as discussed here will substantially minimize the air pollution index. In fact, the EPA will soon be instituting an air pollution indexing system that will be factored in all future construction permits. At that time all major industrial projects will be required to meet and adhere to the air pollution standards and offset excess energy consumption by means of solar or renewable energy resources.

ENERGY ESCALATION COST PROJECTION

According to an Energy Information Administration data source published in 1999, California is among the top-10 energy consumers in the world, and this state alone consumes just as much energy as Brazil or the United Kingdom. Since the entire global crude oil reserves are estimated to last about 30 to 80 years and over 50 percent of the nation's energy is imported from abroad, it is inevitable that in the near future energy costs will undoubtedly surpass historical cost escalation averaging projections.

It is estimated that the cost of nonrenewable energy will, within the next decade, be increased by approximately 4 to 5 percent by producers. When compounded with a general inflation rate of 3 percent, the average energy cost over the next decade could be expected to rise at a rate of about 7 percent per year. This cost increase does not

Figure 7.10 Architectural solar power parking stalls. *Photo courtesy of Atlantis Energy.*

Figure 7.11 Solar power canopy rendering. *Courtesy of Integrated Solar Technologies.*

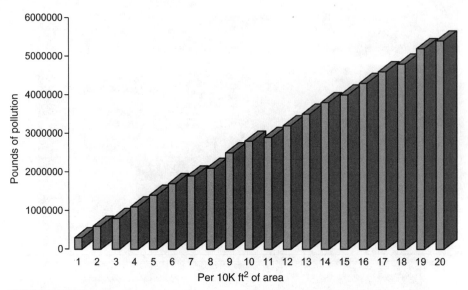

Figure 7.12 Commercial energy pollution graph per 10,000 ft^2, pounds of CO$_2$ emission of area.

take into account other inflation factors such as regional conflicts, embargoes, and natural catastrophes.

In view of the fact that solar power cogeneration systems require nearly zero maintenance and are more reliable than any human-made power generation devices (the systems have an actual life span of 35 to 40 years and are guaranteed by the manufacturers for a period of 25 years), it is my opinion that in a near-perfect geographic setting such as Hemet, their integration into the mainstream of the architectural design will not only enhance the design aesthetics, but will generate considerable savings and mitigate adverse effects on the ecology and from global warming.

As indicated in the solar power cogeneration study, the average installed cost per watt of electric energy is approximately 40 percent of the total installed cost. The unit cost encompasses all turnkey cost components such as engineering design documentation, solar power components, PV support structures, electrical hardware, inverters, integration labor, and labor training.

PROJECT COST ANALYSIS AND STATE-SPONSORED REBATE FUND STATUS

Structures in the preceding costs include roof-mount support frames and simple carport canopies only. Special architectural monuments if required may necessitate some incremental cost adjustment.

As per the California Energy Commission, all solar power cogeneration program rebate applications if applied for by December of 2004 will be subject to a 40 percent subsidy. Rebate allotments are strictly dependent on the amount of funding available at the time of application and are granted on a first come, first serve basis.

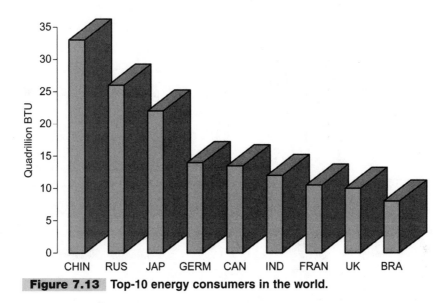

Figure 7.13 Top-10 energy consumers in the world.

SYSTEM MAINTENANCE AND OPERATIONAL COSTS

As mentioned earlier, solar power systems have a near-zero maintenance requirement. However, to prevent marginal degradation in output performance from dust accumulation, solar arrays require a biyearly rinsing with a regular water hose. Since solar power arrays are completely modular, system expansion, module replacement, and troubleshooting are simple and require no special maintenance skills. All electronic dc-to-ac inverters are modular and can be replaced with a minimum of downtime.

An optional (and relatively inexpensive) computerized system-monitoring console can provide a real-time performance status of the entire solar power cogeneration system. A software-based supervisory program featured in the monitoring system can also provide instantaneous indication of a solar array performance and malfunction.

CONCLUSION

Even though the initial investment of a solar power cogeneration system requires a large capital investment, the long-term financial and ecological advantages are so significant that their deployment in the existing project should be given special consideration.

A solar power cogeneration system, if applied as per the recommendations reviewed here, will provide a considerable energy expenditure savings over the life span of the recreation facility and provide a hedge against unavoidable energy cost escalation.

SPECIAL NOTE

In view of the depletion of existing CEC rebate funds, it is recommended that application for the rebate program be initiated at the earliest possible time. Furthermore,

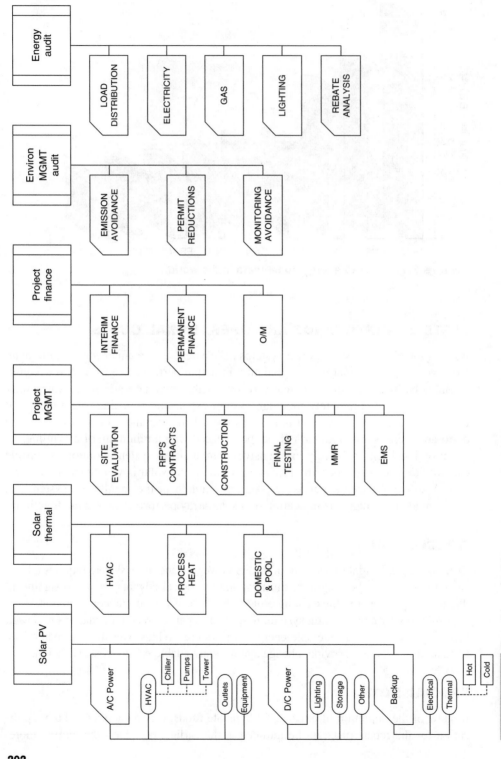

Figure 7.14 Energy project team. *Courtesy of EnviroTech.*

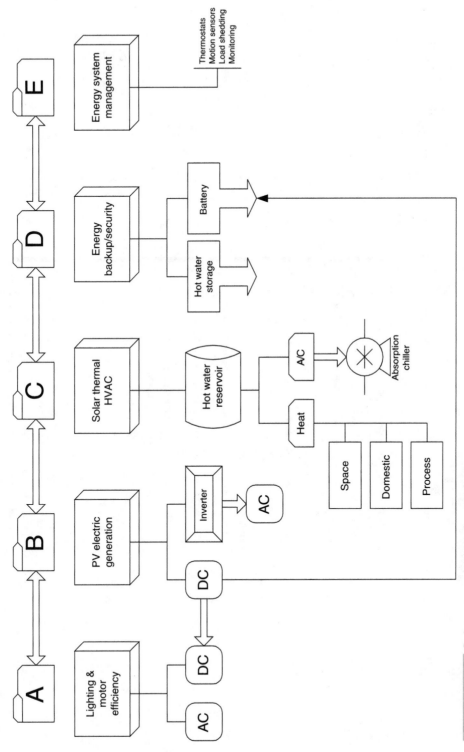

Figure 7.15 Solar energy integration project. *Courtesy of EnviroTech.*

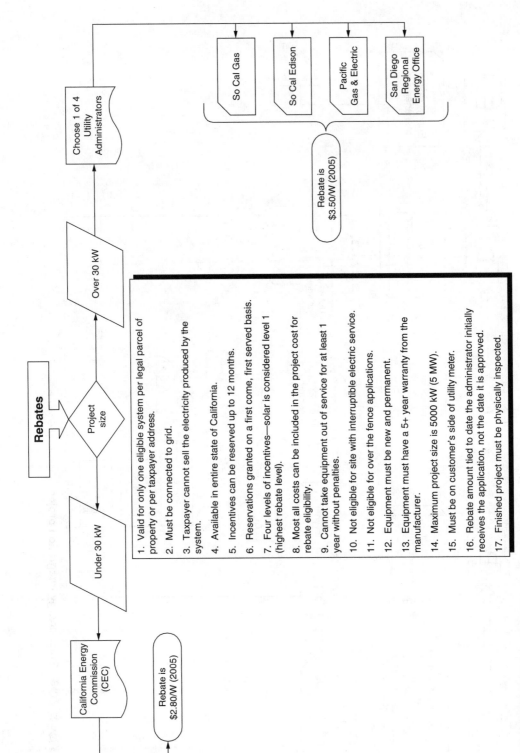

Rebates

Project size

Under 30 kW

Over 30 kW

California Energy Commission (CEC)

Rebate is $2.80/W (2005)

Choose 1 of 4 Utility Administrators

So Cal Gas

So Cal Edison

Pacific Gas & Electric

San Diego Regional Energy Office

Rebate is $3.50/W (2005)

1. Valid for only one eligible system per legal parcel of property or per taxpayer address.
2. Must be connected to grid.
3. Taxpayer cannot sell the electricity produced by the system.
4. Available in entire state of California.
5. Incentives can be reserved up to 12 months.
6. Reservations granted on a first come, first served basis.
7. Four levels of incentives—solar is considered level 1 (highest rebate level).
8. Most all costs can be included in the project cost for rebate eligibility.
9. Cannot take equipment out of service for at least 1 year without penalties.
10. Not eligible for site with interruptible electric service.
11. Not eligible for over the fence applications.
12. Equipment must be new and permanent.
13. Equipment must have a 5+ year warranty from the manufacturer.
14. Maximum project size is 5000 kW (5 MW).
15. Must be on customer's side of utility meter.
16. Rebate amount tied to date the administrator initially receives the application, not the date it is approved.
17. Finished project must be physically inspected.

Figure 7.16 Energy rebate process. *Courtesy of EnviroTech.*

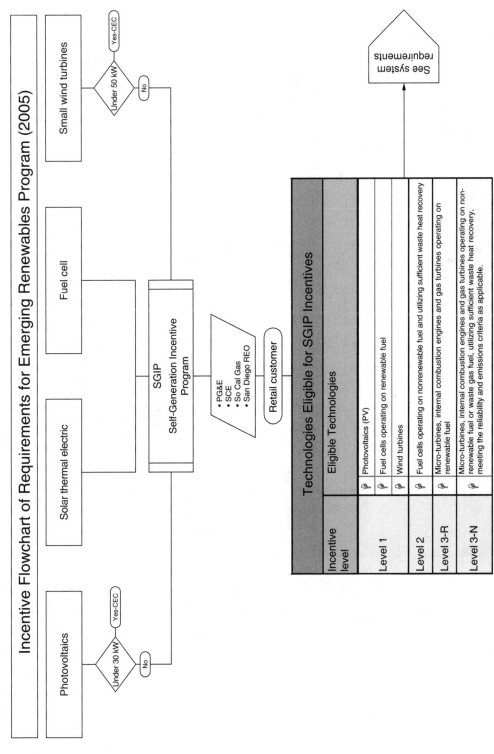

Figure 7.17 Incentive program flow chart. *Courtesy of EnviroTech.*

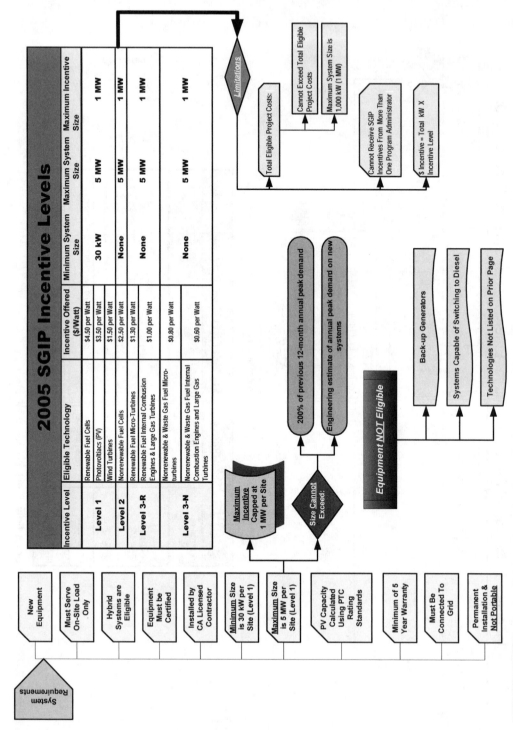

2005 SGIP Incentive Levels

Incentive Level	Eligible Technology	Incentive Offered ($/Watt)	Minimum System Size	Maximum System Size	Maximum Incentive Size
Level 1	Renewable Fuel Cells	$4.50 per Watt	30 kW	5 MW	1 MW
	Photovoltaics (PV)	$3.50 per Watt			
	Wind Turbines	$1.50 per Watt			
Level 2	Nonrenewable Fuel Cells	$2.50 per Watt	None	5 MW	1 MW
Level 3-R	Renewable Fuel Micro-Turbines	$1.30 per Watt	None	5 MW	1 MW
	Renewable Fuel Internal Combusion Engines & Large Gas Turbines	$1.00 per Watt			
Level 3-N	Nonrenewable & Waste Gas Fuel Micro-turbines	$0.80 per Watt	None	5 MW	1 MW
	Nonrenewable & Waste Gas Fuel Internal Combustion Engines and Large Gas Turbines	$0.60 per Watt			

System Requirements

- New Equipment
- Must Serve On-Site Load Only
- Hybrid Systems are Eligible
- Equipment Must be Certified
- Installed by CA Licensed Contractor
- Minimum Size is 30 kW per Site (Level 1)
- Maximum Size is 5 MW per Site (Level 1)
- PV Capacity Calculated Using PTC Rating Standards
- Minimum of 5 Year Warranty
- Must Be Connected To Grid
- Permanent Installation & Not Portable

Maximum Incentive Capped at 1 MW per Site

Size Cannot Exceed:
- 200% of previous 12-month annual peak demand
- Engineering estimate of annual peak demand on new systems

Equipment NOT Eligible
- Back-up Generators
- Systems Capable of Switching to Diesel
- Technologies Not Listed on Prior Page

Limitations

Total Eligible Project Costs:
- Cannot Exceed Total Eligible Project Costs
- Maximum System Size is 1,000 kW (1 MW)

- Cannot Receive SGIP Incentives From More Than One Program Administrator
- $ Incentive = Total kW X Incentive Level

Figure 7.18 **Renewable energy incentive levels.** *Courtesy of EnviroTech.*

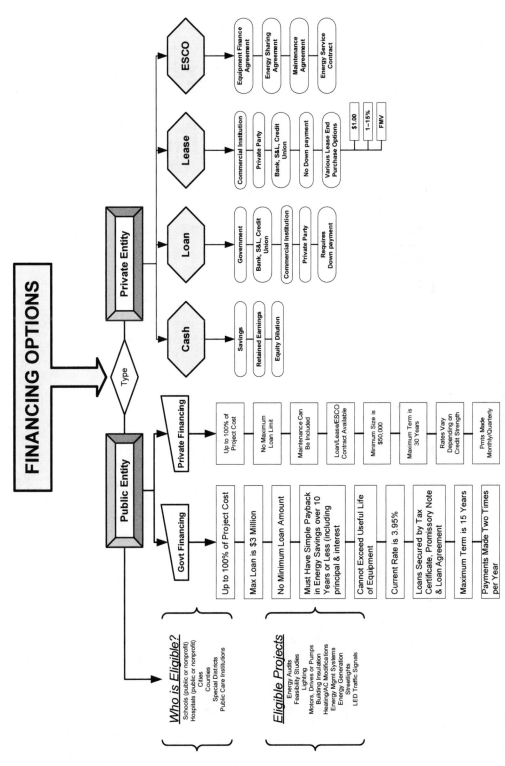

Figure 7.19 Renewable energy financial options. *Courtesy of EnviroTech.*

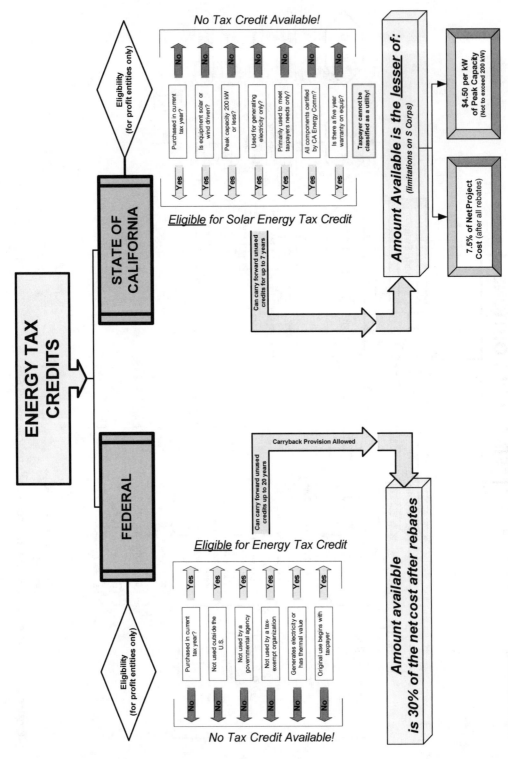

Figure 7.20 Energy tax credit system process. *Courtesy of EnviroTech.*

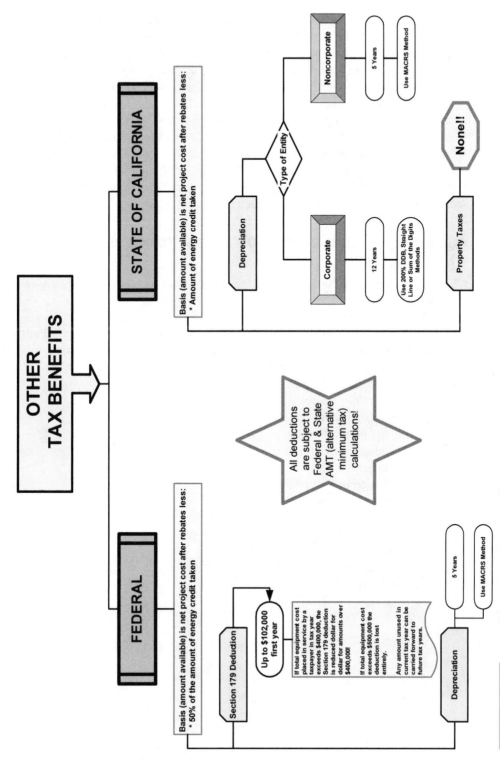

Figure 7.21 Renewable energy tax benefit process. *Courtesy of EnviroTech.*

because of the design integration of the solar power system with the service grid, the decision to proceed with the program must be made at the commencement of the construction design document stage.

As indicated in this report, electrical service planning of the overall campus must also be given serious consideration; otherwise, the present electrical system will be overloaded without possibility of expansion, which will necessitate a costly replacement of switchgear equipment and infrastructure underground conduit works. It is therefore recommended that the overall service demand requirements of the campus be reviewed with the local Romoland, California, SCE office service planning personnel (Mr. Sy Granillo) to ensure that future needs of the project are met.

We would also suggest that Valley Wide Recreation management inquire about site-specific, special low-lighting tariffs which may result in additional savings.

SUMMARY OF ADVANTAGES OF SOLAR POWER SYSTEM

1 State tax rebate—$2.80 to $2.50 per ac watt (subject to availability of CEC rebate funds)
2 Peak power shaving
3 Increased R-value—about 10 percent less air-conditioning power use. The "R" value denotes standard insulation rating class
4 Accelerated capital equipment depreciation—5 years
5 Federal tax credit—10 percent
6 Low interest rate—5 percent for 5 years
7 Increases property value
8 No additional property tax on property
9 Extremely long equipment life
10 Minimal maintenance
11 High system component reliability
12 Modular system expandability
13 Ecologically friendly
14 Excellent hedge against increased fuel cost
15 Yearly net metering credit
16 Substantial savings over the life span of the equipment
17 Mortgage rollover

8

PASSIVE SOLAR HEATING
TECHNOLOGIES

Introduction

In this chapter we will review the basic principles of passive solar energy and applications. The term *passive* implied that solar power energy is harvested by direct exposure of fluids, such as water or a fluid medium, that absorb the heat energy and subsequently convert the energy to steam or vapor, which in turn is used to drive turbines or provide evaporation energy in refrigerating and cooling equipment.

Solar power is the sun's energy without which life as we know it on our planet would cease to exist. Solar energy has been known and used by humankind throughout the ages. As we all know, concentrating solar rays with a magnifying glass can provide intense heat energy that can burn wood or heat water to a boiling point. As discussed later, recent technological developments of this simple principle are currently being used to harness solar energy and provide an abundance of electric power.

Historically the principles of heating water to a boiling point were well known by the French, who in 1888 used solar power to drive printing machinery (see Figure 8.1).

Please refer to Appendix D for a detailed solar power historical time line.

Passive Solar Water Heating

The simplest form of harvesting energy is accomplished by exposing fluid-filled pipes to the sun's rays. Modern technology passive solar panels used for heating water for pools and general household use are constructed from a combination of magnifying glasses and fluid-filled pipes. In some instances pipes carry special heat-absorbing fluids, such as bromide, that heat up quite rapidly. In other instances, water is heated and circulated by small

Figure 8.1 Historical use of passive solar power to run the printing press.

pumps. In most instances pipes are painted black and are laid on a silver-colored reflective base that further concentrates the solar energy. Another purpose of silver backboards is to prevent heat transmission to the roofs or support structures. Figure 8.2 shows diagram of a passive solar panel components.

Pool Heating

Over the years, a wide variety of pool heating panel types have been introduced. Each has its intrinsic advantages and disadvantages. Four primary types of solar pool collector design classifications are as follows:

1 Rigid black plastic panels (polypropylene)
2 Rubber mat or other plastic or rubber formulations

Energy from the sun

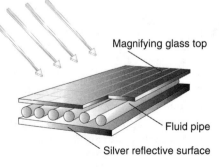

Magnifying glass top

Fluid pipe

Silver reflective surface

Figure 8.2 Passive solar water heating panel.

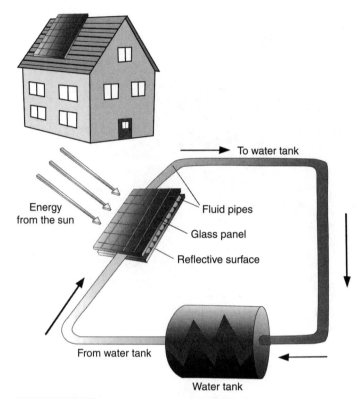

Energy from the sun

To water tank

Fluid pipes

Glass panel

Reflective surface

From water tank

Water tank

Figure 8.3 Household passive solar water heating panel.

3 Tube and fin metal panels with a copper or aluminum fin attached to copper tubing
4 Plastic-type systems

Plastic panels   This technology makes use of modular panels with panel dimensions ranging about 4 ft wide and 8, 10, or 12 ft in length. Individual panels are coupled together to achieve the desired surface area. The principal advantages of this type of technology are lightness of product, chemical inertness, and high efficiency. The panels are also durable and can be mounted on racks. The products are available in a glazed version to accommodate for windy areas and colder climates.

The disadvantage of this technology is its numerous system surface attachments, which can limit mounting locations.

Rubber mat   These systems are made up of parallel pipes, called headers, which are manufactured from extruded lengths of tubing that have stretching mats between the tubes. The length and width of the mats are adjustable and are typically custom fit for each application, as seen in Figure 8.5.

The advantage of this technology is that due to the flexibility of the product, it can be installed and adapted to roof obstructions, like vent pipes. Installations require

Figure 8.4 Passive solar power water heater in industrial use. *Courtesy of Department of Energy.*

Figure 8.5 Rubber mat roof-mount solar water heater. *Courtesy of UMA/HELICOL.*

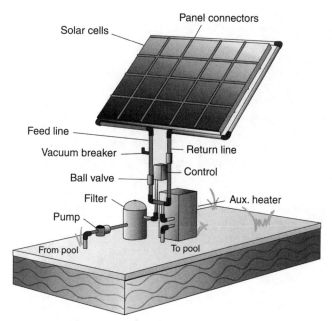

Figure 8.6 **Roof-mount solar water heating system diagram.**

few, if any, roof penetrations and are considered highly efficient. Because of the expandability of the product, the headers are less subject to freeze expansion damage. The main disadvantage of the system is that the mats are glued to the roof and can be difficult to remove without damaging either the roof or the solar panel. The installation also cannot be applied in rack-type installations.

Metal panels   These classes of products are constructed from copper waterways that are attached to either copper or aluminum fins. The fins collect the solar radiation and conduct it into the waterways. Advantages of these classes of product include their rigidity and durability of construction. Like rubber mats, glazed versions of these panels are also available for application in windy areas and cold climates.

A significant disadvantage of this type of technology is that it requires significantly more surface area, has low efficiency, and has no manufacturer's warranty.

Plastic pipe systems   In this technology, plastic pipes are connected in parallel or are configured in a circular pattern. The main advantage of the system is that installation can be done inexpensively and could easily be used as an overhead "trellis" for above-deck installations, as shown in Figure 8.7.

The main disadvantage of this type of installation is that it requires significantly larger surface area than other systems, and, like metal panel, does not carry a manufacturer's warranty.

Figure 8.7 **Residential roof-mount solar pool heater.** *Courtesy of UMA/HELICOL.*

PANEL SELECTION

One of the most important considerations when selecting a pool heating system is the amount of panel surface area that is required to heat the pool. The relationship of the solar collector area to the swimming pool surface area must be adequate in order to ensure that your pool achieves the temperatures you expect, generally in the high 70s to the low 80s in degrees Fahrenheit at a minimum, during the swimming season. The percentage of solar panel surface area to pool surface area varies with geographic location and is affected by factors, such as local microclimates, solar collector orientation, pool shading, and desired heating season.

It is very important to keep in mind that solar energy is a very dilute energy source. Only a limited amount of useful heat falls on each square foot of panel. Consequently, whatever type of solar system is used, a large panel area is needed to collect adequate amount of energy.

In southern California, Texas, and Arizona, where there is abundant sunshine and warm temperatures, the swimming season stretches from April or May to September or October. To heat a pool during this period, it is necessary to install enough solar collectors to equal a minimum of 70 percent of the surface area of the swimming pool, when the solar panels are facing south.

Generally, it is desirable to mount the panels on a southerly exposure; however, an orientation within 45 degrees of south will not significantly decrease performance as long as shading is avoided. A due-west exposure will work well if the square footage of the solar collector is increased to compensate. However, a due-east exposure is generally to be avoided, unless significantly more solar collectors are used.

Figure 8.8 An industrial solar liquid heating panel installation.
Courtesy of Solargenix.

As the orientation moves away from ideal, sizing should increase to 80 to 100 percent, or more for west or southeast orientations. If climatic conditions are less favorable, such as near the ocean, even more coverage may be required. In general, it is always recommended to exceed the minimums to offset changing weather patterns. However, there is a point of diminishing returns, where more panels won't add significantly to the pool heating function. Table 8.1 shows economics for a typical pool heating installation.

HELICOL SOLAR COLLECTOR SIZING EXAMPLE

Divide the solar collector area needed by the total square feet of individual collectors as shown here. The following example demonstrates the use of the insolation chart:

- Calculate the solar panel requirement for a 14 ft × 28 ft pool located in Las Vegas, Nevada.
- The pool surface area is 14 ft × 28 ft = 392 ft².
- Las Vegas is located in zone 5 which has a 0.52 multiplier.
- The collector area is 392 ft × 0.52 ft = 203.8 ft².
- The approximate number of panels required using Helicol HC-40 panels is 5.1, or 5 panels.

Sizing is an art, as well as a science. There are so many factors that affect swimming pool heat losses, that no one has yet come up with the "perfect" model or sizing

TABLE 8.1 TYPICAL POOL HEATING INSTALLATION ECONOMICS	
Pool size	7500 ft^2
Pool depth	5 ft
Total purchase price	$180,000
First year energy savings	$30,575 (fuel cost per therm is $1.00)
10-year energy savings	$38,505 (estimated)
Expected investment payback	5.3 years
Yearly average return: internal rate of return	21%
Size of each panel	12.5 ft^2
Number of panels	144
Total required area	7200 ft^2
Panel tilt	2°
Panel orientation	180°

calculation. Your company may already have a sizing guide, or sizing method, that works well for installations in your particular area and you may not want to use the sizing method outlined in this chapter. Be sure to find out from your sales manager what sizing method or calculation you should use to determine the proper sizing for your systems.

This chapter outlines a sizing method that can be applied to any geographical area. If you follow the guidelines detailed in this chapter and have a thorough grasp of your geographical factors, you should be able to properly size all your solar system proposals with a reasonable degree of accuracy and confidence.

For the average pool water, temperatures ranging from 76° to 80°F are usually considered comfortable. In the northern states, however, 72°F is considered warm, and in the southern states, the swimmers usually want the temperature to be 82°F.

SOLAR WATER HEATING SYSTEM SIZING GUIDE

This guideline should be used as a general example for calculating solar water heating systems. It is assumed that the pool will be covered when the nighttime temperatures drop below 60°F. If you heat a pool, you should use a solar blanket. Not to do so is much like heating a house without a roof—the heat just goes right out the top! Use of a cover retains more than two-thirds of the collected heat needed to maintain a comfortable swimming temperature.

The key to properly sizing a system is to take into account all the environmental and physical factors that pertain to your area in general and the prospect's home in particular. There are 10 questions that you need to have answered to size a system. They are as follows:

1 How many months of the year is the pool used?
2 Taking into account the geographic location, how long can the season reasonably be extended?
3 Will there be a backup heating system? If so, what kind?
4 Does the pool have a screen enclosure?
5 Will a blanket be used?
6 Is there a solar window?
7 Is wind going to be a problem?
8 Is shading going to be a problem? If so, how many hours a day?
9 What direction and at what angle will the collectors be mounted?
10 What is the surface area of the pool?

Some of these questions will be answered as part of your pool heating survey. The rest will be determined by measurement and inspection.

The following guideline will give a *sizing factor* that represents how many square feet of solar collector area is needed in relation to the pool's surface area. Once determined, this factor is multiplied times the pool surface area. The resulting answer is divided by the selected collector area to determine the number of collectors required.

1 To begin, you will want to determine a sizing factor for optimum conditions in your geographic location. It is recommended to contact the local weather bureau and ask for the *mean daily solar radiation* (Langley reading) for the coldest month of the desired swimming season. Using the following table, determine the starting sizing factor by the corresponding Langley reading.

LANGLEY READING	SIZING FACTOR
200	1.05
250	0.96
300	0.85
350	0.75
400	0.67
450	0.60
500	0.55
550	0.51
600	0.48

2 For optimum efficiency, solar collectors should face south. If you are unable to face your system south, multiply the sizing factor by the following amount:

East facing   1.25
West facing   1.15

Increase this figure if you have a roof with a pitch equal to or greater than $^6/_{12}$. Decrease this figure if you have a roof with a pitch equal to or less than a $^4/_{12}$.

3 If the pool is shaded, you need to multiply the sizing factor by the following amounts:

25% shaded   1
50% shaded   1.25
75% shaded   1.50
100% shaded   1.75

As a general rule of thumb, if there is a screen enclosure, multiply the sizing factor by 1.25. If the pool is indoors, multiply the sizing factor by 2.00.

4 In the northern states, the best collector angle is the latitude minus 10 degrees. This gradually changes as you move south until it reaches the latitude plus 10 degrees in southern Florida. For each 10-degree variance from the optimum angle, multiply the sizing factor by the amount indicated in the following table:

DEVIATION	FIGURE
±10°	1.05
±20°	1.10
±30°	1.20

As a general rule of thumb, if collectors are laid flat, multiply the sizing factor by 1.10.

5 For collector area, use the following:

HELICOL COLLECTOR NO.	AREA (FT²)
HC-50	50
HC-40	40
HC-30	30

An example   The following example is used to illustrate sizing factor determination and the number of collectors needed for a 15 ft × 30 ft rectangular pool, partially shaded, flat roof.

1 Langley reading for coldest month of swimming is 500, so the sizing factor is 0.55.

2 Collectors are going to face south.

3 Pool is shaded by 25 percent, so multiply the sizing factor by 1.1 or 10% increase.

4 Collectors are going to go on a flat roof, so multiply by another 1.1: the final sizing factor is then (0.55 × 1.1 × 1.1) = 0.67.

5 Surface area of the pool is 450 ft².

6 Multiply the surface area by the sizing factor of 0.67. The collector area needed is 302 ft².

7 HC-40s are going to be used on the project, so divide by 40 ft². The number of HC-40 collectors needed is 8.

This guide is an approximation only. Wind speeds, humidity levels, desired pool temperature, and other factors can also affect proper solar pool system sizing. If the prospect does not want to use a cover, you may have to double or even triple the solar coverage to achieve the desired swimming seasons.

The choke of the collector size and system configuration is dependent on the designer. The roof space as well as the associated cost must be considered. Installation of smaller collectors will be a great deal more difficult than that of larger ones. None of these rules are concrete, and the determinations require the designer's best judgment.

SOME USEFUL SUGGESTED PRACTICES

Use of common sense when investing in solar pool heating is very important. The first-time buyer should consider the following:

- Buy only from a licensed contractor, and check on the contractor's experience and reputation.
- Be aware that several factors should be considered when evaluating various system configurations. More solar panels, generally, means your pool will be warmer.
- Use a pool cover, if possible.
- Make sure the system is sized properly. An inadequately sized system is guaranteed dissatisfaction.
- Beware of outrageous claims, such as "90-degree pool temperatures in December with no backup heater." No solar heating system can achieve such a performance.
- The contractor should produce evidence of adequate worker's compensation and liability insurance.
- Insurance certificates should be directly from the insurance company and not the contractor.
- Check the contractor's referrals before buying.
- Get a written description of the system, including the number of solar panels, the size of the panels, and the make and model number.
- Get a complete operation and maintenance manual and startup demonstration.
- The price should not be the most important factor. But it should also not be dramatically different from prices of competing bidders for similar equipment.
- Be sure the contractor obtains a building permit, if required.

Concentrator Solar Technologies

Concentrating solar power (CSP) technologies concentrate solar energy to produce high-temperature heat, which is then converted into electricity. The three most advanced CSP technologies currently in use are parabolic troughs (PT), central receivers (CR), and dish engines (DE). The CSP power plant is today considered to be one of the most efficient. CSP can readily substitute solar heat for fossil fuels, fully or partially, to reduce emissions and provide additional power at peak times. Dish engines are better suited for distributed power, from 10 kW to10 MW, while parabolic troughs and central receivers are suited for larger central power plants, 30 to 200 MW and higher.

The solar resource for generating power from concentrating solar power systems is very plentiful, and sufficient electric power for the entire country could be generated by covering only about 9 percent of the state of Nevada, a plot of land 100 mi^2 with parabolic trough systems.

The amount of power generated by a concentrating solar power plant depends on the amount of direct sunlight. Like photovoltaic concentrators, these technologies use only a direct beam of sunlight to concentrate the thermal energy of the sun.

The southwestern United States potentially offers an excellent opportunity for developing concentrating solar power technologies. It is well known that peak power demand, generated as a result of air-conditioning systems, can be offset by solar electric-generating system plants, which operate for nearly 100 percent of the on-peak hours of Southern California Edison.

Concentrating solar power systems can be of sizes from 2 to 10 kW or could be large enough to supply grid-connected power of up to 200 MW. Some existing systems use thermal storage during cloudy periods and are combined with natural gas resulting in hybrid power plants that provide grid-connected dispatchable power. Solar power–driven electric generator conversion efficiencies make concentrating technologies a viable renewable energy resource in the Southwest. The U.S. Congress recently requested that the Department of Energy develop a plan for installing 1000 MW of concentrating solar power in the Southwest over the next 5 years. Concentrating solar power technologies are also considered an excellent source for providing thermal energy for commercial and industrial processes.

BENEFITS

CSP technologies incorporating storage do not burn any fossil fuels and produce zero greenhouse gas (NO$_x$ and SO$_x$) emissions. They have also proven to be reliable. For the past decade the San Diego Electric and Gas Service (SEGS) plants have operated successfully in the southern California desert, providing enough power for 100,000 homes. Plants with cost-effective storage or natural gas hybridization can deliver power to the utility grid whenever that power is needed, not just when the sun is shining. Existing CSP plants produce power now for around 11 cents/kWh (including both capital and operating costs) with projected costs dropping below 4 cents/kWh within the next 20 years as technology refinements and economies of scale are implemented.

Because CSP uses relatively conventional technologies and materials (glass, concrete, steel, and standard utility-scale turbines), production capacity can be scaled up to several hundred megawatts per year rapidly.

EMISSIONS

The emissions benefits of CSP technologies depend on many factors including whether the plants have their own storage capacity or are hybridized with other electricity or heat production technologies. CSP technologies with storage, as stated before, produce zero emissions, and hybrid technologies can reduce emissions by 50 percent or more.

PARABOLIC THROUGH HEATING SYSTEM TECHNOLOGIES

In this technology, a large field of parabolic systems, which are secured on a single-axis solar tracking support, are installed in a modular parallel row configuration aligned in the northsouth horizontal direction. Each of the solar parabolic collectors track the movement from east to west during daytime hours and focus the sun's rays onto a linear receiver tubing which circulates a heat transfer fluid (HTF). The heated fluid is in turn passed through a series of heat exchanger chambers where the heat is transferred to superheated vapor that drives steam turbines. Upon propelling the turbine the spent steam is condensed and is returned to the heat exchanger via condensate pumps.

At present the technology has been successfully applied in thermal electric power generation. A 354-MW solar power–generated electric plant installed in 1984 in California's Mojave Desert has been in operation with remarkable success.

SOLAR TOWER TECHNOLOGY

Another solar concentrator technology that generates electric power from the sun focuses concentrated solar radiation on a tower-mounted heat exchanger. The system

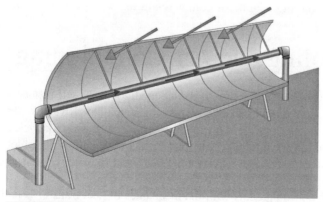

Figure 8.9 Solar parabolic heater system diagram.

Figure 8.10 **Solar parabolic heating system installation.** *Courtesy of Solargenix.*

basically is configured from thousands of sun-tracking mirrors, commonly referred to as heliostats, which reflect the sun's rays onto the tower.

The receiver contains a fluid that once heated in a similar manner to the trough parabolic system transfers the absorbed heat in the heat exchanger to produce steam that then drives a turbine to produce electricity.

Power generated from this technology produces up to 400 MW of electricity. The heat transfer fluid, usually a molten liquid salt, can be raised to 550°F. The HTF is stored in an insulated storage tank and used in the absence of solar ray harvesting.

Recently a solar pilot plant located in southern California, called Solar Two, which uses nitrate salt technology, has been producing 10 MW of grid-connected electricity with a sufficient thermal storage tank to maintain power production for 3 hours, which has rendered the technology viable for commercial use.

Solar Cooling and Air Conditioning

Most of us associate cooling, refrigeration, and air conditioning as self-contained electro-mechanical devices, connected to an electric power source, that provide conditioned air for spaces in which we live as well as refrigerate our food stuff and groceries.

Technically speaking, technology that makes the refrigeration possible is based upon a basic fundamental concept of physics called heat transfer. Cold is essentially the absence of heat; likewise darkness is the absence of light.

The branch of physics that deals with the mechanics of heat transfer is called thermodynamics. There are two principal universal laws of thermodynamics. The first law concerns the conservation of energy, which states that energy neither can be created nor destroyed; however, it can be converted from one type to another. The second law of thermodynamics deals with the equalization and transfer of energy from a higher state to a lower one. Simply stated, energy is always transferred from a higher potential or state to a lower one, until two energy sources achieve exact equilibrium. Heat is essentially defined as a form of energy created as a result of the transformation of another form of energy, a common example of which is when two solid bodies are rubbed together, which results in friction heat. In general, heat is energy in a transfer state, because it does not stay in any specific position and constantly moves from a warm object to a colder one, until such time, as per the second law of thermodynamics, both bodies reach heat equilibrium.

It should be noted that volume, size, and mass of objects are completely irrelevant in the heat transfer process; only the state of heat energy levels are factors in the energy balance equation. With this principle in mind, heat energy flows from a small object to a much larger mass. For example, a hot cup of coffee will transfer heat to one's hand. The rate of travel of heat is directly proportional to the difference in temperature between the two objects.

Heat travels in three forms, namely as radiation, conduction, and convection. As radiation, heat is transferred in waveform similar to radio waves, microwaves, or light.

For example, the sun transfers its energy to Earth by rays or radiation. In conduction, heat energy flows from one medium or substance to another by physical contact. Convection on the other hand is the flow of heat between air, gas, liquid, or a fluid medium. Installed solar heat collector panels are shown in Figure 8.13. These types of heat collecting panels are deployed in hybrid chiller systems when the heat of the circulating liquid is further increased by auxiliary natural gas burners. In refrigeration the basic principles are based on the second law of thermodynamics, that is, transfer or removal of heat from a higher energy medium to a lower one by means of convection.

TEMPERATURE

Temperature is a scale for measuring heat intensity by the directional flow of energy. Water freezes at 32°F (0°C) and boils at 212°F (100°C). Temperature scales are simply temperature differences between freezing and boiling water temperatures measured at sea level. As mentioned earlier, based on the second law of thermodynamics, heat transfer or measurement of temperature is not dependent on the quantity of heat.

MOLECULAR AGITATION

Depending on the state of heat energy, most substances in general can exist in vapor, liquid, and solid states. As an example, depending on the heat energy level, water can exist as solid ice when it is frozen, as a liquid at room temperature (between the two boundary temperatures of 32°F and 212°F), and as a vapor form when heated above the boiling temperature of 212°F.

Figure 8.11 Solar heat collector panel used in hybrid chillers.
Courtesy of Solargenix.

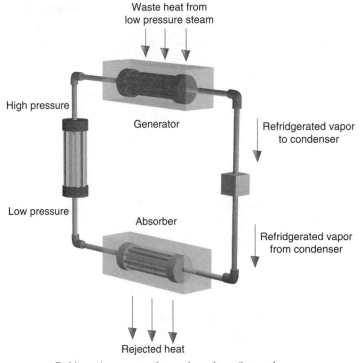

Refrigeration evaporation and condensation cycle

Figure 8.12 **Second law of thermodynamics.** *Courtesy of Vector Delta Design Group, Inc.*

Figure 8.13 Solar heat collector panel installation in Arizona Desert. *Courtesy of Solargenix.*

Steam will condense back to the water state if heat energy is removed (cooled) from it. Water will change into a solid state (ice) when sufficient heat energy is removed from it. The processes can be reversed when heat energy is introduced.

The state of change is related to the fact that in various substances, depending upon the presence or absence of heat energy, a phenomenon referred to as atomic thermal agitation causes expansion and contraction of molecules. Close contraction of molecules forms solids, and larger separations transform matter into the liquid and gaseous states. A hybrid absorption chiller evaporator is shown in Figure 8.16. In borderline energy conditions, beyond the solid and gaseous states, excess lack and surplus of energy are referred to, respectively, as supercooled or superheated states.

PRINCIPLES OF REFRIGERATION

Refrigeration is accomplished by two distinct processes. In one process, referred to as the compression cycle, a medium, such as freon gas, is first given heat energy by compression, which turns the gas into liquid. Then in a subsequent cycle, energy is removed from the liquid, in the form of evaporation or gas expansion, which disperses the gas molecules and turns the surrounding chamber into a cold environment.

A circulating medium of energy-absorbing liquid such as water or air when circulated within the so-called evaporation chamber gives up its heat energy to the expanded gas. The cold water or air is in turn circulated by means of pumps into environments that have higher ambient heat energy levels. The circulated cold air in turn exchanges or passes the cold air into the ambient space through radiator tubes or fins, thus lowering the energy of the environment.

Temperature control is realized by opening and closing of cold media circulating tube valves or air duct control vanes, modulated by a local temperature-sensing device, such as a thermostat or a set point control mechanism.

COOLING TECHNOLOGIES

There are two types of refrigeration technologies currently in use, namely electric vapor-compression (freon gas) and heat-driven absorption cooling.

Absorption cooling chillers are operated by steam, hot water, fossil fuel burners, or combinations of all. There are two types of absorption chillers, one uses lithium bromide (LiBr) as an energy conversion medium and water as a refrigerant. In this type of technology the lowest temperature achieved is limited to 40°F.

Another absorption chiller technology uses ammonia as the energy conversion medium and a mix of ammonia and water as the refrigerant. The maximum limit of temperature for this technology is 20°F. Both of these technologies have been around for about 100 years.

The basic principle of absorption chillers is based on gasification of the LiBr or ammonia. Gasification takes place when either of the media is exposed to heat. Heat could be derived from fossil fuel gas burners, hot water obtained from geothermal energy, passive solar water heaters, or microturbine generators, which use landfill gases to produce electricity and heat energy.

COEFFICIENT OF PERFORMANCE

The energy efficiency of an air-conditioning system is defined by a coefficient of performance (COP), which is defined as the ratio of cooling energy to the energy supplied to the unit. A ton of cooling energy is 12,000 British thermal units per hour (Btu/h), which as defined in olden days is the energy required to remove heat from space to ambient air obtained through melting a ton of ice. One ton or 12,000 Btu is equal to 3514 W of electric power.

Based on these definitions, the cooling coefficient of performance of an air-conditioning unit, which requires 1500 W of electric power would have 12,000 Btu/h (3514/1500) = 2.343. Obviously, a lower electric energy requirement will increase the COP rating which brings us to the conclusion that the lower the amount of energy input, the better the efficiency.

SOLAR POWER COOLING AND AIR CONDITIONING

A combination use of passive solar and natural gas-fired media evaporation has given rise to a generation of hybrid absorption chillers that can produce a large tonnage of cooling energy by the use of solar or geothermal heated water. A class of absorption chillers, which commonly used LiBr and have been commercially available for some time, use natural and solar power as the main sources of energy. A 1000-ton absorption chiller can reduce electric energy consumption by an average of 1 MW, or 1 million watts, which will have a very significant impact on reducing the electric power consumption and resulting environmental pollutions as described in earlier chapters.

TABLE 8.2   TYPICAL BUILDING COOLING CAPACITIES

SPACE	SIZE	COOLING TONS
Medium office	50,000 ft^2	100–150
Hospital	150,000 ft^2	400–600
Hotel	250,000 ft^2	400–500
High school	50,000 ft^2	100–400
Retail store	160,000 ft^2	170–400

DESICCANT EVAPORATORS

Another solar power cooling technology makes use of a solar desiccant evaporator air-conditioning technology, which reduces outside air humidity and passes it through an ultraefficient evaporative cooling system. This cooling process, which uses an indirect evaporative process, minimizes the air humidity, and this makes use of the technology quite effectively in coastal and humid areas. Typical building cooling capacities are shown in Table 8.2.

Direct Solar Power Generation

The following project undertaken by Solargenix Energy makes use of special parabolic reflectors that concentrate the solar energy rays into circular pipes located at the focal center of the parabola. Concentrated reflection of energy elevates temperature of the circulating mineral liquid oil within the pipes, raising the temperature to such levels that allow considerable steam generation via special heat exchangers that drive power turbines. The following abstract reflects a viable electrical power generation in Arizona.

RED ROCK, ARIZ.—APS today broke ground on Arizona's first commercial solar trough power plant and the first such facility constructed in the United States since 1990.

Located at the company's Saguaro Power Plant in Red Rock, about 30 miles north of Tucson, the APS Saguaro solar trough generating station will have a one megawatt (MW) generating capacity, enough to provide for the energy needs of approximately 200 average-size homes. APS has contracted with Solargenix Energy to construct and provide the solar thermal technology for the plant, which is expected to come online in April 2005. Solargenix, formerly Duke Solar, is based out of Raleigh, North Carolina. Solargenix has partnered with Ormat who will provide the engine to convert the solar heat, collected by the Solargenix solar collectors, into electricity.

"The APS Saguaro Solar Trough Power Plant presents a unique opportunity to further expand our renewable energy portfolio," said Peter Johnston, manager of Technology Development for APS. "We are committed to developing clean renewable energy sources today that will fuel tomorrow's economy. We believe solar-trough technology can be part of a renewable solution.

The company's solar-trough technology uses parabolic-shaped reflectors (or mirrors) to concentrate the sun's rays to heat a mineral oil between 250 and 570 degrees. The fluid then enters the Ormat engine passing first through a heat exchanger to vaporize a secondary working fluid. The vapor is used to spin a turbine, making electricity. It is then condensed back into a liquid before being vaporized once again.

Historically, solar-trough technology has required tens of megawatts of plant installation to produce steam from water to turn generation turbines. The significant first cost of multi-megawatt power plants had precluded their use in the APS solar portfolio. This solar trough system combines the relatively low cost of parabolic solar trough thermal technology with the commercially available, smaller turbines usually associated with low temperature geothermal generation plants, such as the Ormat unit being used for this project.

In addition to generating electricity for APS customers, the solar trough plant will help APS meet the goals of the Arizona Corporation Commission's Environmental Portfolio Standard, which requires APS to generate 1.1 percent of its energy through renewable sources—60 percent through solar—by 2007. APS owns and operates approximately 4.5 megawatts of photovoltaic solar generation around the state and has partnered on a 3-megawatt biomass plant in Eager, which came online in February, and a 15-megawatt wind farm to be constructed near St. Johns. APS, Arizona's largest and longest-serving electricity utility, serves about 902,000 customers in 11 of the state's counties. With headquarters in Phoenix, APS is the largest subsidiary of Pinnacle West Capital Corp.

Innovations in Passive Solar Power Technology

The following are a few innovations under development by Energy Innovations, a subsidiary of Idea Labs, located in Pasadena, California.

SUNFLOWER 250

The experimental solar power tracking concentrator shown in Figure 8.18 consists of a 3-m^2 platform equipped with adjustable motorized reflective mirrors that concentrate solar rays on to focally located solar cells which produce electricity. The prototype units are currently undergoing testing in Pasadena, California, and throughout various geographic locations. Initial tests have showed very promising performance. Manufactured units are expected to be available within the next couple of years.

Manufactured units would be ground mounted on stands and will be connected via internal wiring. The undercarriage would be shielded by a wind guard around the perimeter. Each unit is expected to produce a peak power rating of 200 W.

The continuous sun tracking mechanism of the design enables the sunflower to produce 30 percent more energy than a traditional flat PV panel of similar rating. At present, development continues to provide wind mitigation solutions.

STIRLING ENGINE SUNFLOWER

The Stirling Engine Sunflower, shown in Figure 8.19, uses a radical concept since it does not use a stationary photovoltaic cell technology; rather it is constructed from lightweight polished aluminum zed plastic reflector petals each being adjusted by a

Figure 8.14 **Sunflower 250 prototypes.** *Photo courtesy of Energy Innovations, Pasadena, CA.*

Figure 8.15 **Stirling Engine Sunflower prototype.** *Photo courtesy of Energy Innovations, Pasadena, CA.*

Figure 8.16 **SunPod prototype.** *Photo courtesy of Energy Innovations, Pasadena, CA.*

microprocessor-based motor controller, which enables the petals to track the sun in an independent fashion. This heat engine is essentially used to produce hot water by concentrating solar rays onto a low-profile water chamber.

At present the technology is being refined to produce higher-efficiency and more cost-effective production models. At present the company is working on a larger-scale model for use in large-scale solar water heating installations.

SUNPOD

The SunPod design shown in Figure 8.20 involves a plastic Fresnel lens that focuses light onto a strip of PV material at the bottom of the tube. This late design requires only one-axis tracking to keep the units pointed toward the sun for producing electricity. This particular design requires rooftop penetration to ensure it is held firmly in place under windy conditions.

U.S. Government Annual Research Expenditure

In the recent past, the U.S. federal government had allocated about $90 million for solar energy research, unfortunately, the solar power fund, at the expense of coal, oil, and fusion energy research, has been reduced by 50 percent.

9

FUEL CELL TECHNOLOGIES

Introduction

In general, fuel cells are battery-like devices that produce electric power by means of electrochemical reactions. Unlike batteries, as long as fuel is supplied, the cells produce electricity with minimal degradation or recharging. Fuel cell degradation, depending on the type of product and technology, may degrade 1 to 4 percent per 1000 hours; as a result over a period of 5 to 7 years, stacks are replaced or rebuilt.

Fuel Cell Technology

The technical definition of a fuel cell is "an energy conversion device that generates electricity and heat by electrochemically blending a gaseous fuel and oxidizing gas using an ion-conducting electrolyte." In simplistic terms, a fuel cell is able to convert chemical energy directly into electric energy without combustion, thereby being able to furnish higher efficiencies than the ion-conducting electrolyte (ICE) method. Without combustion this has another advantage; fuel cells have lower carbon dioxide (CO_2) emissions than fossil fuels for the same power output. In principle, a fuel cell operates like a battery; however, unlike a battery; a fuel cell does not run down or require recharging and, in theory, will keep running as long as it is supplied with fuel. Fuel cells typically use hydrogen as the gaseous fuel and oxygen as the oxidant in the electrochemical reaction. The two reactions make up what is called redox (oxidation-reduction) of the fuel cell; the result is electricity, heat, and a by-product of water.

A fuel cell is essentially an electrochemical sandwich with different layers. Cells can be joined together, called cell stacking, to produce more electric power. The method employed to connect fuel cells in series is with bipolar plates that perform the function to collect and conduct the current from the anode of one cell to the cathode of the next. Fuel cells are almost silent; whatever sound is generated is due in part to auxiliary equipment like pumps, a blower to cool the stack, or a humidifier.

Fuel cells are an important technology with worldwide applications that will change the course of commercialization and pave the way for a cleaner environment in the future and decreased dependence on petroleum-based fuels.

At present, fuel cell research and development has not reached maturity and cannot fully compete with established power generation technologies; however, because of their efficient energy conversion and extremely low emissions, the cells represent a viable and promising alternative as a primary or standby source of electricity.

Fuel cells have a great potential to promote energy diversity and provide a transition to renewable energy sources. Hydrogen is the most abundant element on Earth and in the universe, representing 75 percent of all the mass of visible matter in stars and galaxies and 90 percent of their molecular mass.

When hydrogen is used as an energy source in a fuel cell, the only emission that is created is water vapor, which can be electrolyzed to produce additional hydrogen.

This continuous cycle of energy production has real potential in conjunction with solar power technology to replace the traditional energy sources, which cause depletion of precious nonrenewable resources and create pollution.

Fuel cells are an environmentally clean, quiet, and highly efficient method for generating electricity and heat from natural gas and other fuels. Fuel cells are an ideal technology for small power plants capable of producing several megawatts of electricity, and because of their very clean emissions and almost silent operation, this technology can be installed in urban areas and next to buildings and homes close to the electric demand source.

A fuel cell resembles a battery in the sense that it uses an electrochemical process to convert chemical energy into electric power. However, a fuel cell differs from batteries in one significant characteristic—it never runs down as long as fuel in the form of hydrogen, natural gas, or gasoline is supplied to the cell.

As with other technologies, a side benefit of the electric energy produced is that waste heat is generated, which can be used for the production of domestic or process hot water, or for space heating purposes.

Short History

The fuel cell was first discovered in 1839 by Sir William Grove, a Welsh judge and scientist. The discovery became dormant up until the 1960s when practical application of the electrochemical conversion used in U.S. space programs paved the way for today's research and product development. In the past, U.S. space programs chose fuel cells to power the Gemini and Apollo spacecrafts and still use the technology to provide electricity and water for the space shuttle.

Nowadays, most industrialized nations of the world such as the United States, Canada, Germany, Holland, Japan, and Italy have extensive nationally sponsored research-and-development programs, which promise to have a significant impact on the world economy and a reduction of global greenhouse gas production.

Basic Operation Principles

A basic fuel cell consists of two electrodes that sandwich an electrolytic membrane. Hydrogen fuel when fed into the *anode* (+) of the fuel cell, with the aid of the electrolytic membrane which acts as a catalyst, reacts chemically with oxygen or air introduced at the cathode (−).

In this chemical reaction, the hydrogen atom is split into a proton and an electron. The proton passes through the electrolyte membrane. The electrons on the other hand travel from the anode to the cathode and give rise to electric current. Electrons at the cathode reunite with the hydrogen protons. Hydrogen, in turn, reacts chemically with the oxygen, producing water molecules and heat.

This electrochemical process makes use of hydrogen as a source of fuel. Some fuel cell systems also include a preprocessing mechanism known as *fuel reformer*, which enables the hydrogen from any form of hydrocarbon-based fuel, such as natural gas, methanol, landfill methane gases, and regular gasoline, to be separated from the main molecules and used in the electrochemical conversion process.

Fuel Reformers

As mentioned earlier, fuel cells use hydrogen as the fuel for electroconversion; however, hydrogen-rich materials such as methanol, ethanol, natural gas, petroleum distillates, liquid propane, and gasified coal can also serve as possible fuel sources.

One method of reformation, known as *endothermic steam reforming*, combines fuels with steam by vaporizing them together at very high temperatures. In this process, hydrogen is separated from the hydrocarbon molecules by use of a special membrane, referred to as a membrane exchange assembly (MEA), or proton exchange membrane (PEM), as shown in Figure 9.1. The main drawback of the steam reforming technique is that the process requires energy consumption.

Another type of fuel reformation is known as *partial oxidation* (POX), which produces carbon dioxide (CO_2) but no particulates, nitrogen monoxide (NO_x), sulfur-monoxide (SO_x) smog-producing agents, or gases.

Types of Fuel Cells

PHOSPHORIC ACID FUEL CELLS

This type of fuel cell uses a phosphoric acid compound as the catalytic conversion membrane. It generates electricity with a conversion efficiency of about 40 percent. The large amount of heat produced by the process (about 400°F) is used for steam production, which through cogeneration is used to produce additional electricity.

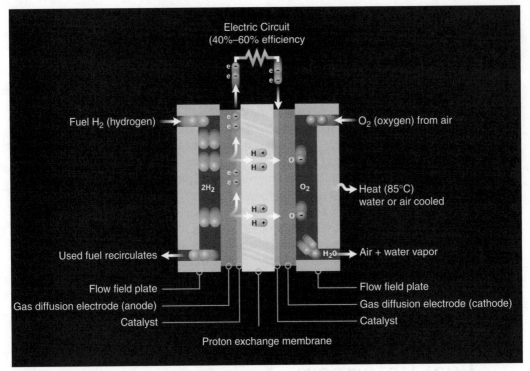

Figure 9.1 **Fuel cell operation diagram.** *Courtesy of Ballard Engineering.*

To date, hundreds of commercially available phosphoric acid fuel cells (PAFC) have been installed throughout the world that are used by power utility companies, airport terminals, hotels, hospitals, municipal waste dumps, offices, and buildings.

Recently, many automotive manufacturers such as Toyota, DaimlerChrysler, and many more, use PEM fuel cells in their hybrid automobiles (using a combination of a fuel cell and an internal combustion engine) and are racing to produce more efficient product systems that will ultimately be used in transportation technology as the principal means of power generation.

PROTON EXCHANGE MEMBRANE FUEL CELLS

Proton exchange membrane fuel cells (PEMFC) operate at low temperatures (about 200°F), have a relatively high energy density, and are capable of changing output power demand at a quick rate. These particular characteristics make the application of these cells in automobiles, where quick startup is paramount, very appropriate.

The proton exchange membrane, in this type of fuel cell, is a thin plastic sheet that allows hydrogen ions to pass through it. This MEA is coated on both sides with dispersed platinum alloy particles, which act as a catalyst. Hydrogen extracted by a fuel reformer is fed to the anode side of the fuel cell, where the catalytic membrane separates the electrons and permits passage of protons to the cathode side.

Figure 9.2 A stationary fuel cell power generator.
Courtesy of Ballard Engineering.

Similar to the PAFC electrochemical process, the hydrogen ions are recombined with the electrons, which subsequently react with the oxygen-producing heat and water molecules. A stationary fuel cell assembly power generator is shown in Figure 9.2.

MOLTEN CARBONATE FUEL CELLS

These fuel cells use molten carbonate as a catalyst to separate the hydrogen electrons from the molecule. The cells operate at very high temperatures (1200°F) and are considered to be very efficient. Heat generated as a by-product of fuel cell operation is recovered and used in heat exchangers that provide heated water for use in many commercial and industrial applications, which substantially augments the overall operational efficiency.

Molten carbonate fuel cells (MCFC) operate on hydrogen, carbon monoxide, natural gas, propane, landfill gases, marine diesel fuels, and coal gasified products. In Italy and Japan, multimegawatt MCFCs have been successfully installed and tested as a stationary stand-alone power generation system. At present there are a number of stationary MCFCs, used for base load electricity and heat energy, in operation in California.

SOLID OXIDE FUEL CELLS

This technology uses a hard ceramic solid oxide material as an electrolytic conversion catalyst, which operates at extremely high temperatures (1800°F). Solid oxide fuel cells (SOFC) are capable of producing several hundred kilowatts of power at an efficiency near or exceeding 60 percent. A special solid oxide fuel cell, known as a tubular solid oxide fuel cell, uses tubes of compressed solid oxide discs resembling metal can tops, which are stacked to 100 cm high.

Another experimental fuel cell technology, referred to as a planar solid oxide fuel cell (SOFC), promises the potential for producing small-scale (3–5 kW) power generators. It is intended for use in stationary power generation, as a power supply in remote areas, and as an auxiliary power for use in vehicles. SOFC fuel cell applications at present are not considered to be suitable for use as a main prime power source.

Recently, SOFC technologies have made significant progress in product development and are being manufactured commercially by a number of companies. At present the main obstacle to use of the technology is the high-temperature production resulting from the chemical reaction of hydrogen and oxygen which must be mitigated. The fuel cell, due to the significant power production capacity and high efficiency, will in the near future be used in motor vehicles and as a source for electric power generation.

ALKALINE FUEL CELLS

Alkaline fuel cells (AFC) use potassium hydroxide in the electrochemical catalytic conversion process and have a conversion efficiency of about 70 percent. In the last few decades, NASA has used alkaline cells to power space missions.

Up until recently, due to the high cost of production, alkaline fuel cells were not available commercially. However, improvements and a cost reduction in fuel cell production have created a new opportunity to commercialize the technology.

DIRECT METHANOL FUEL CELLS

This type of fuel cell uses a similar proton exchange electroconversion process as discussed earlier and uses a polymer membrane as the electrolyte; however, without a reformer mechanism, the anode acts as a catalyst that separates the hydrogen from the liquid. It should be noted that until recently, due to the high cost of production, alkaline fuel cells have not been available commercially; however, improvements and cost reduction in fuel cell production have created a new opportunity to commercialize the technology. Direct methanol fuel cells (DMFC) operate at a relatively low temperature (120°F–200°F) and have about 40 percent efficiency. At more elevated temperatures, these fuel cells operate at a higher conversion efficiency.

In the United States, future commercialization of DMFC technology will depend upon genetically modified corn, which can be fermented at low temperatures. Such a breakthrough could indeed make the technology a viable source of electric energy production, which could significantly contribute to reducing air pollution and minimizing the expensive import of crude oil.

REGENERATIVE FUEL CELLS

This technology, which is still in its early research-and-development stage, is a closed-loop electroconversion system which uses solar-powered electrolysis to separate hydrogen and oxygen molecules from water. Hydrogen and oxygen are fed into a fuel cell, which in addition to generating electricity produces heat and water. At present NASA is undertaking research using the technology for space applications and as a stationary auxiliary power unit.

Benefits of Fuel Cell Technology

It is estimated that the fuel cell market within the next decade could exceed $20 billion worldwide. In addition to an expanding market for alternative electric energy, a significant percentage of the millions of vehicles produced annually throughout the world, which use internal combustion, will be converted into hybrid technology. It is also believed that the demand for fuel cells by the transportation industry within the next decade will increase annually by an additional $15 billion. In the United States, passenger vehicles alone consume over 6 million barrels of oil everyday, which represents 85 percent of our oil imports.

If only 25 percent of vehicles could operate with fuel cells, oil imports could be reduced by 1.8 billion barrels a day or about 650 billion barrels a year, which produces an unfavorable balance of trade and a considerable amount of air pollution. Furthermore, if each vehicle produced in the future was designed to operate on fuel cells, this country's electric power generation capacity would increase by 200 percent.

The U.S. Department of Energy estimates that if only 10 percent of the nation's vehicles were powered by fuel cells, the yearly import of crude oil would be reduced by about 13 percent or 800,000 barrels. The production of greenhouse gases, such as carbon dioxide, would be reduced by 60 tons and air pollution particulates would be reduced by one million tons.

Impact of Fuel Cells on the Global Economy

In view of wide-ranging applications of technology in markets, such as for steel production, electric power generation, vehicles, and transportation industries, fuel cells could have a significant impact on the global market economy since they could provide employment for tens of thousands of high-quality jobs.

It is estimated that each 1000 MW of fuel cell energy production would create 5000 new jobs, and if only 25 percent of cars in the nation were to use the technology, jobs created would exceed 1,000,000, which would have a significant positive impact on the U.S. gross national product.

Examples of Fuel Cell Demonstrations

Ballard participates in a number of demonstration programs designed to showcase fuel technology around the world as part of the commercialization process.

CALIFORNIA FUEL CELL PARTNERSHIP

The California Fuel Cell Partnership (CaFCP), South Coast Air Quality Management District, and a collaboration of auto manufacturers, fuel suppliers, fuel cell manufacturers, and state government agencies. To demonstrate and test fuel cell vehicles under every-day conditions, the CaFCP, headquartered in Sacramento, California, investigates and demonstrates the viability of site infrastructure technology, promotes public awareness of premembrane fuel cell–powered vehicles, and explores the path to commercialization by identifying potential problems and solutions.

VANCOUVER FUEL CELL VEHICLE PROGRAM

The Vancouver Fuel Cell Vehicle Program (VFCVP) is a 5-year, $9 million joint initiative between Canada, Ford Motor Company, the government of Canada, Technology Early Action Measures, and the government of British Columbia.

This project will demonstrate five Ballard-powered fuel cell vehicles in real-world conditions on British Colombia's lower mainland and is the first fleet demonstration of fuel cell vehicles in Canada. Demonstrating third-generation Ford fuel cell vehicles, this project will provide valuable information on performance and reliability that can be applied toward the evolution of fuel cell vehicles to the commercial marketplace in the transition to a hydrogen economy.

Ballard will have its own fuel cell for the duration of the program. Other vehicle users include the city of Vancouver, the government of British Columbia, Fuel Cells Canada, and the (National Research Council (NRC).

SANTA CLARA VALLEY TRANSPORTATION AUTHORITY

The Santa Clara Valley Transportation Authority (VTA) contracted with Gillig Corporation and Ballard Power Systems of Burnaby, Canada, to build three hydrogen-powered, zero-emission cell buses (ZEBs) for use in regular transit service. Air Products & Chemicals, Inc., will supply hydrogen, which is converted to hydrogen gas at VTA's fueling station at Cerone. An additional project is the Sunline Transit Agency in Thousand Palms, California, which has an extended history of use of alternative fuel vehicles and buses and is also a testing facility for many vehicle companies that conduct a variety of tests in the California desert.

EUROPEAN FUEL CELL BUS PROJECT

Ballard heavy-duty fuel cell engines are inside 30 Mercedes buses running in revenue transit service in 10 European cities in yearly demonstration programs, which include nine cities and the Ecological City Transport System Reykjavik, Iceland. The European Union has led the way in zero-emission fuel cell technology.

SUSTAINABLE TRANSPORT ENERGY FOR PERTH

Three Ballard fuel cell-powered buses commenced operation in Perth, Australia, on September 27, 2004. As of November, the buses were working more than 8 hours per day and 5 days per week. The buses had traveled more than 8000 kilometers (km) and had operated for 400 hours.

10

WIND ENERGY TECHNOLOGIES

Introduction

This chapter describes the operation of wind turbine technology and provides an overview of its application. Some of the material can also be found at the educational Web sites of the American Wind Energy Association (www.awea.org), the U.S. Department of Renewable Energy Laboratories Turbine Manufacturers Association (www.nrel.gov/wind).

Wind Power Energy

The contents of this chapter are based on a wind power tutorial that can be found on the American Wind Energy Association Web site. It was prepared in cooperation with the U.S. Department of Energy and the National Renewable Laboratory.

In reality, wind energy is a converted form of solar energy. The sun's radiation heats different parts of Earth at different rates, most notably during the day and night but also when different surfaces (for example, water and land) absorb or reflect at different rates.

This causes portions of the atmosphere to warm differently. Hot air rises, reducing the atmospheric pressure at Earth's surface, and cooler air is drawn in to replace it. The result is wind. Air has mass, and when it is in motion, it contains the energy of that motion, *kinetic energy*. Some portion of that energy can be converted into other forms—mechanical force or electricity—which we can use to perform work.

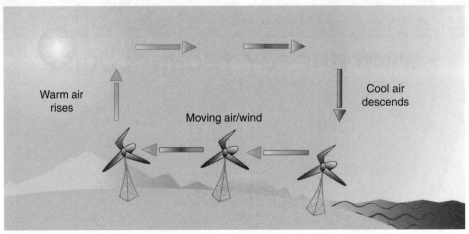

Figure 10.1 Graphic representation of wind kinetic energy at work.

Basics of Wind Power Turbine Operation

A wind energy system transforms the kinetic energy of wind into mechanical or electric energy that can be harnessed for practical use. Mechanical energy is most commonly used for pumping water in rural or remote locations. The farm windmill, still seen in many rural areas of the United States, is a mechanical wind pump, but it can also be used for many other purposes (e.g., grinding grain and sawing). Wind electric turbines generate electricity for homes, businesses, and state utilities.

There are two basic designs of wind electric turbines, vertical-axis and horizontal-axis (propeller style) machines. Horizontal-axis wind turbines are most common today, constituting nearly all the utility-scale (100-kW capacity and larger) turbines in the global market. Turbine system components include the following:

- A blade or a rotor (which in some installations could span over 50 ft), which converts the wind energy into rotational shaft energy
- An enclosure referred to as nacelle, which contains the drive mechanism consisting of a gearbox and an electric generator
- A support structure such as a tower that supports the nacelle
- Electric equipment and components, such as controls and interconnection equipment

There are also types of turbines that directly drive the generator and do not require a gearbox. The amount of electricity that wind turbines generate depends on the turbine's capacity or power rating.

The electricity generated by a utility-scale wind turbine is normally collected and fed into utility power lines, where it is mixed with electricity from other power plants and delivered to utility customers.

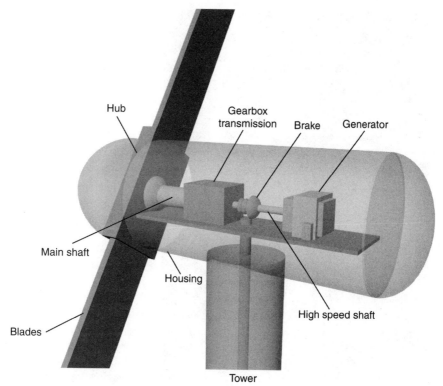

Figure 10.2 Wind turbine mechanisms.

Energy Generation Capacity of a Wind Turbine

The ability to generate electricity is measured in watts. Watts are very small units, so the terms kilowatt (kW, or 1000 W), megawatt (MW, 1 million W), and gigawatt (pronounced "jig-a-watt," GW, 1 billion W) are most commonly used to describe the capacity of generating units like wind turbines or other power plants.

Electricity production and consumption are most commonly measured in kilowatt-hours (kWh). A kilowatt-hour means 1 kW of electricity is produced or consumed for 1 hour. One 50-W lightbulb left on for 20 hours consumes 1 kilowatt-hour (kWh) of electricity (50 W × 20 h = 1000 Wh = 1 kWh). The output of a wind turbine depends on the turbine's size and the wind's speed through the rotor. Wind turbines being manufactured now have power ratings ranging from 250 W to 1.8 MW (MW).

For example, a 10-kW wind turbine, shown in Figure 10.3, can generate about 10,000 kWh annually at a site with wind speeds averaging 12 mi/h or about enough to power a typical household. A 1.8-MW turbine can produce more than 5.2 million kWh in a year, enough to power more than 500 households. The average U.S. household consumes about 10,000 kWh of electricity each year.

Figure 10.3 A small capacity wind turbine installation. *Photo courtesy of Berg Electric.*

A practical example of a project is a 250-kW turbine installed at the elementary school in Spirit Lake, Iowa. It provides an average of 350,000 kWh of electricity per year, more than is necessary for the 53,000-ft² school. Excess electricity is fed into the local utility system, which earned the school $25,000 in the turbine's first 5 years of operation. The school uses electricity from the utility at times when the wind does not blow. This project has been so successful that the Spirit Lake school district has since installed a second turbine with a capacity of 750 kW.

Wind speed is a crucial element in projecting turbine performance, and a site's wind speed is measured through wind resource assessment prior to a wind system's construction. Generally, an annual average wind speed greater than 4 m/s (9 mi/h) is required for small wind electric turbines (less wind is required for water-pumping operations).

Utility-scale wind power plants require minimum average wind speeds of 6 m/s (13 mi/h). The power available in the wind is proportional to the cube of its speed, which means that doubling the wind speed increases the available power by a factor of 8. Thus, a turbine operating at a site with an average wind speed of 12 mi/h could in theory generate about 33 percent more electricity than one at an 11-mi/h site, because the cube of 12 (1768) is 33 percent larger than the cube of 11 (1331). In the real world, the turbine will not produce quite that much more electricity, but it will still generate much more than the 9 percent difference in wind speed.

The important thing to understand is that what seems like a small difference in wind speed can mean a large difference in available energy and in electricity produced, and therefore, a large difference in the cost of the electricity generated. Also, there is little energy to be harvested at very low wind speeds; 6-mi/h winds contain less than one-eighth the energy of 12-mi/h winds.

Construction of Wind Turbines

Utility-scale wind turbines for land-based wind farms come in various sizes, with rotor diameters ranging from about 50 to 90 m and with towers of roughly the same size. A 90-m machine, definitely at the large end of the scale, with a 90-m tower would have a total height from the tower base to the tip of the rotor of approximately 135 m (442 ft). Offshore turbine designs now under development have rotors that have a 110-m diameter. It is easier to transport large rotor blades by ship than by land. Wind turbines intended for residential or small business use are much smaller. Most have rotor diameters of 8 m or less and would be mounted on towers of 40 m in height or less.

Most manufacturers of utility-scale turbines offer machines in the 700 kW to 1.8 MW range. Ten 700-kW units would make a 7-MW wind plant; while ten 1.8-MW machines would make an 18-MW facility. In the future, machines of larger size will be available, although they will probably be installed offshore, where larger transportation and construction equipment can be used. Units larger than 4 MW in capacity are now under development.

One megawatt of wind energy can generate between 2.4 million and 3 million kWh annually. The towers are mostly tubular and made of steel. The blades are made of fiberglass-reinforced polyester or wood epoxy. A wind farm is shown in Figure 10.4.

Figure 10.4 Wind turbine installation in Palm Springs, California. *Courtesy of AWEMA.*

Wind Turbine Energy Economics

The *energy payback time* is a term used to measure the net energy value of a wind turbine. In other words it determines how long the plant has to operate to generate the amount of electricity that was required for its manufacture and construction.

Several studies have looked at this question, over the years, and have concluded that wind energy has one of the shortest energy payback times of any energy technology. A wind turbine typically takes only a few years (3–8 years, depending on the average wind speed at its site) to pay back the energy needed for its fabrication, installation, operation, and retirement.

Since you can't count on the wind blowing, what does a utility gain by adding 100 MW of wind to its portfolio of generating plants? Does it gain anything? Or should it also add 100 MW of fueled generation capacity to allow for the times when the wind is calm?

First, it needs to be understood that the bulk of the value of any supply resource is in the energy that the resource produces, not the capacity it adds to a utility system. In general, utilities use fairly complicated computer models to determine the value in added capacity that each new generating plant adds to the system.

According to some models, the capacity value of a new wind plant is approximately equal to its capacity factor. Thus, adding a 100-MW wind plant with an average capacity factor of 35 percent to the system is approximately the same as adding 35 MW of conventional fueled generation capacity.

The exact answer depends on, among other factors, the correlation between the time that the wind blows and the time that the utility sees peak demand. Thus wind farms whose output is highest in the spring months or early morning hours will generally have a lower capacity value than wind farms whose output is high on hot summer evenings.

Since wind is a variable energy source, its growing use presents problems for utility system managers. At current levels of use, however, this issue is still some distance from being a problem for most utility systems.

A conventional utility power plant uses fuel, so it will normally run much of the time unless it is idled by equipment problems or for maintenance. A capacity factor of 40 to 80 percent is typical for conventional plants.

A wind plant is fueled by the wind, which blows steadily at times and not at all at other times. Although modern utility-scale wind turbines typically operate 65 to 80 percent of the time, they often run at less than full capacity. Therefore, a capacity factor of 25 to 40 percent is common; although, they may achieve higher capacity factors during windy weeks or months.

The most electricity per dollar of investment is gained by using a larger generator and accepting the fact that the capacity factor will be lower as a result. Wind turbines are fundamentally different from fueled power plants in this respect.

If a wind turbine's capacity factor is 33 percent, it doesn't mean it is only running one-third of the time. A wind turbine at a typical location in the midwestern United States runs about 65 to 80 percent of the time. Much of the time it will be generating at less than full capacity.

AVAILABILITY FACTOR

The *availability factor*, or just *availability*, is a measurement of the reliability of a wind turbine or other power plant. It refers to the percentage of time that a plant is ready to generate, that is, when it is not out of service or under maintenance or repairs. Modern wind turbines have an availability of more than 98 percent, higher than most other types of power plants. After two decades of constant engineering refinement, today's wind machines are highly reliable.

WIND TURBINE POWER GENERATION CAPACITY

Utilities must maintain enough power plant capacity to meet expected customer electricity demand at all times, plus an additional reserve margin. All other things being equal, utilities generally prefer plants that can generate as needed (that is, conventional plants) to those that cannot (such as wind plants).

However, despite the fact that wind is variable and sometimes does not blow at all, wind plants do increase the overall statistical probability that a utility system will be able to meet demand requirements.

A rough rule is that the capacity value of adding a wind plant to a utility system is about the same as the wind plant's capacity factor multiplied by its capacity, as shown earlier. Thus, a 100-MW wind plant with a capacity factor of 35 percent would be similar in capacity value to a 35-MW conventional generator. For example, in 2001 the Colorado Public Utility Commission found the capacity of a proposed 162-MW wind plant in eastern Colorado (with a 30 percent capacity factor) to be approximately 48 MW.

Figure 10.5 A typical wind turbine installation. *Courtesy of AWEMA.*

The exact amount of capacity value that a given wind project provides depends on a number of factors, including average wind speeds at the site and the match between wind patterns and utility load (demand) requirements. It also depends on how geographically dispersed wind plants on a utility system are and how well-connected the utility is with neighboring systems that may also have wind generators. The broader the wind plants are scattered geographically, the greater the chance that some of them will be producing power at any given time.

Wind Turbine Energy Supply Potential for the United States

Wind energy could supply about 20 percent of the nation's electricity, according to Battelle Pacific Northwest Laboratory, a federal research laboratory. Wind energy resources used for generating electricity can be found in nearly every state.

However, the U.S. wind resources are even greater. North Dakota alone is theoretically capable of producing enough wind-generated power to meet more than one-third of the U.S. electricity demand. The theoretical potentials of the windiest states are shown in Table 10.1.

Present projections show that wind power can provide at least up to a fifth of a system's electricity, and the figure could probably be higher. Wind power currently provides nearly 25 percent of the electricity demand in the northern German state of Schleswig-Holstein. In western Denmark, wind supplies 100 percent of the electricity that is used during some hours on windy winter nights.

CONSISTENCY OF SUPPORT POLICY

Over the past 5 years, the federal production tax credit has been extended twice, but each time Congress allowed the credit to expire before acting and then only approved short durations. The credit expired again on December 31, 2003, and as of March 2004 had still not been renewed. These expiration and extension cycles inflict a high cost on the industry, cause large layoffs, and hold up investments. Long-term, consistent policy support would help unleash the industry's pent-up potential.

Transmission Line Access

Transmission line operators typically charge generators large penalty fees if they fail to deliver electricity when it is scheduled to be transmitted. The purpose of these penalty fees is to punish generators and deter them from using transmission scheduling as a "gaming" technique to gain advantage against competitors, and the fees are therefore not related to whether the system operator actually loses money as a result of the generator's action. But because wind is variable, wind plant owners cannot

TABLE 10.1   TOP-20 WIND ENERGY–PRODUCING STATES

STATE	BILLIONS OF KILOWATT-HOURS
North Dakota	1210
Texas	1190
Kansas	1070
South Dakota	1030
Montana	1020
Nebraska	868
Wyoming	747
Oklahoma	725
Minnesota	657
Colorado	481
New Mexico	435
Idaho	73
Michigan	65
New York	62
Illinois	61
California	59
Wisconsin	58
Maine	56
Missouri	52
Iowa	51

guarantee delivery of electricity for transmission at a scheduled time. Wind energy needs a new penalty system that recognizes the different nature of wind plants and allows them to compete on a fair basis.

The electrical transmission system in the plains, which cover the central third of the United States, need to be extensively redesigned and redeveloped. At present, this system consists mostly of small distribution lines. Instead, a series of new high-voltage transmission lines are needed to transmit electricity from wind plants to population centers. Such a redevelopment will be expensive, but it will also benefit consumers and national security, by making the electrical transmission system more reliable and by reducing shortages and the price volatility of natural gas. Transmission will be a key issue for the wind industry's future development over the next two decades. Wind energy potential as measured by annual energy potential is in the billions of kilowatt-hours.

World Wind Power Production Capacity

As of the end of 2003, there were over 39,000 MW of generating capacity operating worldwide, producing some 90 billion kWh each year—as much as 9 million average American households use, or as much as a dozen large nuclear power plants could generate. Yet this is but a tiny fraction of the potential of wind.

New Transmission Lines

According to the U.S. Department of Energy, the world's winds could theoretically supply the equivalent of 5800 quadrillion BTUs (quads) of energy each year—more than 15 times current world energy demand. (A quad is equal to about 172 million barrels of oil or 45 million tons of coal.) The potential of wind to improve the quality of life in the world's developing countries, where more than 2 billion people live with no electricity or prospect of utility service in the foreseeable future, is vast.

Wind Force 12, a study performed by Denmark's BTM Consult for the European Wind Energy Association and Greenpeace, found that by the year 2020, wind could provide 12 percent of world electricity supplies, meeting the needs of 600 million average European households.

Denmark is revisiting and currently rewriting its wind policy. The degree to which that means the United States should reexamine its own policy revolves around the degree to which our situation is similar to Denmark's. In fact, a brief analysis of some major differences suggests that there are strong reasons for continuing to support wind development in the United States rather than backing away from it.

Wind supplies 20 percent of the national electricity demand in Denmark. Although the United States has nearly twice as much installed wind equipment as Denmark, wind generates only 0.4 percent of our electricity, far below the 10 percent threshold identified by most analysts as the point at which wind's variability becomes a significant issue for utility system operators.

Denmark is also so small geographically (half the size of Indiana) that high winds can cause many of its wind plants to shut down almost at once. In the United States, wind plants are much more geographically dispersed (from California to New York to Texas) and do not all experience the same wind conditions at the same time.

Rapid development of wind and new small-scale power plants within the past 5 years have brought Denmark to the point where power produced by so-called nondispatchable resources in the country's west exceeds 100 percent of the demand in the region. At many times, this excess generation leaves the country scrambling to increase electricity export capabilities to handle the surplus. This situation is essentially unimaginable in the United States.

Denmark's approach encourages community involvement but places particular stress on low-capacity distribution networks (at the "end of the line" on transmission systems). In the United States, our larger wind plants require advance transmission planning, but feed into main transmission lines and do not affect the customer distribution network.

Growing Use of Wind Energy for Utility Systems

Denmark's situation should not cause concern in the United States. Denmark's problem is that its wind turbine system has been too successful, too quickly for a small country, and Denmark must now take steps to manage that success. It is unfortunate that the United States has not dealt with its energy problems so decisively. At current levels of use, this issue is still some distance from being a problem on most utility systems.

Up to the point where wind generates about 10 percent of the electricity that the system is delivering in a given hour of the day, there is not an issue. There is enough flexibility built into the system for reserve backup, varying loads, and so forth that there is effectively little difference between a 10 percent wind system and a system with 0 percent wind. Variations introduced by wind are much smaller than routine variations in load (customer demand).

At the point where wind is generating 10 to 20 percent of the electricity that the system is delivering in a given hour, wind variation is an issue that needs to be addressed, but that can probably be resolved with wind forecasting (which is fairly accurate in the time frame of interest to utility system operators), system software adjustments, and other changes.

Once wind is generating more than about 20 percent of the electricity that the system is delivering in a given hour, the system operator begins to incur significant additional expense because of the need to procure additional equipment that is solely related to the system's increased variability.

These figures assume that the utility system has an average amount of resources that are complementary to wind's variability (e.g., hydroelectric dams) and an average amount of load that can vary quickly (e.g., electric arc furnace steel mills). Actual utility systems can vary quite widely in their ability to handle as-available output resources like wind farms. However, as wholesale electricity markets grow, fewer, larger utility systems are emerging. Therefore, over time, more and more utility systems will look like an "average" system. Since wind is a variable energy source, doesn't it cost utilities extra to accommodate on a system that mostly uses fuel-powered plants with predictable outputs?

However, as the previous answer suggests, the added cost is modest. Three major studies of utility systems with less than 10 percent of their electricity supplied by wind have found the extra or ancillary costs of integrating it to be less than 0.2 cents/kWh. Two major studies of systems with 20 percent or more electricity supplied by wind have found the added cost to be 0.3 to 0.6 cents/kWh.

Wind Farm Study Cases in the United Kingdom

The following case studies were reported by British Wind Energy of the United Kingdom. The majority of systems presented are owned and operated by private energy provider organizations.

NOVAR WIND FARM, NOSSHIRE, SCOTLAND

This wind farm is owned and operated by National Wind Power Ltd. The wind farm, which covers 300 hectares (ha) of land, has 34 turbines that cover approximately 1 percent of the land space. The installation was completed in October 1997 and each of its 34 turbines produce 500 kWh of power with a total power output of 17 MW. The land which is located 600 m above sea level also serves as cattle grazing land.

HAGSHAW HILL WIND FARM

This wind farm is located in Lanarkshire, Scotland, and is the first wind farm installation in the country; it was commissioned in 1995. The project consists of 26 turbines which produce a total of 15.6 MW of electricity. The wind farm is owned and operated by Scottish Power.

WINDY STANDARD WIND FARM

This project is located in Galloway on the hills above the Carsphaim Forest and is the largest wind farm installation in Scotland. The project was commissioned in 1996 and consists of 36 wind turbines which produce a total of 21.6 MW of electricity. The project is owned and operated by the National Wind Power Ltd., UK. The wind farm covers 350 ha of land, with the turbines covering approximately 1 percent of the total land area.

BLYTH OFFSHORE WIND FARM

This is the first offshore wind farm in the United Kingdom. The project is located in the Blyth Harbour, Northumberland, in England and was commissioned in December 2000. It consists of two turbines which produce 3.8 MW of electricity. The project is owned and operated by Blyth Offshore Wind Ltd. The wind turbines are installed 1 km from the shore in an area anchored on submerged bedrock with a water depth of approximately 6 m and tidal wave heights of 8 m.

Figure 10.6 Blyth offshore wind turbine installation.
Photo courtesy of Blyth Offshore wind Ltd., UK.

Wind Power Calculations

The following wind power calculation is based on the annual electricity consumption for residential dwellings that use gas heating.

The maximum energy produced by a wind farm is as follows:

Power produced = total turbine output capacity (kWh) × 8760 h/yr × site wind power production multiplier (0.35)

To find out the number of dwellings a project can support, the result is divided by 3880 W for gas-heated housing and 10,100 W for electric-heated residential units.

When applied to the Novar project referenced earlier, the preceding calculation results in the following:

Annual power production = 17,000 × 8760 × 0.35 = 52,122,000 W

Gas-heated dwellings supported = 52,122,000/3880 = 13,433 dwellings

Electric-heated dwellings supported = 52,122,000/10,100 = 5160 dwellings

Advantages and Disadvantages of Wind Power

ADVANTAGES

■ The wind energy is available globally, the technology is mature, and energy production is relatively inexpensive.

■ After the initial installation capital, the wind turbines require relatively minimal maintenance and produce energy without producing any atmospheric pollution.

■ Even though wind turbines are very tall, they occupy a minimal footprint, so the occupied land can be used for agricultural and cattle rearing purposes.

■ Wind farms offer an economic alternative in rural locations where grid utility service installations are not readily available.

■ Wind turbines are one of the best means to provide electric energy in third-world countries.

■ In most installations, electric power produced by wind turbines is more competitive than grid-supplied power.

DISADVANTAGES

■ To produce power, wind turbines require minimum wind speeds of 7 mi/h, which means that power produced by windmills must be supplemented with an alternative source of electric power.

■ Because of the possibility of accidental bird collisions with wind turbine blades and disturbance of marine life in offshore installations, some animal protection groups and environmentalists object to wind energy production.

■ Because of the imposing structure heights of windmills, most people in urban areas object to the landscape disfigurement.

■ During the process of manufacturing there is some pollution produced.

■ To harvest dependable amounts of energy, a large number of turbines must be installed in sufficient quantities in rural locations that have appropriate atmospheric wind current conditions to ensure a higher probability of power production. Wind turbine output production must often be supplemented with auxiliary gas turbines or other energy production means to provide some degree of power production stability.

11

OCEAN ENERGY TECHNOLOGIES

Introduction

This chapter addresses a number of technologies that are used to harvest oceanic power, which manifests in many different forms such as waves, tides, and ocean undercurrents.

Tidal Power

Tidal movement occurs due to the twice-daily variation in sea level caused primarily by the moon's pull, and to some extent the sun's pull, on Earth's oceans. Tidal power has been in use for milling grains since the eleventh century in Britain and France.

TIDAL PHYSICS EFFECTS OF THE TERRESTRIAL CENTRIFUGAL FORCE

The interaction of the moon and Earth causes the oceans facing the moon to bulge out toward the moon. In the mean time oceans on the opposite side of the globe are partly shielded from the moon's gravitational effect by Earth, but the centrifugal force created by Earth's rotation results in the oceans bulging out a little away from the moon. This is known as the lunar tide. This process is further complicated by the gravitational effect of the sun which causes a similar bulging of waters on the facing and opposing sides of Earth. This is also known as the solar tide.

Because the sun and moon are not in fixed positions in the celestial sphere, but change positions with respect to each other, they influence the tidal range, which is the difference between low and high tide. For example, when the moon and the sun are in the same plane as Earth, the overall tidal range results from the superposition of the tidal ranges due to the lunar and solar tides. This results in the maximum tidal range, which is also known as the spring tides. On the contrary when the moon and the sun are at right angles to each other, lower tidal differences are experienced, which results in what is called neap tides.

TIDAL POWER GENERATION

The generation of electricity from tides is similar to hydroelectric generation, with the exception that water flows in and out of the turbines in both directions. Therefore generators are designed to produce power when the rotor is turned in either direction.

A tidal power station known as an ebb generating system, involves construction of a dam, known as a barrage, across an estuary. A number of sluice gates on the barrage allow the tidal basin to fill on the incoming high tides and to exit through the turbine system on the outgoing tide referred to as an ebb tide. As mentioned earlier, special generators are used to produce electricity from both incoming and ebb tides

Tidal power is a technology that captures the energy contained in a moving water mass due to tides and converts it into electricity. The types of tidal energies extracted are kinetic energy, which results from currents arising between ebbing and surging tides, and potential energy, which results from the difference in height or the head differential height between high and low tides. The potential energy contained in a volume of water is determined by the formula $E = xmg$ where x is the height of the tide, m is the mass of water, and g is the acceleration due to gravitational energy.

Power generation from the kinetic energy of tidal currents has proven to be more pragmatic and much more feasible today than building ocean-based dams or barrages. At present many coastal sites worldwide are being evaluated for their suitability for harvesting tidal current energy.

The technology for extracting the tidal potential energy involves construction of barrages and tidal lagoons. The barrage technology traps water at a certain level inside a basin, thus making use of the created water head differential height that result when the water level outside of the basin or lagoon changes relative to the water level inside. The differential head static energy is converted into dynamic energy when the sluice gates open and the flow is used to drive turbines.

Tidal power is considered a renewable resource since the tidal phenomenon is caused by the orbital mechanics of the solar system which is inexhaustible for the foreseeable future.

The efficiency of tidal power generation largely depends on the amplitude of the tidal swell, which can rise up to 33 ft. These swells manifest as tidal waves which are funneled into rivers, fjords, and estuaries. Amplitudes of up to 56 ft have been observed in the Bay of Fundy, Canada. Superimposition of tidal waves gives rise to tidal resonance which amplifies the tidal waves.

Tidal energy generators are in general installed in locations where high-amplitude tides occur. Some of the significant worldwide installation locations include the former USSR, the United States, Canada, Australia, Korea, and the United Kingdom as discussed subsequently. Smaller-scale tidal power plants are also in operation in Norway.

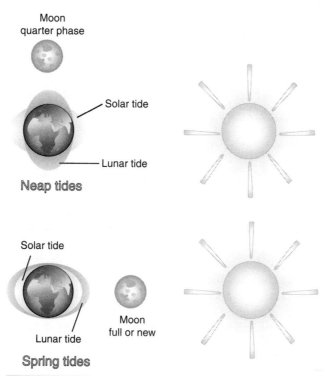

Figure 11.1 Effects of the moon and sun on the ocean tides.

Barrages

Tidal barrages, as mentioned earlier, are used to trap water in basins by means of water-trapping mechanical doors referred to as sluice gates. These gates resemble flap gates, which are raised vertically. The basic elements of a barrage are caissons, embankments, sluice gates, turbines, and ship locks. The sluices, turbines, and ship locks are housed in compartments within large concrete blocks called caissons. Embankments are used to seal the basins where they are not sealed by caissons.

INSTALLED CAPACITY

The total rated capacity of the turbines is specially optimized to meet the performance requirements of each barrage construction. In small-capacity installations where a barrage is capable of producing only a small amount of energy, the turbines will be operated at a low-power production capacity over a long period of time.

In large-capacity reservoirs where the basins can be drained very quickly in every cycle, power generation is at a very high rate, but lasts only for short periods of time, which makes it somewhat incompatible for grid interconnection. Fast pond water drainage, causing a quick drop in the water level, could also have negative effects on the environment.

MODES OF OPERATION

Ebb generation   In tidal ebb the basin is filled through the sluices when turbines are set in freewheeling mode up until high tide. When the reservoir is full, the sluice gates and turbine gates are kept closed until the sea level becomes low enough to create a sufficient head across the barrage. Then the sluice gates are lifted and the turbines generate electric power until the head becomes too low. This cycle is repeated over and over. This process is known as ebb generation, or outflow generation, because power generation occurs as the tide ebbs.

Flood generation   In this power generation process the basin is emptied through the sluice gates and the turbines generate electricity at tide flood. This process is relatively less efficient than ebb generation, since the volume contained in the upper half of the basin where ebb generation systems operate is greater than the volume of the lower half where flood generation occurs.

Another factor that contributes to the lower efficiency is the fact that the main water flow to the basin is from a river and that as the tide rises; the difference in the height of the basin and the sea side of the barrage becomes less than desirable.

Bidirectional flow power generation   Certain turbines as discussed subsequently are designed in a manner that makes them capable of power generation at tide ebbs and floods.

Pumping   Turbines can also be powered in reverse by excess energy in the grid to increase the water level in the basin at high tide for ebb generation and two-way generation. The energy is returned to the grid during generation.

Two-basin schemes   In this type of construction, tidal power generation plants have two basins where one is filled at high tide and the other is emptied at low tide. Turbines are placed between the two basins. In general two-basin configurations have added advantages over the normal schemes described earlier since the power generation time can be adjusted with a certain flexibility that allows for continuous electric power production. Because of the larger construction and capital investment, two-basin schemes are more expensive to build; however, there are geographic locations that are well suited for this scheme.

Tidal power bulb turbines   In general, tidal turbines are constructed to meet the specific requirements of a barrage design. The La Rance tidal plant near St Malo on the Brittany coast in France uses a bulb turbine (Figure 11.2). In this type of system, water flows around the turbine, making access for maintenance difficult, as the water must be prevented from flowing past the turbine.

Rim turbine   Another type of mechanism is the rim turbine (Figure 11.3). An example is the Straflo turbine at Annapolis Royal in Nova Scotia, Canada. These turbines do not have maintenance access problems because the generator is mounted in the barrage,

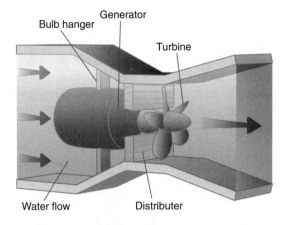

Figure 11.2 Bulb turbine diagram.

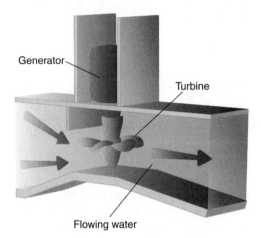

Figure 11.3 Rim turbine diagram.

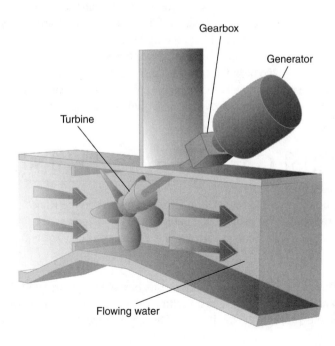

Figure 11.4 Tubular turbine diagram.

at right angles to the turbine blades. This however makes it difficult to regulate the performance of these turbines, and they are unsuitable for use in pumping applications.

Another proposed turbine construction technique involves the use of a design where the blades are connected to a long shaft and are oriented at an angle so that the generator can be mounted on top of the barrage.

Types of Tidal Energy

Currently there are two developing technologies for harvesting the ocean's energy: one uses tidal barrages and the other, tidal streams.

TIDAL FENCES

Tidal fences are composed of individual, vertical axis turbines that are mounted within the fence structure, known as a caisson. They can be thought of as giant turnstiles completely blocking a channel and forcing the water to go through them.

TIDAL BARRAGE

A barrage or a dam is built across an estuary or bay that has an adequate tidal range, which is usually in excess of 5 m. The purpose of the dam or barrage is to let water flow through it into the basin as the tide comes in. The barrage has gates that allow the water to pass through. The gates are closed when the tide stops coming in, trapping the water within the basin or estuary and creating a hydrostatic head. As the tide recedes, the barrage gates that are channeled through turbines are opened, and the hydrostatic head causes the water to come through these gates, driving the turbines and generating power. Power can be generated in both directions through the turbines; however, this usually affects efficiency and the economics of the power plant. Essentially the turbine technology is similar to hydropower. Construction of barrages in general requires extensive civil engineering design.

Current Tidal Generation Technologies

The world's largest tidal power barrage station was constructed on the Rance estuary in France nearly 40 years ago. The project generates 240 MW of power. The Annapolis Royal Barrage, located in Nova Scotia, Canada, was constructed in 1984 and produces 18 MW of power.

La Rance Tidal Barrage   The construction of this barrage began in 1960. The system consists of a dam 330 m long and a basin with a tidal range of 8 m; it incorporates a lock to allow for the passage of small craft. During construction, two temporary

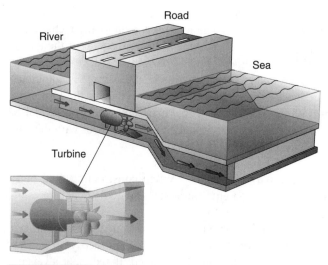

Figure 11.5 La Rance tidal barrage graphics.

dams were built on either side of the barrage to ensure that it would be dry; this was for safety and convenience. The work was completed in 1967 when twenty-four 5.4-m turbines, rated at 10 MW, were connected to the 225-kV French transmission network. Figure 11.5 is a diagrammatic presentation of the La Rance tidal barrage.

This barrage uses bulb turbines, described earlier, which were developed by Electricite de France and allow for power generation on both ebbs of the tide.

Bulb-type turbines have been popular with hydropower systems and have been used on mainland Europe in dams on the Rhine and Rhone rivers. An estimate of power generated from the La Rance power station, which provides electricity to a large majority of homes in Brittany, is calculated as follows. The turbines are rated at 10 MW; therefore, they have a total capacity of 240 MW.

Maximum electricity generated per annum (kWh)
= 240,000 × 8760 (hours in a year) = 2,102,400,000 kWh

Wave energy is like most other forms of renewable energy in that it cannot be relied upon 100 percent of the time, so the value derived from the preceding equation will almost certainly never be generated in a year. The capacity factor is used to estimate the percentage of the maximum that will actually be generated in a year. A capacity factor of approximately 40 percent is assumed for Scottish waters.

Electricity generated per annum (kWh) = 2,102,400,000 × 40%
= 840,960,000 kWh

To determine the number of homes this quantity of electricity can provide for in a year, the average annual household consumption is used. This is assumed to be 4377 kWh/year, and the number of homes supplied is then 840,960,000/4377 = 192,131.

Barrage economics   The capital required to start construction of a barrage is quite significant and has been the main obstacle in deployment of the technology, which is associated with long payback periods. In general, advancement of the technology has always been subsidized by government funding or large organizations getting involved with tidal power. However, once the construction of the barrage is complete, there are few associated maintenance and operational costs. In general, the turbines only need replacing once every 30 years. The key to the economic success of tidal barrages is optimum design, which is one that can produce the most power with the smallest barrage possible.

Environmental concerns of barrages   When constructing tidal barrages, it is important to take into consideration environmental and ecological effects on the local area, which may be different for each location.

The change in water level and possible flooding could affect the vegetation around the coast, therefore having an impact on the aquatic and shoreline ecosystems. The quality of the water in the basin or estuary could also be affected. If sediment levels change, this could affect the turbidity of the water and therefore could affect the animals that live in it and depend upon it, such as fish and birds. Fish would undoubtedly be affected unless a provision was made for them to pass through the barrages. Not all these changes adversely affect the environment, since some may result in the growth of different species of plants and creatures that may flourish in an area where they are not normally found.

Concerns over the environmental effects of barrage tidal plants present at the La Rance tidal power station have been recently overcome by development of technologies that result in newly designed tidal turbines known as tidal mills.

Social implications of barrages   The building of a tidal barrage can have many social consequences on the surrounding area. An example is the world's largest tidal barrage, La Rance in France, which took over 5 years to construct. The barrage can be used as a road or rail link, providing a time-saving way to cross the bay or estuary. Bays can also be used as recreation facilities or tourist attractions.

Other Tidal Technologies

STINGRAY TIDAL ENERGY UK

An alternate way to harness tidal energy is applied in the Stingray system. This technology consists of a parallel linkage that holds large hydroplanes to the flow of the tides. The angle of these hydroplanes to the flow of the tide is varied causing them to move up and down. This motion is used to extend a cylinder, which produces high-pressure oil that drives a hydraulic motor which in turn drives an electric generator.

This concept has received recognition for its potential from the UK Department of Trade and Industry. Initially the project was funded by the Department of Industrial Technology, Water and Power Technologies Panel, United Kingdom. The prototype system was to be designed, built, and installed in 2002.

ANNAPOLIS ROYAL, BAY OF FUNDY, CANADA

Bay of Fundy tidal power plant located in Canada is recognized as the world's geographic location for heist tides. North Atlantic waters are funneled up this deep inlet between the Canadian provinces of New Brunswick and Nova Scotia, and swirl through the narrowest parts of the bay with impressive speed and power.

So when in the 1980s the Canadian government decided to explore the feasibility of tidal power, the Bay of Fundy was selected as a natural site for an experimental station. The site chosen at Annapolis Royal had been closed off by a causeway, which was built to control tidal flow further up the Annapolis River.

The power plant was designed to generate power by water flowing through the sluice gates which in turn could activate a turbine and generate power on its way down the Bay of Fundy and out into the sea.

Plant engineering Twice every day, as the tide rises, the sluice gates are opened to let water flow up into the lower part of the Annapolis River, which now serves as the head pond for the power station. Just before high tide, the gates are closed, leaving only a narrow passage for fish to pass through. Now all the operators have to do is wait for the tide to turn and the water level on the seaward side to drop. When there is enough difference between the water levels on the two sides, the operators begin to let water flow through the giant turbine, slowly at first to get it turning, and then at full strength.

Once the huge 25-m-diameter wheel is up to its operating speed of 50 revolutions per minute (r/min), the station starts to generate power. At peak power it supplies a very respectable 20 MW to the Canadian grid.

Once the tide has subsided and the water level is equalized with the sea level, the turbine slows to a stop, before the whole cycle begins again.

THE LIMPET ISLAY

Limpet stands for "land installed marine powered energy transformer." The project was sponsored and developed by Wavegen, an investment company in collaboration with Queen's University in Belfast. It is the result of research on the island where a demonstration plant was built capable of generating 75 kW of power. Figure 11.6 shown below depicts graphic operational principle of the Limpet tidal turbine.

The concept involves a wave chamber that is constructed on the shore. The waves cause the air in the chamber to rise and decompress resulting in a rush of air that drives a Wells turbine which generates power. Therefore, the turbine has to be capable of turning regardless of the direction of the airflow; this turbine was designed by Professor Wells of Queen's University.

This device is currently generating 5 MW of power which it supplies to the grid on the Island of Islay off the west coast. This is an example of how this technology can be used to meet small-scale local needs. Wavegen has said that at present the answer lies not in huge operating plants but in small ones such as these, which can concentrate on meeting local or regional needs.

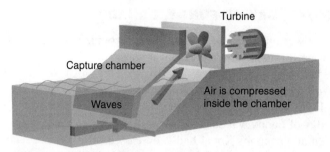

Figure 11.6 Graphic presentation of Limpet tidal
turbine installation. *www.darvill.clara.net.*

A device that is rated 500 kW according to the following calculations provides power to 400 households.

$$\text{Maximum power generated per annum} = 500 \text{ kW} \times 8760 \text{ h} = 4,380,000 \text{ kWh}$$

$$\text{Assuming a capacity factor of 40\%, annual electricity generated} = 4,380,000 \times 40\% = 1,752,000 \text{ kWh}$$

$$\text{Assuming a 4377-kWh power demand per year, the number of homes receiving power} = 1,752,000/4377 = 400$$

PUMPING

The turbines in the barrage can also be used to pump extra water into the basin at periods of low demand. This usually coincides with cheap electricity prices, generally at night, when demand is low. Establishments that provide tidal power in general buy inexpensive electric energy from grids usually during low demand such as night hours and pump water in the basins, and then generate power at times of high demand when prices are high. This practice is commonly used in hydroelectric power providers and is known as storage.

TIDAL POWER SYSTEM OUTPUT

The tidal power schemes described earlier are not considered to be a constant source of energy since within a 24-hour period the power production is limited to a maximum of 6 to 12 hours. Since the tidal cycle is based on the period of revolution of the moon and electricity demand is based on the period of revolution of the sun, the energy production cycle is not long enough to satisfy the power demand cycle, and as result grid connection becomes somewhat incompatible.

ENVIRONMENTAL CONSIDERATIONS

Impact on aquatic habitat Construction of barrage systems in estuaries has a considerable negative impact on aquatic life within the basin, such as fish and shellfish.

With proper ecological design considerations, on the other hand, large holding basins referred to as lagoons could be used for fish or lobster farming, which could be a benefit to local economic viability.

Turbidity Turbidity is the amount of silt and sedimentary matter in the water. An increase in silt and sedimentary matter results in a smaller volume of water being exchanged between the basin and the sea. Suspended particulates in the water also prevent sunlight from penetrating the water which affects aquatic ecosystem conditions.

Salinity Salinity is a result of less water exchange with the sea. If the average salinity inside the basin decreases, this also affects the ecosystem. Lagoons when properly designed can readily prevent this problem.

Sediment movement In general, rivers flowing through estuaries provide a high volume of sedimentation which eventually is moved to the sea. Introduction of a barrage into an estuary frequently results in sediment accumulation within the barrage which affects the ecosystem and creates a significant maintenance and drainage dilemma.

Pollutants Biodegradable pollutants, such as sewage, increase in concentration and cause bacteria growth within the basin, which in turn has human and ecosystem health consequences.

Fish In some instances fish pass through turbines safely; however, when the gates are closed, fish when attempting to escape get sucked through turbine fins. Mitigation measures such as screens or fish-friendly turbine designs seldom prevent fish mortality. The recent research on sonic guidance technology may hopefully provide a solution to this ongoing problem.

ECONOMIC CONSIDERATIONS

Tidal power projects are both labor and material intensive and require a very high capital expenditure; however, they are relatively inexpensive to run. Most tidal power projects do not produce returns on investment for decades; as such it is very difficult to attract investors. To date, worldwide projects have been supported by governments that have the required financial resources to support such undertakings.

Tidal Streams

Tidal streams are fast-flowing volumes of water caused by the motion of tides. These usually occur in shallow depths of seas where a natural construct forces the water to speed up. The technology involved is similar to wind energy; however, there are differences.

Water is 800 times denser than air and has a much slower flow rate; this means that the turbine experiences much larger forces and moments. This results in turbine designs that have much smaller diameters. The turbines used either generate power on

Figure 11.7 **Stream energy turbine test.** *Courtesy of A.J. Consulting, United Kingdom.*

both ebbs of the tide or are able to withstand severe structural strain resulting from the currents. The technology is still in the development stage but has potential as a reliable, predictable renewable energy source.

Tidal stream technology has the advantage over tidal barrages when comparing ecological issues. The technology is less intrusive than offshore turbines or tidal barrages, which create hazards to navigation or shipping. Figures 11.7 and 11.8 are

Figure 11.8 **Artist's rendering of stream energy turbine.** *Courtesy of A.J. Consulting, United Kingdom.*

graphic presentation of tidal stream designed by J.A. Consulting engineering UK that promotes tidal stream energy project concepts.

TIDAL STREAM TURBINES

Tidal turbines don't pose a problem for navigation and shipping and require the use of much less material in construction. They are also less harmful to the environment. They function best in areas where the water velocity is 2 to 2.5 m/s. Above this level, the turbine experiences heavy structural loads, and not enough generation takes place below this level.

ECONOMICS OF TIDAL STREAM TURBINES

Tidal stream technology, likewise, is in the developmental stage. It holds significant potential to harvest renewable energy. The cost of utilizing tidal streams is, in general, site specific and depends on the type of technology used.

Once the tidal stream turbines are installed, electricity is produced with no fuel costs and with complete predictability. The only operational cost will be turbine maintenance, which will be dependent on the type of technology used.

There are only a few places around the earth where this tide change occurs. Some power plants are already operating using this idea. One plant in France makes enough energy from tides to power 240,000 homes.

Marine Current Turbines

Marine current turbines are, in principle, like submerged windmills. They are installed in the sea at places with high tidal current velocities, to take out energy from the huge volumes of flowing water, as shown in Figure 11.9. These flows have the major advantage of being an energy resource as predictable as the tides that cause them, unlike wind or wave energy, which responds to the more random quirks of the weather system.

The technology under development by the Marine Current Turbine Ltd. (MCT) consists of twin axial flow rotors of 15 to 20 m in diameter, each now driving a generator via a gearbox much like a hydroelectric turbine or a wind turbine, as shown in Figure 11.10. The twin power units of each system are mounted on winglike extensions on either side of a tubular steel monopile some 3 m in diameter, which is set into a hole drilled into the seabed from a jack-up barge.

The technology for placing monopiles at sea has been well-developed by Seacore Ltd., a specialist offshore engineering company, which is cooperating with MCT in this work. The patented design of the turbine allows it to be installed and maintained without any use of costly underwater operations.

The artist's impression in Figure 11.10 shows the propeller of a turbine raised for maintenance from a small workboat. The turbine is connected to the shore by a marine cable lying on the seabed, which emerges from the base of the pile.

Figure 11.9   Artist's rendering of marine current turbines. *Courtesy of MCT, United Kingdom.*

The submerged turbines, which will generally be rated at from 500 to 1000 kW each depending on the local flow pattern and peak velocity, will be grouped in arrays or "farms" under the sea, at places with high currents, in much the same way that wind turbines in a wind farm are set out in rows to catch the wind. The main difference is that marine current turbines of a given power rating are smaller, because water is 800 times denser than air, and they can be packed closer together because tidal streams are normally bidirectional whereas wind tends to be multidirectional. Also the technology has a "low profile" and involves negligible environmental impact.

Environmental impact analyses completed by independent consultants have confirmed that the technology does not offer any serious threat to fish or marine mammals.

The rotors turn slowly at 10 to 20 r/min; a ship propeller by comparison typically runs 10 times as fast, and moreover these rotors stay in one place, whereas some ships move much faster than sea creatures can swim. There is no significant risk of leakage of noxious substances, and the risk of impact from the rotor blades is extremely small, bearing in mind that the flow spirals in a helical path through the rotor and that nature has adapted marine creatures so that they do not collide with obstructions (marine mammals generally have sophisticated sonar vision). Another advantage of this technology is that it is modular, so small batches of machines can be installed with only a small period between investment in the technology and the time when revenue starts to flow. This is in contrast to large hydroelectric schemes, tidal barrages, nuclear power stations, or other projects involving major civil engineering, where the lead time between investment and gaining a return can be many years.

It is expected that turbines will generally be installed in batches of about 10 to 20 machines. Many of the potential sites so far investigated are large enough to accommodate many hundreds of turbines. As a site is developed, the marginal cost of adding more turbines and of maintaining them decreases, so there is a considerable improvement in the economies as the project grows.

Marine Current Technologies, United Kingdom   Marine Current Turbines Ltd is currently in the process of starting a program of tidal turbine development through the research-and-development (R&D) and demonstration phases, to commercial manufacture. An initial grant of 1 million euros has been received from the European Commission toward R&D costs, and this has been followed by a grant toward the cost of the first phase of work from the UK government, Department of Trade and Industry (DTI) worth 960,000 euros. The German partners also received a grant worth approximately 150,000 euros, from the German government.

The company's plan is to complete the initial R&D phase by 2006 and to start commercial installations at that time. It is planned that some 300 MW of installations will be completed by 2010, and, after that, there is a far larger growth potential from a market literally oceanic in size.

Figure 11.10   **Marine current pile-mounted turbine installation.**
Courtesy of MCT, United Kingdom.

Some Interesting Oceanic Technologies

OCEANIC WAVE POWER

Oceanic wave power technology harvests electric power from the dynamics of wave movements. The basic operational principles of this technology are based upon an oscillating movement of water columns as shown in Figure 11.11. As waves enter the lower chamber, the pressure resulting from air compression forces the propeller to rotate which then turns an electric generator shaft. The impeller mechanism is prevented from corrosion since the housing where the blades are mounted does not get in contact with the seawater.

THE SALTER DUCK AND WAVE PENDULUM

These technologies are quite recent. The slater duck devices shown in Figure 11.13 harness the ocean wave energy by a pendulum mechanism construction which resembles a floating dock decoy. When moved up and down by waves, a pendulum movement within drives a turbine within each floating unit. The floating devices, anchored to the seabed are arranged in special patterns that take maximum advantage of the wave movements.

The wave pendulum power generator rendering shown in Figure 11.13 consists of a paddle mechanism, which is displaced back and forth by ocean waves which in turn drives an electric power generator.

POWER-GENERATING SEA BUOY

This technology is a U.S.-patented mechanism that employs pendulum movement mechanisms similar to technologies previously discussed and develops power from the upward and downward movement of sea waves.

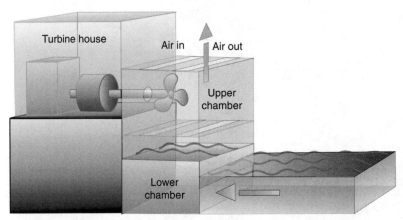

Figure 11.11 Rendering of oceanic air column wave power generator.

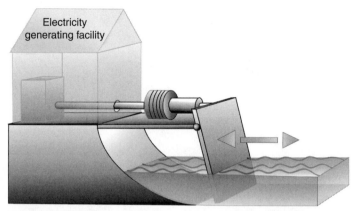

Figure 11.12 Rendering of oceanic wave pendulum power generator.

As the Salter Duck moves with the motion of the waves, the pendulum inside swings forward and backward generating electricity

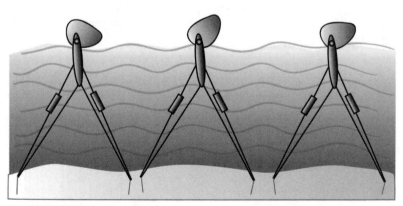

Figure 11.13 Rendering of oceanic wave Slater Duck power generator.

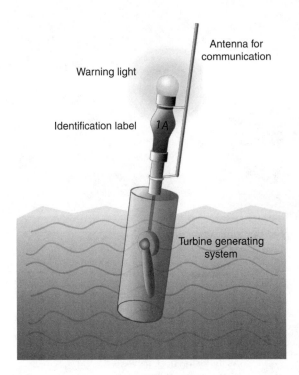

Antenna for
communication

Warning light

Identification label   1A

Turbine generating
system

Figure 11.14 Rendering of
oceanic buoy power generator.

The system depicted in Figure 11.4 is used to provide power for buoys and devices that monitor ocean parameters such as wave height, water temperature, and wave directional movement, and seismic activity of the ocean floor for tsunami warnings. In most instances the data monitored are transmitted to a remote data-monitoring center via a satellite.

12

GEOTHERMAL ENERGY

Introduction

The term *geothermal* is a composition of two Greek words, *geo* meaning "earth" and *therm* meaning "heat." The combined word means heat generated from the earth.

Earth's center is formed from molten iron located about 4000 mi from its crust. The estimated temperature of Earth's core is about 5000°C, the heat which conducts outward from the center heats up the outer layers of rock referred to as the mantel. When the mantel melts and is spewed out of the crust, it is called magma.

Rainwater seepage through geological cracks and faults becomes superheated and emerges as geysers and hot springs, and sometimes water is trapped in underground voids which become geothermal reservoirs. Geothermal energy as a technology involves the production of electric energy by using these hot water reservoirs. The process involves drilling wells as deep as 2 mi that reach the geothermal reservoirs, where the hot water is brought to the surface as steam and is heated up to 250°F if so required. The steam in turn is used to drive electric generator turbines.

In areas where the water is not so hot as to become steam, hot water is circulated through commercial, industrial, and residential projects for space heating and drying process. After being used, the spent water is circulated back to the reservoir and the process is repeated.

Geothermal Resources

Geothermal resources are in general manifested in a wide variety of forms including the following:

Hot water reservoirs are exothermally heated underground water reservoirs.

Natural steam reservoirs which manifest as ground steams.

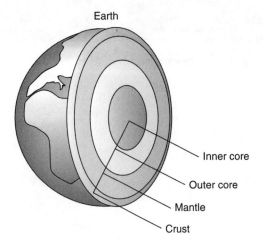

Earth

Inner core

Outer core

Mantle

Crust

Figure 12.1 **Earth's core.**

Geopressured reservoirs are underground salt water (brine) which are extremely pressure saturated with natural gas as a result of the weight of the overlaying land mass.

Geothermal gradients are human-made drilled shafts which allow access to dry hot rocks for heat energy extraction. In general, the thermal gradient or temperature rise for every kilometer of drilled depth is 30°C. Holes drilled to 20,000 ft can therefore reach a temperature of 190°C which provides the potential for an enormous commercial energy extraction.

Molten magma, which is produced as a result of volcano activities, has a temperature of 2000°C, which is not suitable for thermal power extraction.

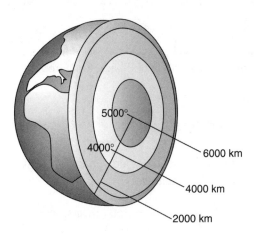

5000°

4000°

6000 km

4000 km

2000 km

Figure 12.1 **Temperatures in Earth's core (in degrees Celsius).**

Geothermal Power Extraction Potential

The following example is used to demonstrate the geothermal power potential calculation of a site that covers 50 km^2 with a thermal crust of 2 km where the temperature gradient is 240°C. At this depth the specific heat of the rock is determined to be 2.5 J/cm^3, and the mean surface temperature is measured to be 15°C. Thermal heat generation is calculated as follows:

$$\text{Volume of rock } V = 50 \text{ km}^2 \times 2 \text{ km} = 100 \text{ km}^3$$

Heat content $Q_h = V \times$ specific heat $\times$ temperature difference between depth gradient and surface temperature (240°C − 15°C = 225°C)

Prior to completing the calculation of the volume, the bedrock must be converted from cubic kilometers to cubic centimeters, which in this case is 10^{15} cm^3:

$$Q_h = 100 \text{ km}^2 \times 10^{15} \text{ cm}^3 \times 2.5 \text{ J/(cm}^3 \times °C) \times 225°C = 5.625 \times 10^{19} \text{ J}$$

Assuming that only 2 percent of the available thermal energy of the geothermal mass could be used to provide power for electricity generation, the question becomes, How many years would it take to produce 1000 MW/year of power?

Taking the useful portion of the mass as being 2 percent, then the overall power generated is calculated as follows:

(Note 1 J/s = 1 W)

Total capacity to generate 1000 MW per year

$$= 50{,}000 \text{ MW} \times \text{year} \times 1{,}000{,}000 \text{ W/MW} \times (3150 \times 10^6 \text{ s/year})$$

$$= 1.575 \times 10^{18} \text{ J/year}$$

Lifetime power production $= (5.625 \times 10^{19})/(1.5\,75 \times 10^{18}) = 35$ years

Types of Geothermal Power Plants

DRY STEAM PLANTS

These classes of plants, an example of which is shown in Figure 12.3, use very hot water and steam, which is heated above 300°F, as found in hot water resources and geysers. The steam is either used directly or is depressurized or "flashed" and purged to eliminate carbon dioxide, nitric oxide, and sulfur, which are usually associated with the process. The clean steam is then used to turn turbines, which drive generators, as shown in Figure 12.4. The pollution resulting from removal of these toxic gases and elements is about 2 percent of what is generated by traditional fossil fuel power plants.

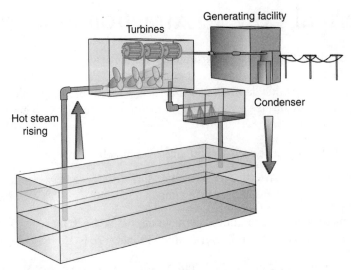

Figure 12.3 Dry steam geothermal steam plant.

BINARY PLANTS

In this technology, geothermal steam is extracted using lower-temperature hot water resources with temperatures that range from 100 to 300°F (superheated vapor). The hot water is passed through heat exchangers, which produce a flow of secondary fluid, such as isobutene or isopentane, which have a lower boiling point.

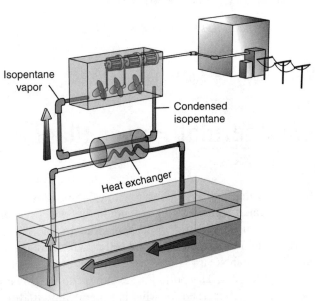

Figure 12.4 Binary geothermal steam plant diagrams.

The secondary fluid vaporizes and turns the turbines that generate electricity. The remaining secondary fluids are then recycled through heat exchangers. Upon completion of the process, the geothermal fluid is condensed and returned to the reservoir. Because binary plants use a self-contained cycle, no pollutants are ever emitted or introduced into the atmosphere.

FLASH STEAM POWER PLANTS

In this geothermal technology the hot water is pumped under extreme pressure to the earth's surface by pumps. When the hot water reaches the surface, the pressure is reduced and as a result the hot water is transformed into steam; this is referred to as "steam blasting." As mentioned earlier the spent water is in turn circulated back to the reservoir and the cycle is repeated.

Potential of Geothermal Power

The Pacific Northwest in the United States has the potential to generate up to 11,000 MW of electricity from geothermal power. A geothermal plant is shown in Figure 12.5. Although estimates of available resources are uncertain, until exploratory work is done, the Northwest Power Planning Council has identified 11 specific areas where it expects there are about 2000 MW to be developed, enough power to serve over 1.3 million homes.

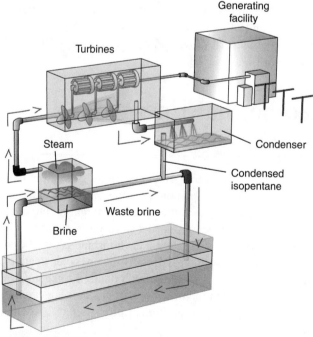

Figure 12.5 Geothermal plant.

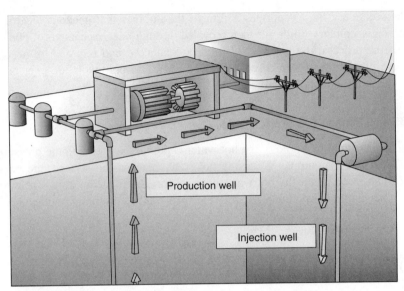

Figure 12.6   Geothermal flash steam power production.

A graph is shown in Figure 12.7. Geothermal areas in the western United States are usually found where there has been recent volcanic activity, such as in the northwest basins in Oregon, Washington, California, Nevada, and Utah. Low-temperature geothermal heating systems have been in operation for decades in Klamath Falls, Oregon; Boise, Idaho; and the Big Island of Hawaii, which generates 25 percent of its power from geothermal systems. Geothermal power production between 1970 and 2000 in the United States is shown in Figure 12.10.

Cost of Geothermal Energy and Economics

It is estimated that the average cost of geothermal electricity generation per kilowatt-hour is about 4.5 to 7 cents. This is comparable to some fossil fuel plants; however, power production does not produce any pollution and, when taking into consideration the pollution abatement cost, the power produced is very competitive. Geothermal plants are built in modular system configurations; each turbine is sized to deliver 25 to 50 MW of electric power. When burning fossil fuels such as coal, steam plants generate substantial amounts of noxious gases and precipitants. In general, geothermal steam carries a lesser amount of contaminants. Geothermal plants are capital-intensive projects, which require no fuel expenditure. Typical projects pay back their capital costs within 15 years.

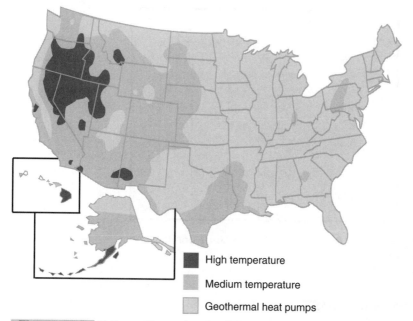

Figure 12.7 **U.S. geothermal potential.** *Courtesy of Southern Methodist University Geothermal Lab.*

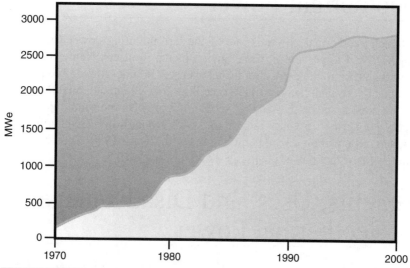

Figure 12.8 Graph showing growth in U.S. geothermal power, 1970–2000.

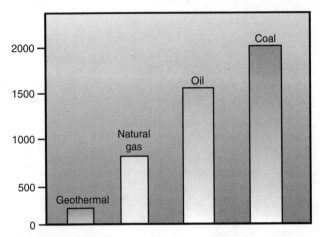

Figure 12.9 Carbon dioxide emissions comparison (lb/MWh).

ECONOMIC COST BENEFITS

Geothermal power, similar to other types of renewable energy resources; maintains benefits in the geographic location of installation, which provides local jobs and contributes royalties and taxes to the county.

Environmental Impact of Geothermal Power

Although geothermal power is one of the less polluting power sources, it must be properly sited to prevent possible environmental impacts. New geothermal systems reinject water into the earth after its heat is used, in order to preserve the resource and to contain gases and heavy metals sometimes found in geothermal fluids.

Care must be taken in planning geothermal projects to ensure that they don't cool nearby hot springs or cause intermixing with groundwater. Geothermal projects can produce some carbon dioxide emissions, but these are 15 to 20 times lower than the cleanest fossil fuel power plants, as shown in Figure 12.9.

Benefits, Uses, and Disadvantages of Geothermal Power

Benefits

- It provides clean and safe energy, using little land space.
- The energy produced is sustainable and renewable.

■ It generates continuous, reliable power.
■ The power produced is cost competitive.
■ It conserves the use of fossil fuels.
■ It reduces energy imports.
■ It benefits local economies.
■ Power plants are modular and can be increased in potential incrementally.

Direct uses

■ Balneology—hot springs, baths and bathing
■ Agriculture—greenhouse and soil warming
■ Aquaculture—fish, prawn, and alligator farms
■ Industrial—product drying and warming
■ Direct heating—residential and industrial

Advantages

■ Although geothermal power produces a minute amount of fumes during the well drilling process, it is environmentally friendly and does not produce harmful pollutants generated by fossil fuel power plants.
■ Geothermal power production is reliable and can produce uninterrupted constant power.
■ Geothermal energy is quite flexible; it can be used to produce electric power, and hot water for heating or industrial drying processes.
■ It should be noted that geothermal power derived from natural steam is 97 percent efficient; 3 percent of the loss of efficiency is attributed to turbine friction.

It should be noted that geothermal energy technology that makes use of water recirculation such as binary systems does in fact withdraw the earth's thermal energy at a considerable rate and therefore cannot be considered a completely renewable source of energy.

Ocean Thermal Energy

Ocean energy principles use the temperature differences in the ocean. The difference in the ocean's surface temperature and deep water temperature is quite significant, and power plants can use this difference in temperature to make energy. A difference of at least 38°F is needed between the warmer surface water and the colder deep ocean water. This type of energy source is called ocean thermal energy conversion (OTEC). It is being used in both Japan and in Hawaii in some demonstration projects.

13

BIOFUELS AND BIOGAS
TECHNOLOGIES

Introduction

Biomass is the stored solar energy in plant and animal tissues and materials in chemical form and is considered to be the most vital resource on Earth. Biomass in addition to providing sustenance to plants and animals is also the most significant source of energy and is used in building the majority of human-made materials such as fabrics, medicines and chemicals, and construction materials. The use of biomass as a source of energy dates back to the discovery of fire by humans. In this section we will review biomass fuels and technologies which provide us with various types of energies used in various forms for fueling cars, generating power, and building computer components.

The Chemical Composition

The chemical composition of biomass is dependent on the various types of tissues found in plant and animal species. In general, plant structures consist of about 25 percent lignin and 75 percent carbohydrates or sugars. The carbohydrate in plants consists of many sugar molecules that are linked together in long chains called polymers. Some of the carbohydrate has a significant content of cellulose and hemicellulose. Lignin is a nonsugar polymer that acts as the mortar for the building blocks of plants. It gives plants their strength and acts like a glue that holds the cellulose fibers together.

Origins of Biomass

Essentially biomass in nature is formed when carbon dioxide from the atmosphere and water from the earth are combined in the photosynthetic process to produce carbohydrates or sugars, which are the building blocks of biomass. Solar energy promotes photosynthesis in plants through a process involving chlorophyll. This process stores the sun's energy in the chemical bonds of the structural components of biomass. Biomass burns and in the process the energy stored in the chemical bonds is extracted whereby oxygen from the atmosphere combines with the carbon in plants to produce carbon dioxide and water. This is a cyclic process since the carbon dioxide is absorbed by the plants which produce new biomass. As discussed in earlier chapters, for millennia human beings have used biomass as a fuel and have consumed plants as food for their nutritional energy in the form of sugar and starch.

Exploitation of fossilized biomass in the form of coal fuel, which is essentially a compacted biomass resulting from very slow chemical transformations over the millennia, is essentially a converted form of sugar polymer that has been transformed into the chemical composition of coal, which is a concentrated source of energy as fuel. All the fossil fuels, such as coal, oil, and natural gas, are ancient biomass formed from plant, tree, and animal remains. Even though fossil fuels contain the same constituents such as hydrogen and carbon as those found in living plants and animals, they are not considered renewable because nature has taken numerous years to create them.

When plants and animals decay, they release most of their chemical constituents back into the atmosphere. On the contrary when not burned or processed, fossil fuels do not affect Earth's atmosphere. Some of the well-known biomass residues used for generating energy include sugarcane; corn fiber; rice straw and hulls and nutshells; sawdust, timber slash, and mill scrap; paper trash and urban yard clippings in municipal waste; energy crops; fast-growing trees like poplars and willows; grasses like switchgrass or elephant grass; and the methane captured from landfills, municipal wastewater treatment, and manure from cattle or poultry.

At present in developed countries biomass is approximately 14 percent of the world's primary energy needs; however, the population living in developing countries use biomass as their most important source of energy. As referenced in previous chapters, with increases in world population and per capita demand for fuels and the depletion of fossil fuel resources, the demand for biomass is expected to increase very rapidly.

Biomass currently provides 40 to 90 percent of the primary source of energy in developing countries and is expected to remain as the global source energy for the foreseeable future.

With advances in biomass energy conversion technologies some industrialized countries such as Sweden and Austria derive 15 percent of their primary energy consumption from biomass. At present the United States only derives 4 percent, or 9000 MW, of electric energy from biomass, which is equivalent to the same energy amount generated from its nuclear power plants. It is estimated that with further advancement of biomass technologies the total power production could supply 20 percent or more of the U.S. energy consumption. In countries with significant agricultural infrastructure, various biomass energy such as ethanol, biogas, or biodiesel could produce sufficient energy to offset the use of oil imports by as much as 50 percent. The following is the global biomass energy distribution potential.

The total mass of living matter including moisture is estimated to be 2000 billion tons.

The total mass in land plants is estimated to be about 1800 billion tons.

The total mass in forests is estimated to be 1600 billion tons.

Per capita terrestrial biomass is estimated to be 400 tons, and the net annual production of terrestrial biomass is estimated to be 400,000 million tons.

Biomass Energy Potential

The potential of bioenergy extraction from biomass is enormous. With advances in the application of this technology it is possible to convert raw biomass into various forms of energy such as electricity, liquid or gaseous fuels, or processed solid fuels which could result in significant social and economic benefits to the world. It is a well-known fact that for the very large majority of the world population the quality of life and health is directly related to the availability of one or another form of energy. It is a proven fact that improvements in infrastructure, health and social advancement, and jobs hinges upon the availability of dispersible energy.

ENERGY VALUE OF BIOMASS

When referring to biomass energy, we consider the energy potential stored in plants, crop and forest residues, and animal wastes. The energy content of biomass, when used as a solid fuel, is usually comparable to that of coal. Stored heat energy values of dry biomass range from 17.5 gigajoules/ton (GJ/ton) for wheat straw and sugar cane to about 20 GJ/ton for wood. Corresponding values for coal and lignite are 30 and 20 GJ/ton, respectively. Generally, freshly harvested biomass contains a considerable amount of moisture which could represent 8 to 20 percent for wheat straw mass, 30 to 60 percent for woods, and 75 to 90 percent for animal manure. On the other hand the moisture content of most coal ranges from 2 to 12 percent. Therefore the energy density content for biomass is lower than that for coal. However, the chemical attributes of biomass have many distinctive advantages; one such attribute is that the ash content of biomass is much lower than that of coal and is relatively free of toxic metals and other harmful contaminants.

When processed, biomass fuels offer a wide diversity of fuel supplies which can be produced and used to generate electricity through direct combustion in modern devices or be processed to produce a wide variety of liquid fuels such as ethanol, biodiesel, or other alcohol fuels used in motor vehicles. Biomass energy undoubtedly can increase the global economic development without contributing to the greenhouse effect since biomass does not contribute to the inversion layer of global warming. This is because the net amount of CO_2 produced from burning biomass fuels is recycled and absorbed by plants and is therefore considered a sustainable energy resource.

TABLE 13.1 ENERGY CONTENT OF VARIOUS BIOMASS AND BIOGAS FUELS

BIOMASS FUEL	CONTENT OF WATER (%)	MJ/kG	kW/kG
Oak tree	20	14.1	3.9
Pine tree	20	13.8	3.8
Straw	15	14.3	3.9
Grain	15	14.2	3.9
Rape oil	NA	37.1	10.3
Hard coal	4	30.0–35.0	8.3
Brown coal	20	10.0–20.0	5.5
Heating oil	NA	42.7	11.9
Biomethanol	NA	19.5	5.4
BIOGAS FUELS		**MJ/(N·M³)**	**kWh/(N·M³)**
Sewer gas		16.0	4.4
Wood gas		5.0	1.4
Biogas from cattle dung		22.0	6.1
Natural gas		31.7	8.8
Hydrogen		10.8	3.0

Benefits of Biomass Energy

Undoubtedly the evolution of biomass technology and the extended use of bioenergy-based fuels, which maximize the use of otherwise discarded agricultural, industrial, and animal waste, can create an industrial infrastructure that could have a major impact on industrial growth in rural areas. The U.S. Department of Agriculture estimates 17,000 jobs can be created per every million gallons of ethanol produced. Likewise production of 5 quadrillion Btu of electricity on 50 million acres of land would increase overall farm income by $12 billion annually. It is important to note that the United States consumes about 90 quadrillion Btu of energy annually. If promoted by state and federal governments, biomass energy generation can provide farmers with stable incomes and a higher standard of living. Rural industrial development and diversification can, in addition to strengthening local economies, contribute to the elevation of local community living standards.

ENVIRONMENTAL BENEFITS

Extended-use biomass technologies and use of bioenergy will provide a degree of ecological balance and climate change, will mitigate acid rain, minimize soil erosion, reduce water pollution, provide a better sanctuary and habitat for wildlife, and help maintain forest health.

Impact of Biomass Energy on Climatic Conditions

The extended use of fossil fuels throughout the centuries as a primary energy resource has contributed toward serious deterioration of the environment and the atmosphere by release of hundreds of millions of tons of greenhouse gases into the atmosphere. Greenhouse gases, which include carbon dioxide (CO_2) and methane (CH_4), have altered Earth's climate, disrupting the entire biosphere which currently supports life as we know it. Biomass energy technologies can help minimize and perhaps reverse this trend. Even though methane and carbon dioxide are the two most significant contributors to global warming, methane, which is relatively short lived in the atmosphere has had a significantly more detrimental effect than carbon dioxide since it is 20 times more potent.

Methane gas trapped and harnessed from landfills, wastewater treatment facilities, and manure lagoons which are vented to the atmosphere could be used to generate electricity or fuel for use in motor vehicles. Biomass energy crops, which have significant amounts of stored carbohydrates in the plants, roots, and stalk sink, have roots that remain in the soil after the harvest which can regenerate carbon on a yearly basis.

ACID RAIN

Acid rain is a result of the combustion of fossil fuels which releases sulfur and nitrogen oxides into the atmosphere. Acid rain has been implicated in the killing of fish and rendering lakes inhabitable for aquatic life. Since biomass contains no sulfur, when it is mixed with coal, referred to as cofiring, it reduces sulfur emissions and prevents acid rain.

SOIL EROSION AND WATER CONTAMINATION

Biomass crops are considered excellent for reducing water pollution since they can be readily cultivated on more marginal lands, in floodplains, and on farmland as interseasonal crops between two adjacent harvest periods. Planted crops stabilize the soil and reduce soil erosion. They also play a significant role in reducing nutrient runoff, which can protect aquatic ecosystems. Shading provided from vegetation growth enhances the habitat for numerous aquatic creatures such as fish prawns and shellfish. Since most bioenergy crops tend to be perennials, they do not have to be planted every year, which means a decreased use of farm machinery and less soil deterioration.

Another property intrinsic to biomass energy is that it reduces water pollution by capturing methane, through anaerobic digestion, from manure lagoons on cattle, hog, and poultry farms. As discussed later, by constructing and utilizing large anaerobic digester lagoons, farmers can reduce odors, capture the methane for energy production, and create either liquid or semisolid soil fertilizers.

Biomass Technologies

The following are technologies that extract the stored energy in biomass fuels.

MICROTURBINE GENERATORS

Microturbines are a third-generation technology that evolved from many years of research and development in civil and military aviation and maritime technologies. The principle of this technology is based on gas turbine engine system engineering which was developed in 1938 in England.

A microturbine's principal function is the utilization of very high internal combustion pressures developed in a chamber to provide rotational power to a shaft. This torque is then transferred to a microturbine generator which generates alternating current.

Typically a microturbine is constructed from a few mechanical components including a recuperator chamber where fuel such as methanol, ethanol, natural gas, petroleum distillates, liquid propane, and gasified coal are vaporized at very high temperatures. In a combustion chamber the vaporized gases are combusted, resulting in extremely high pressures and rotational torque produced by turbine blades. Frictional losses from the rotational torque are minimized by an air-bearing mechanism which provides an airfoil around the output shaft. The generator design also involves an intricate water or liquid cooling and filtration system that allows the system to operate at lower temperatures. Efficiencies of microturbines range from 21 to 31 percent. Higher system efficiencies are attained by using the waste heat generated from the exhaust gases. Microturbines require natural gas at a high pressure of at least 75 lb/in^2, which is achieved by use of an on-site compressor that somewhat reduces the system efficiency and increases operational costs.

Microturbine sizes range from 25 to 1,00 kW and are designed as stand-alone modular power-generating units that are capable of tandem synchronized or combined operation that can generate the desired electric power output.

A microturbine, shown in Figure 13.1, is a compact turbine generator that delivers electricity close to the point where it is needed. It operates on a variety of gaseous and liquid fuels and is designed to operate with low-Btu landfill gases (LFGs). Microturbine technology use became commercialized in 1998.

Microturbines in general serve as primary, emergency, or standby power, which adds capacity and reduces the grid consumption bottleneck for peak shaving purposes. The units deliver energy cost savings, while supplying clean, reliable power with low maintenance needs. Microturbines are very compact and are packaged in compartments the size of a refrigerator, with power generation capacities of 30 to 60 kW of electricity, enough to power a small business.

Maximum thermal efficiencies of microturbines are achieved when the exhaust is used to recover generated heat, referred to as cogeneration. The generation capacity of the units is literally unlimited when they are running in parallel.

Like a jet engine, microturbines mix fuel with air to create combustion, as shown in Figure 13.3. This combustion turns a magnet generator, a compressor, and turbine wheels on a revolutionary single-shaft, air-bearing design at high speed with no need

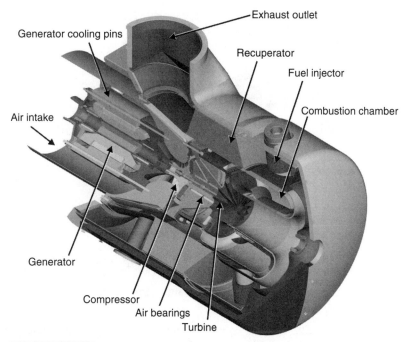

Figure 13.1 Microturbine system components. *Graphic courtesy of Capstone Turbine.*

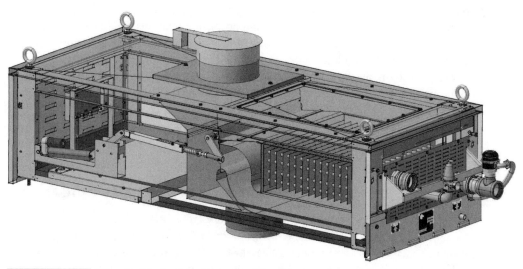

Figure 13.2 Microturbine pictorial assembly diagram. *Courtesy of Capstone Microturbine.*

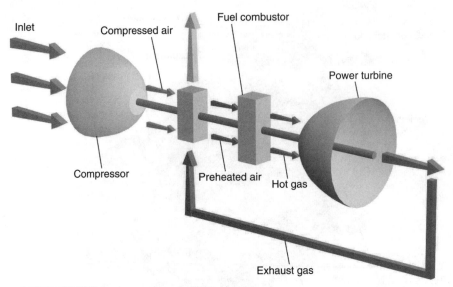

Inlet

Compressed air

Fuel combustor

Power turbine

Compressor

Preheated air   Hot gas

Exhaust gas

Figure 13.3 **Microturbine operation diagram.** *Graphic courtesy of Capstone Turbine.*

for additional lubricants, oils, or coolants. The result is a highly efficient, reliable, clean combustion generator with very low NO$_x$ emissions that, unlike diesel generators, can operate around the clock without restrictions. Unlike combined cycle gas turbines, these power systems use no water.

Microturbines are low-emission generators that are used in applications where combined electricity and heat generation could be used simultaneously. For example, a typical microturbine manufactured by Capstone can provide up to 29 kW of power and 85 kW of heat for combined heat and power applications. The technology makes use of solid-state power electronics that allow synchronized tandem coupling of 2- to 20-unit stand-alone modules with no external hardware except computer cables.

These systems also incorporate circuits that allow for automatic grid and stand-alone switching, a heat recovery unit, up to100-unit networking, remote monitoring and dispatch, and other functionalities. The major functional components of microturbine systems include a compressor, combustor, turbine, and permanent-magnet generator. The rotating components are mounted on a single shaft, supported by air bearings. Figure 13.4 shown below depicts a microturbine functional process.

LANDFILL AND WASTEWATER TREATMENT PLANT BIOGAS

Methane gas is also commonly generated in very large volumes as a by-product of the biological degradation of organic waste or the result of many industrial processes. Although methane is an excellent waste-gas fuel for microturbines, like the one shown in Figure 13.5, it is also an especially potent greenhouse gas. Landfill gases are created when organic waste in a municipal solid waste landfill decomposes. This gas consists of about 50% methane (CH$_4$), the primary component of natural gas, about 50% carbon dioxide (CO$_2$), and a small amount of nonmethane organic compounds.

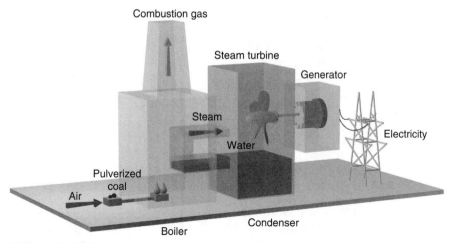

Figure 13.4 A microturbine functional process diagram.

Instead of being allowed to escape into the air, LFG can be captured, converted, and used as an energy source. Capturing and using LFG helps to reduce odors and other hazards associated with LFG emissions and helps prevent methane from migrating into the atmosphere and contributing to local smog and global climate change.

Landfill gas is a readily available, local, and renewable energy source that offsets the need for nonrenewable resources, such as coal and oil. In fact, LFG is the only renewable energy source that, when used, directly prevents atmospheric pollution.

Figure 13.5 Microturbine system assembly. *Courtesy of Capstone Microturbine.*

Landfill gas can be converted and used in many ways: to generate electricity, heat, or steam; as an alternative vehicle fuel for fleets like school buses, taxis, and mail trucks; or in niche applications like microturbines, fuel cells, and greenhouses.

Of the 6000 landfills across the United States, there are about 340 energy projects currently in operation. However, the EPA estimates that as many as 500 additional landfills could cost effectively have their methane turned into an energy source producing electricity to power 1 million homes across the United States. This is equivalent to removing the greenhouse gas emissions that would be generated in 1 year by 13 million cars.

HOT WATER AND HOT AIR

With a hot-water heat exchanger fully integrated into an exhaust system, useful heat is easily extracted from microturbines. Microturbines when in operation produce significant amount of hot exhaust air, which used to heat water through a heat exchange mechanism this is referred to as heat cogeneration. This built-in cogeneration ability allows overall system efficiencies to approach 80 percent, depending on the temperature of the inlet water.

Facilities can use this heat in a variety of ways. In suitable climates, the hot water can reduce building heating fuel use by providing space heating. Since the water heat exchanger is rated for potable water use, the system can supply domestic hot water directly. Figures 13.7 and 13.8 show microturbine installations in landfill projects.

Figure 13.6 A microturbine equipment assembly. *Courtesy of Capstone Microturbine.*

Figure 13.7 Microturbine in landfill installation. *Courtesy of Capstone Microturbine.*

Figure 13.8 LFG can be captured and used as an energy source. This example is the Cal Poly San Luis Obispo lagoon digester.
Courtesy of Capstone Microturbine.

The hot water can also be used in conjunction with other heat-driven devices such as absorption chillers or desiccant wheels used for dehumidification. In the latter case, the heat provided by the microturbine helps regenerate the desiccant wheel by driving out the captured moisture. Of course, the 400°F (200°C) exhaust from the microturbine can also be used directly in some applications. Since the gas turbine engine emissions are low, the exhaust is relatively clean. And because of the design of the engine, the exhaust still contains plenty of oxygen and can support follow-up burners to further raise temperature.

Economic Benefits of Using Landfill Gas

Landfill gas projects are a win-win opportunity for all parties involved, whether they are the landfill owners or operators, the local utility, the local government, or the surrounding community. Even before LFG projects produce profits from the sale or use of electricity, they produce a related benefit for communities. Landfill gas projects involve engineers, construction firms, equipment venders, and utilities or end users of the power produced. Much of this cost is spent locally for drilling, piping, construction, and operational personnel, providing additional economic benefits to the community through increased employment and local sales. Once the LFG system is installed, the captured gas can be sold for use as heat or fuel or be converted and sold on the energy market as renewable "green power." In so doing, the community can turn a financial liability into an asset.

Environmental Benefits of Using Landfill Gas

Converting LFG to energy offsets the need for nonrenewable resources, such as coal and oil and reduces emissions of air pollutants that contribute to local smog and acid rain. In addition LFG can improve the global climate changes discussed earlier.

Case Study of a Successful Application of Microgenerator Technology

The following application reflects use of Capstone Micro Turbine application in a landfill gas recovery project.

The HOD Landfill, located within the Village of Antioch in Lake County in northeastern Illinois, is a superfund site consisting of approximately 51 acres of landfill

Figure 13.9 **A microturbine equipment assembly.** *Courtesy of Capstone Microturbine.*

area. On September 28, 1998, the EPA issued a record of decision (ROD) for the site requiring that specific landfill closure activities take place. The remedial action (in response to the ROD), which was completed in January 2001, included the installation of a landfill gas collection system with 35 dual gas and leachate extraction wells. This system collects approximately 300 ft^3/min of landfill gas. In 2001, RMT, Inc., an environmental engineering contractor who undertook project installation, and the Antioch Community School District began exploring the option of using this landfill gas to generate electricity and heat for the local high school.

The design and construction of the energy system posed a number of challenges, including resolving local easement issues, meeting local utility requirements, connecting to the existing school heating system, crossing under a railroad, and meeting the EPA's operational requirements. One-half mile of piping was installed to transfer approximately 200 ft^3/min of cleaned and compressed landfill gas to the school grounds, where 12 Capstone microturbines are located in a separate building. The 12 microturbines produce 360 kW of electricity and, together with the recovered heat, meet the majority of the energy requirements for the 262,000-ft^2 school. The system began operating in September 2003.

This use of landfill gas proved beneficial to all parties involved. It provides energy at a low cost for the high school; clean, complete combustion of waste gas; decreased emissions to the environment by reducing the need for traditional electric generation sources; public relations opportunities for the school and community as being the first school district in the United States to get electricity and heat from landfill gas; and educational opportunities in physics, chemistry, economics, and environmental management for Antioch Community High School (ACHS) students, as a result of this state-of-the-art gas-to-energy system being located at the school.

The design of the energy system included tying into the existing gas collection system at the landfill, installing a gas conditioning and compression system and transferring the gas a half-mile to the school grounds for combustion in the microturbines to generate electricity and heat for the school.

A diagrammatic layout of the landfill gas-to-energy system is shown in Figure 13.10. RMT staff worked with the local government, school officials, and the EPA, in addition to leading the design efforts and managing the construction activities throughout the project. RMT also provided public relations assistance to ACHS by attending Antioch Village Board meetings to describe the project and to answer any questions from concerned citizens and Village Board members. Potential options evaluated included using the LFG to produce electricity, for use in the school's existing boilers, and for use in a combined heat and power system through these evaluations. It was determined that the only economically viable option was to produce electricity and heat for the school.

In 2002 ACHS applied for, and received, a $550,000 grant from the Illinois Department of Commerce and Community Affairs to be used for the development of the LFG combined heat and power project. Shortly after this, RMT and ACHS entered

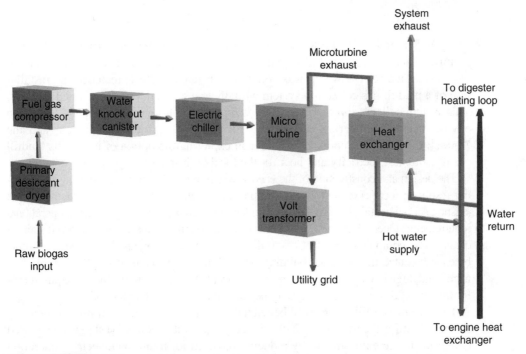

Figure 13.10 **LFG gas recovery process diagram.** *Graphic courtesy of Capstone Turbine.*

into an agreement to turn the landfill gas into the primary energy source for the high school. The overall cost of this project, including design, permits, and construction was approximately $1.9 million.

RMT was the designer and general contractor on the project. The team was responsible for designing the system, administering contracts, coordinating access rights and railroad access, obtaining all appropriate permits, creating a health and safety plan, managing construction, and coordinating utility connections.

PROJECT DESIGN

This project included 12 Capstone microturbines used to turn landfill gas into the primary energy source for the 262,000-ft^2 ACHS. This is the first landfill gas project in the United States that is owned by a school and provides heat and power requirements for the institution.

The collection system at the HOD Landfill, which includes 35 landfill gas extraction wells, a blower, and a flare, must remain operational to control landfill gas migration. Therefore, the construction of the new cogeneration system required connection to the existing system to allow for excess LFG to be used.

The gas pipeline to the microturbines located at the school was made of high-density polyethylene (HDPE SDR 9) pipe 4 inches in diameter and a half-mile long. It was installed 4 to 12 ft below ground, running from the HOD Landfill to the microturbines at the school. The use of horizontal drilling techniques allowed the pipe to cross beneath a stream, a road, public utilities, athletic fields, and a railroad, with minimal disturbance of the ground surface. This was extremely important for the community and the school athletic programs.

Twelve Capstone microturbines are located at the school to provide the electricity and heat from the LFG. Each Capstone microturbine fueled by the landfill gas produces up to 30 kW of three-phase electricity at 480 V, using 12 to 16 ft^3/min of landfill gas for a total of 360 kW of electricity—enough to power the equivalent of approximately 120 homes. The microturbine system incorporates a combustor, a turbine, and a generator. The rotating components are mounted on a single shaft supported by air bearings that rotate at up to 96,000 r/min. The generator is cooled by airflow into the gas turbine. Built-in relay protection (overvoltage and undervoltage, and overfrequency and underfrequency) automatically trips off the microturbines in the event of a utility system outage or a power quality disturbance. Excess electricity not used by ACHS is sold to Commonwealth Edison. A 12-turbine system was selected to provide a system that will remain functional as LFG production from HOD Landfill decreases.

The project's design and construction can be a model for other communities that are interested in the beneficial reuse of nearby landfill gas resources. It is an example of how to deal with the numerous community concerns related to developing an alternative energy system based on landfill gas. Determining suitable equipment for the system design, construction, and operation, while considering local community needs and requirements, is critical to a successful project. Figure 13.11 shows photograph of the Lopez Canyon Landfill in Los Angeles. Figure 13.12 is a generalized graphic diagram of a landfill gas recovery.

Figure 13.11 Gas compression component installation at the Lopez Canyon Landfill in Los Angeles. *Courtesy of Capstone Microturbine.*

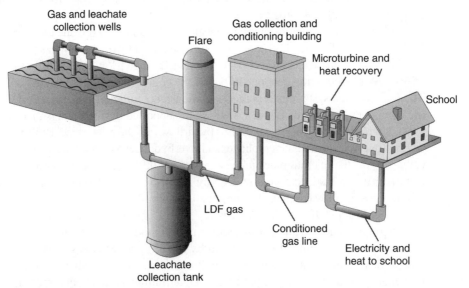

Figure 13.12 Landfill gas configuration diagram. *Courtesy of Capstone Microturbine and Alliant Energy.*

ELECTRIC POWER GENERATION

This project is a prime example of how innovative partnerships and programs can take a liability and turn it into a benefit. This solution has created a win-win situation for all involved, including HOD Landfill, ACHS, the Village of Antioch, the State of Illinois, Commonwealth Edison, and the EPA. Each key player is seeing significant benefits of the energy system: low energy costs for the high school; use of waste heat for internal use in the high school; clean and complete combustion of waste gas; decreased emissions to the environment through reduced need for traditional electric generation sources; reduction in greenhouse gas emissions; and educational opportunities in physics, chemistry, economics, and environmental management as a result of this on-campus, state-of-the art gas-to-energy system.

Cal Poly Biogas Case Study

The following is an interesting case study that demonstrates the application of a microturbine used in farming applications. The study was conducted by Cal Poly University, San Louis Obispo, in California. The study was conducted to convert animal waste to electricity and heat by use of microturbine technology. The project site selected was a dairy farm that was populated with 300 cows and animals. Animals were kept in a covered barn where a large percentage of the manure was initially deposited on the concrete floor which was periodically flushed out by water and collected in a lagoon that has a volume of 19,000 m³. A small percentage of the manure was deposited in corrals and was collected seasonally.

Methane production technologies, which use gases generated from covered lagoons using anaerobic digesters, had previously shown a certain degree of success. Essentially the choice of the digester technology by and large depends on the specifics of the animal waste characteristics. One type of technology makes use of a packed bed and an upflow anaerobic sludge blanket digester for soluble organic waste. Waste is collected in a lagoon covered with 1-mm-thick reinforced polypropylene which is sealed and tied down by weights. Digested gas is allowed to exit through a manifold. Figure 13.13 is a photograph of a floating lagoon at Cal Poly Dairy

Based on this model, the Cal Poly project employed a microturbine that used the digested gases generated by a 14,000-m³ (4 million gal) earthen lagoon. With the simple use of pumps and piping, diluted dairy manure wastewater is transferred to the lagoon.

Digested biogas from the lagoon is compressed and dried in a desiccant tank which is then used by a Capstone 30-kW, 440-V microturbine which generates grid-connected electricity. The system also used a Unifin heat recovery system from the microturbine generator exhaust which is used for space and water. Figure 13.14 is a generalized biogas process block diagram.

Figure 13.13 **Floating lagoon at the Cal Poly Dairy.** *Courtesy of Capstone Microturbine.*

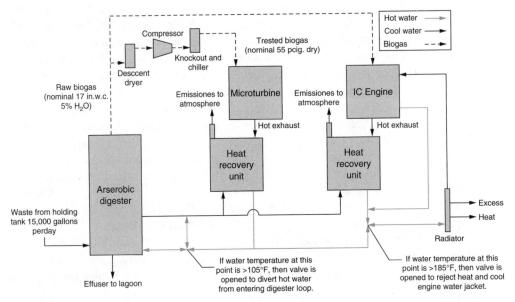

Figure 13.14 **Generalized biogas process block diagram.** *Courtesy of Capstone Microturbine.*

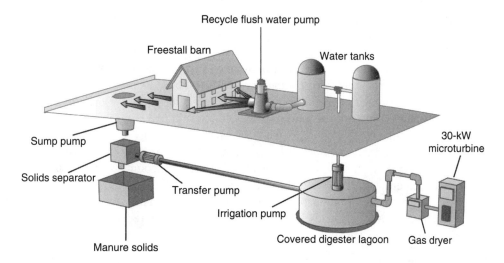

Recycled water is pumped up from the digester logoon to the water tanks.
From here the water pump is used to flush animal solids on the ground
from the barn toward the sump pump.

Figure 13.15 Diagramatic presentation of biogas production in a dairy farm.

CAPITAL COST AND ELECTRICITY BENEFITS

The cost of this methane recovery system for the lagoon construction, flexible cover, piping, gas handling, microturbine system, and associated labor and engineering was approximately $225,000. Based on the measured biogas production and the rated efficiency of the completed methane recovery system, it is estimated to produce about 170,000 kWh of electricity and 77,000 kJ of hot water annually which is estimated to be worth approximately $16,000.

In this case study it was found that the 30-kW turbine with an input of 13 m^3/h of biogas produces 28 kW of electricity which translates into an efficiency of 27 percent. Figure 13.15 represents a diagrammatic presentation of an on site biogas production system deployed in a dairy farm

This project was in part supported by the California State University Agricultural Research Initiative and was also funded by the Western Regional Biomass Energy Program which was matched in part by contributions from Capstone Microturbine.

Biomass Energy

For thousands of years, human beings have used biomass energy in the form of burning wood for heat and cooking food, which generated a relatively small amount of carbon dioxide. Since trees and plants for growth remove carbon dioxide from the atmosphere, and replenish oxygen constantly, the net effect of excess gas production on the ecology was kept in balance.

With the advancement of societies and the growing sophistication of lifestyles, use of diverse nonrenewable forms of biomass such as coal and fossil fuels shifted the balance of carbon dioxide generation and absorption. The net effect of this, as discussed earlier, resulted in greenhouse effects and other air pollution and contaminants, which challenge our very existence.

Fortunately, human intelligence has opened a window of opportunity that allows for the use of various forms of biomass, such as plants; agricultural waste; forestry and lumber residue; organic components of residential, commercial, and industrial waste; and even landfill fumes as bioenergy sources. The use of these forms of biomass not only reduces the atmospheric pollution but also enables creation of biodegradable products, biofuels, and biopower. These substantially reduce the need for use of nonrenewable fossil fuels.

Biofuels Production

PYROLYSIS

Pyrolysis is a method of processing a base fuel to produce a more efficient one. The basic process involves heating the original fuel material which is often pulverized or shredded in the near absence of air, at temperature ranges of about 300 to 500°C, until the volatile matter has been evaporated. The residue is then charcoal, a fuel which has about twice the energy density of the original which burns at a much higher temperature. For many centuries, and in much of the world still today, charcoal is produced by pyrolysis of wood. Depending on the moisture content and the efficiency of the process, the reduction ratio of the primary fuel to pyrolyzed product is about 4 to 1; in other words four tons of wood is required to produce 1 ton of charcoal. Figure 13.16 shows a sugar cane biogas

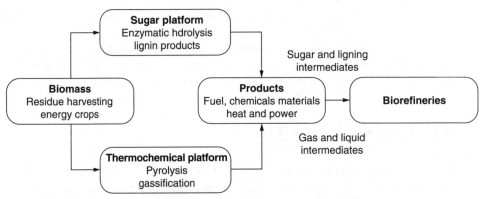

Figure 13.16 **Sugar cane generalized biogas production process block diagram.**

Graph courtesy of DOE/NERL.

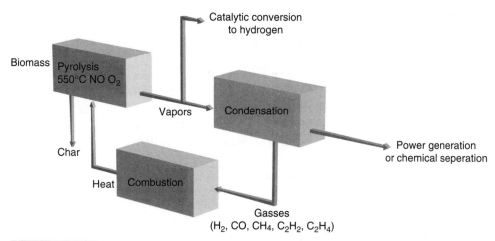

Figure 13.17 **Biogas liquefaction via block diagram via pyrolysis process** *Graph courtesy of DOE/NERL.*

production process block diagram. Figure 13.17 is biogas liquefaction via block diagram via Pyrolysis process

Pyrolysis can also be carried out in the presence of a small quantity of oxygen which is also known as gasification. One of the most useful by-products is methane, which is a suitable fuel that can be compressed and used as liquid fuel or can be used in high-efficiency gas fired turbines to generate electricity.

With more advanced pyrolysis processes, volatile gases are collected in distillation columns similar to those of gas refineries where, by the use of temperature gradients, various compositions of gases such as methane and ethane and even heavier petroleum oils can be extracted. Upon the removal of contaminants such as sulfuric acid, distilled liquid products are used as potential fuel.. In a process called fast pyrolysis, plant material, such as wood or coconut shells, can be converted at temperatures of 800 to 900°C up to 60 percent into a gas rich in hydrogen and carbon monoxide.

Fast pyrolysis therefore can be considered to be an excellent contender to conventional gasification methods. The technology at the present is experimental and has yet to be developed as a treatment for biomass on a commercial scale. At present, fast pyrolysis technology has not been advanced for use with conventional biomass.

GASIFICATION PROCESS

The gasification process is based upon extraction of a flammable gas mixture of hydrogen, carbon monoxide, methane, and other nonflammable by-products from coal and wood. The process involves partially burning and partially heating the biomass in the presence of charcoal. When produced, gases are compressed and liquefied and used as a substitute for gasoline. Such fuels generally reduce the power output of the car by about 40 percent.

SYNTHETIC FUELS

This gasification process uses oxygen rather than air, which produces a gas consisting mainly of H_2, CO, and CO_2, and when the CO_2 is removed, the mixture left is called synthesis gas. The most significant property of synthetic gas is that almost any hydrocarbon compound may be synthesized from its molecules. The reaction of H_2 and CO with the use of a catalyst is another method of producing methane gas. Another possible item produced from the gasification process is methanol (CH_3OH), a liquid hydrocarbon which has an energy caloric value of 23 GJ/ton.

Methanol production consists of highly sophisticated chemical processes under very high temperatures and pressure conditions. Use of methanol as a substitute for gasoline has in recent years made the fuel a significant commodity.

ETHANOL

Ethanol is a very high energy liquid fuel that is used as a direct substitute for gasoline in cars. The fuel that is produced from fermentation of the sugar solution left over in the sugar cane or sugar beet harvest has been successfully produced in large quantities in Brazil.

As mentioned earlier, feedstock used in fermentation includes crushed sugar beets or fruit. Sugars can also be manufactured from vegetable starches and cellulose by a pulping and cooking process or can be derived by a process which involves milling and treatment of cellulose with hot acids.

After approximately 30 hours of fermentation, the brew that contains 6 to 10% alcohol is removed from the liquid by distillation. Fermentation is an anaerobic biological process in which sugars are converted to alcohol by the action of microorganisms, usually by yeast. The resulting alcohol is ethanol (C_2H_3OH), which is used in internal combustion engines, either directly by specially modified engine carburetors or mixed with gasoline. The latter is referred to as gasohol.

The fermentation process of a biomass depends on the ease with which it can be converted to sugars. The best-known source of ethanol is sugar cane or the molasses remaining after the cane juice has been extracted. Other plants that can be readily fermented and have a high content of starch include potatoes, corn, and other grains. The fermentation process involves conversion of the starch to sugar. Biomass carbohydrates such as cellulose, which do not readily ferment, are pulverized to fine dust and broken down to sugars by acid or enzymes.

The energy content of the final products produced from these processes contains about 30 GJ/ton. The fermentation process requires an enormous amount of heat, which is usually derived by burning crop residues left over from sugar cane bagasse or maize stalks and cobs.

Biofuels are renewable energy sources derived from biomass, such as ethanol or wood alcohol, by a fermentation process. All plant and vegetation material carbohydrates, in the form of starch, sugar, and cellulose, like beer and wine, under proper conditions are fermented and converted into ethanol.

Another process of producing methanol is gasification, which involves vaporizing the biomass at very high temperatures. Upon vaporization, biomass gases are purged of impurities, by use of special catalytic media. Ethanol is commonly mixed with gasoline and diesel fuel to reduce vehicular smog emissions. Vehicles designed with flexible fuel type engines not only can use methanol as additives but are also capable of using animal fat, vegetable oil, or recycled cooking greases. Biomass-based alcohols used as pollution-reducing additives are commercially sold as MTBE (methyl tertiary butyl ether) and ETBE (ethyl tertiary ether). Each year in the United States, we blend 1.5 billion gal of ethanol with gasoline to reduce air pollution and improve vehicle performance. Most mixed fuels use 10% ethanol and 90% gasoline, which works best in trucks and cars. A mixed gasoline and ethanol fuel is referred to as an E85 grade fuel.

BIOPOWER

Biopower, or biomass power, uses the direct firing or burning of biomass residue, such as feedstock and forestry residue, to produce steam. The generated steam from the boilers is in turn used to turn steam turbines that convert rotational mechanical energy into electricity. The steam generated is also used to heat buildings and industrial and commercial facilities.

GASIFICATION SYSTEM

In this process the biomass is exposed to extremely high temperatures, which cause the hydrocarbon chains to break apart in an oxygen-starved environment. Under such conditions, biomass is converted and broken into a mixture of carbon monoxide, hydrogen, and methane. The gas fuels generated are used to run a jet engine–like gas turbine that is coupled to an electric generator.

BIOMASS DECAY

When decayed, biomass produces methane gas, which can be used as an energy source. The decaying organic matter in landfills release considerable amounts of methane gas that is harvested by drilling wells.

Partially perforated pipes are strategically placed in a number of locations throughout the landfill, which are connected to a common header pipe that feeds the collected gas to a chamber. The gases collected are compressed and cleansed by a purging process, which uses water steam to capture corrosive gases, such as hydrogen sulfides. Compressed methane is used either as direct burning fuel or for microturbines and fuel cells to generate electricity and cogenerated heat.

BIOPRODUCTS

Interesting uses of biomass are processes that convert them into any product that is presently made from petroleum-based nonrenewable fossil fuel. When converting

Figure 13.18 Biomass plastic products.

biomass into biofuels, scientists have developed a technique for releasing the sugars, starches, and cellulose that make up the basic structure of plants and vegetation.

Biomasses that have carbon monoxide and hydrogen as their building blocks are used to produce a large variety of products, such as plastics, antifreeze, glues, food sweeteners, food coloring, toothpaste, gelatin, photographic film, synthetic fabrics, textiles, and hundreds of products that are completely biodegradable and recyclable.

Chemicals used to produce various renewable materials derived from biomass are referred to as "green chemicals."

Some Interesting Facts about Bioenergy

According to the CEC, California produces over 60 million dry tons of biomass each year, only 5 percent of which is burned to generate electricity. If all the biomass generated were used, the state could generate 2000 MW of electricity, which would be sufficient to supply electricity for about 2 million homes.

Today biomass energy provides about 3 to 4 percent of energy in the United States. Hundreds of U.S. power plants use biomass resources to generate about 56 billion kWh of electricity each year.

Combined biopower plants in the United States generate a combined capacity of 10.3 GW of power, which is equal to 1.4 percent of the nation's total electric-generating capacity. With improvements in biopower technology, by the year 2020, the power capacity could be increased to about 7 percent or 55 GW. It should be noted that biomass fuel, similar to hydroelectric power, is available on a continuous basis.

Biodiesel

Biodiesel is recently being considered one of the most promising alternatives to petroleum-based diesel fuel. Biodiesel, which has the appearance of a yellowish cooking oil, is made from renewable vegetable oils or animal fats. Its chemical composition consists of a combination of long chains of fatty acids known as monoalkyl esters. The process of producing biodiesel is called trans-esterification, which involves removal of esters or fatty acids from the base oil. Upon completion of the process, biodiesel becomes combustible with properties similar to those of petroleum diesel fuel and can replace it in most applications. At present biodiesel is used as an additive to diesel fuel, which improves the lubricity of pure petroleum diesel fuel. With the inevitable cost escalation of crude oil, use of biodiesel has proven to be a viable candidate to partially mitigate a shortage of fossil fuel which at present is the world's primary transport energy source. Since biodiesel is a renewable fuel resource, it can replace petroleum diesel in current engines and can be produced in large quantities, transported and sold using today's commercial infrastructure. Biodiesel use and production has in recent years increased rapidly, and it is being used in Europe. In some provinces of France, farmers extract biodiesel from sunflower seed oil which is used to fuel tractors and agricultural combines. In the United States and Asia, biodiesel sales and use is gradually growing and a number of fuel stations are making biodiesel available to consumers. Recently a growing number of large transport fleets have started to use biodiesel as an additive to diesel fuel.

Biodiesel has a flash point of 150 and is not as readily ignited as petroleum diesel which has a flash point of 64. It is also much less combustible than gasoline which has a flash point of 45. Biodiesel is in fact considered to be a nonflammable liquid by the Occupational Safety and Health Administration (OSHA); however, it burns if heated to a high enough temperature which makes it a much safer fuel in cases of vehicular accidents.

One significant property of biodiesel fuel is that it gels and solidifies at much lower temperatures than petroleum diesel. The gelling characteristic much depends on the property of the base feedstock which biodiesel is made of. To overcome this problem, the fuel holding reservoirs are equipped with thermostatically controlled thermal elements which maintain the biodiesel at an appropriate temperature level.

One of the most significant properties of biodiesel is that unlike petroleum diesel, it is completely biodegradable and nontoxic, which can significantly reduce toxic gas emissions when burned.

A process of the trans-esterification, which reduces methyl esters from the feedstock, is the addition of methanol to the base oil which produces glycerol as a by-product.

At present, because of the economic scale of production, biodiesel is somewhat more expensive to produce than petroleum diesel, which is one of the main reasons why the fuel has not found widespread usage; however, as the cost of crude oil escalates and pollution abatement and emission of exhaust from automobiles becomes more stringent, biodiesel could become a viable substitute petroleum-based diesel fuel. Presently worldwide production of vegetable oil and animal fat is not enough to produce large amounts of biodiesel to replace liquid fossil fuel use. Another issue related to worldwide large-scale production of biodiesel is that it will require a large increase in farming of the feedstock, which will result in overfertilization, increased pesticide use, and land use conversion that would be needed to produce the additional vegetable oil.

ENVIRONMENTAL EFFECTS

The environmental benefits of biodiesel include reduction of emissions of carbon monoxide (CO) by approximately 50 percent and of carbon dioxide by 78.45 percent on a net life-cycle basis since the carbon in biodiesel emissions is produced by vegetation and is reabsorbed by plants.

Biodiesel use also eliminates sulfur emissions present in conventional diesel fuels because biodiesel extracted from animal and vegetable fats does not include sulfur. Biodiesel fuel use does produce emissions; however, the emissions can be reduced through the use of catalytic converters.

BIODIESEL POTENTIAL ENERGY REQUIREMENT

The U.S. Department of Energy's 2004 figures for annual transportation fuel and home heating oil use in the United States was estimated to be about 230,000 million U.S. gal. The amount of biodiesel needed to meet this demand could not be met with waste vegetable oil and animal fats. At present the United States' estimated production capacity of vegetable oil is approximately 23,600 million lb or 3000 million U.S. gal. The estimated production of animal fat is 11,638 million lb.

Currently the U.S. projection of plant production required to produce biodiesel includes the following (measured in U.S. gallons per acre):

Soybean—40 to 50
Rapeseed—110 to 145
Mustard—140
Jatropha —175
Palm oil—650
Algae—10,000 to 20,000

The reason for low soybean production is related to the fact that they are not an efficient producing crop when used solely for the production of biodiesel; however, the plant product is extensively used in food products.

In Europe, rapeseed oil is the preferred feedstock in biodiesel production. In India and Southeast Asia, the Jatropha tree, which is grown also for watershed protection, is considered to be the most efficient fuel source. Another plant specially bred for biodiesel fuel production is mustard plant varieties which can produce a high yield of oil. The production technology of algae harvest and oil production technology for biodiesel has as of yet not been commercialized.

WORLD PRODUCTION OF BIODIESEL

United States   Even though biodiesel is somewhat more expensive than conventional diesel fuel, it is commercially available in most oilseed-producing states in the United States. To promote the use of biodiesel, as of 2003, tax credits have been made available in the United States. It is estimated that in 2004, 30 million U.S. gal of biodiesel were sold in the United States. As environmental pollution abatement measures become more stringent, the U.S. biodiesel market is estimated to grow to 1 or 2 billion U.S. gal by the year 2010. The present cost of biodiesel fuel in the United States is about $1.85 per U.S. gallon which has made it competitive with diesel fuel.

A pilot project in Harbor, Alaska, is currently producing biodiesel from fish oil derived from local fish processing plants. The project is being developed in a joint collaboration with the University of Alaska Fairbanks.

Brazil   In the past decade, Brazil has commercialized ethanol use in automobiles and has recently completed construction of a commercial biodiesel refinery which was inaugurated in March 2005. The refinery is capable of producing 3.2 million U.S. gal of biodiesel fuel per year. The feedstocks used consist of a variety of sunflower seeds, soybeans, and castor beans.

Canada   Ocean Nutrition of Mulgrave, Nova, produces 6 million gal of fatty acid ethyl esters annually as a by-product of its omega-3 fatty acid processing. The by-product is used by Halifax-based Wilson Fuels as a fuel blended for use in transportation and heating fuel.

Fluidized Bed Boilers

The main principle of fluidized bed boilers is based upon a mixture of limestone and coal where the limestone acts as a sponge that absorbs sulfur and pollutants. As coal burns in a fluidized bed boiler, it releases sulfur; however, the tumbling action of limestone around the coal captures the sulfur by a chemical process when the sulfur gases are changed into a dry calcium sulfate powder that can be removed from the boiler. Calcium sulfate product is processed in the production of dry wall and wallboard used in building construction.

Another attribute of this technology is that fluidized bed boilers burn the biomass and fossil fuels in a cooler state at about 1400°F as opposed to the 3000°F required by conventional boilers, which results in production of nitric oxide (NO_x) which is formed as a result of the reaction of nitrogen molecules with the air. As a result, fluidized bed boilers can burn very dirty coal and remove 90 percent of the sulfur and nitrogen pollutants. Fluidized bed boilers can burn all biomass fuels such as wood, ground-up railroad ties, and any organic waste.

Today, very large fluidized bed boilers operate in more than 300 installations throughout the United States. The Clean Coal Technology Program has been conducting experimental tests in Colorado, Ohio, and Florida. A newly designed type of fluidized bed boiler encases the entire boiler inside a large pressure vessel, much like a pressure cooker, which is as large as a single-story building. Burning coal in a pressurized fluidized bed boiler produces a high-pressure stream of combustion gases that are used to generate steam which runs electric power turbines. Pressurized fluidized bed boilers are 50 percent more efficient than conventional boilers. Because of the efficiency of energy conversion, pressurized fluidized bed boilers reduce the amount of carbon dioxide and greenhouse gases released during coal burning.

WOOD BURNING

Wood waste used as fuel in boilers varies with the type of bark from paper mill, wood wastes from lumber mills, from plywood factories and dismantled building materials. Moisture contents and the configuration of wood waste differ with the source of generation.

Takuma Corporation has developed several types of combustion systems which match specific characteristics of the fuel material. Numerous types of designs and technologies are incorporated into each power plant to achieve efficient power generation and a stable power supply.

For example, palm fruits yield edible oil which is an important part of tropical agriculture business in Malaysia and Indonesia. Fruit residues from the palm oil extraction process are a valuable boiler fuel. In such a process the steam from the boiler is used for generating electric power for the plant, which is in turn used to provide heat for the oil extraction and refining processes.

In another installation, corn which is also widely grown in tropical, temperate, and some cold zones, in addition to its use as food and animal feed, is used to make corn starch from the grain and alcohol from the cob. The grain and cob residues left over from the process are used as fuel to provide process heat in the factory.

14

HYDROELECTRIC POWER

Hydroelectric Power Generation

As discussed in earlier chapters the dynamics of water energy is a result of the constant movement and transformation of the global cycle which results from the effect of solar energy. The cycle consists of the evaporation of oceans and seawaters which forms clouds which in turn create precipitation in the form of rain or snow, and eventually the cycle is completed when water in rivers flows back to the ocean. The energy of this water cycle is tapped by a wide variety of technologies described in earlier chapters, including hydropower stations.

Ancient civilizations used waterwheels to relieve humans of some forms of manual labor. Water power was used by the Greeks; around 4000 BC, they used hydropower to turn waterwheels for grinding wheat into flour. With the invention of the water turbine in the early 1800s hydroelectric power technology was soon advanced to produce electricity.

The main advantage of hydroelectric power is that it is renewable, generates no atmospheric pollution during operation, and has relatively very low operation and maintenance costs. Another positive attribute of hydroelectric projects is that dams and reservoirs that hold the water can be used as recreational facilities. Disadvantages associated with hydroelectric power generation include high initial capital cost and potential site-specific negative environmental and ecological impacts, which will be discussed here in detail.

Hydroelectric Power Plants

Among the variety of renewable energy resources, hydroelectric power is the most desirable for utility systems and has a long, successful proven track record. Power generated from hydroelectric plants can exceed 10 GW. On the European continent,

Norway generates 98 percent of its electric energy from hydroelectric power. It is estimated that to date only about 10 percent of the world's hydroelectric potential resources have been exploited; the remaining untapped potential is in Africa and Asia.

The world's total installed hydropower capacity is currently about 630,000 MW. The annual worldwide power production is estimated to be 2200 billion kWh, which means that the power plants are running at 40 percent of their rated power production capacity. Figure 14.1 depicts a hydroelectric power generating diagram

The largest hydroelectric complex in the world, Itaipu, is located on the Parana River, which flows between Paraguay and Brazil. The Itaipu hydroelectric power generation complex has 18 turbines which collectively produce 12,600 MW of electricity.

At present, a number of large hydroelectric power dams are being built throughout Asia. In 1999 China completed its 3300-MW Ertan hydroelectric power station which has six turbines that generate a total of 550 MW.

The Indian government has approved construction of 12 large-scale hydroelectric power station projects which would add 3700 MW to the countries electric power generation capacity.

One of the world's largest hydroelectric projects is currently under construction in China. This project, which has a capacity of 18.2 GW and is named Three Gorges Dam, has entered the second phase of a three-phase construction. Upon completion of phase 3 in 2009 the power station will be providing full power generation. The estimated construction cost of the dam is projected to be about 25 billion U.S. dollars. Upon its final completion, the Three Gorges Dam will extend 2 km across the Yangtze River, be 200 m tall, and create a 550-km-long reservoir.

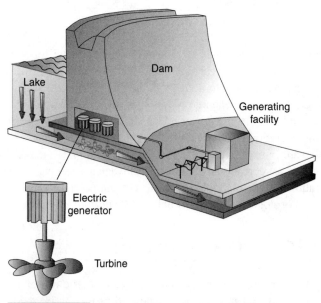

Figure 14.1 **Hydroelectric power generating diagram.**

Construction of the dam to date has created serious environmental and social problems including water pollution along the Yangtze River with numerous pollutants from mining operations, factories, and human settlements that used to be washed out to sea by the strong currents of the river.

In coming years silt in the river is expected to be deposited at the upstream end of the dam which will inevitably clog the major tributaries. According to recent reports an estimated 2 million people have been resettled and 1300 archaeological sites will to be moved or flooded. Construction of the dam has also resulted in destruction of natural habitats of several endangered species and rare plants.

Hydroelectric Power Potential

There are two significant parameters for hydroelectric power plant potential, namely, the amount of water flow per time unit and the vertical height or head that water can be made to fall. In some instances the water head may be attributed to the natural site topography, or it may be created artificially by constructing a dam. Water accumulation in a dam is dependent on the intensity, distribution, and duration of rainfall, as well as direct evaporation, transpiration, ground infiltration, and the field moisture capacity of the basin or reservoir soil.

The potential power derived from the water fall depends upon the height and the rate of flow. The available hydropower is calculated from the following formula:

$$P = 9.8 \times q \times h$$

where P = available power (kW), q = water flow rate (m^3/s), and h = water head (m).

The calculated power derived by this formula does not include power losses resulting from friction, and turbine efficiency. The overall power generation efficiency of a hydropower generating plant could be as high as 90 percent or more.

Hydroelectric power potential example   This example illustrates the power production potential of a hydroelectric power plant with a head of 300 m and an average flow of 1200 m^3/s, with the assumption that the dam dike covers an area of 2000 km^2.

$$P = 9.8 \times q \times h = 9.8 \times 1200 \times 300 = 3520 \text{ MW}$$

Since the flow is 1200 m^3/s, a drop of 1 m in water level corresponds to

$$2000 \times 1,000,000 = 2,000,000,000$$

The time it takes for the volume of water to pass through the turbine is

$$t = 2,000,000,000/1200 = 463 \text{ h} = 19.3 \text{ days}$$

The theoretical equivalent fossil fuel oil operated thermal power generating station when compared to a hydroelectric plant with a power generating capacity of 5000 Wh would require approximately 20 million barrels of oil per day. Estimated amount of pollution created by fossil fuel is conservatively estimated at 6 pound per Kw/hr., then the total generated hourly pollution created by the thermal power will be in the neighborhood of 30,000 lb.

Environmental Effects of Hydroelectric Power

Some negative factors and setbacks associated with the construction of hydroelectric projects are related to the initial capital investment and irreversible ecological and environmental damages that are associated with dam construction and operation. Figure 14.2 is a depiction of a hydroelectric power dam diagram.

A watercourse is an ecological system that can be seriously affected when disturbed by human intervention. As an example, changes in water flow may affect the quality of the water and the production of fish downstream. Dams and barriers are known to alter ecological conditions of aquatic life. Construction of artificial lakes and dam reservoirs often prevents migration of downstream fish to upstream habitats. Figure 14.3 is a diagrammatic presentation of a multi-stage hydroelectric power generating system.

Environmental changes resulting from construction of dams affects the aquatic ecology along entire stretches of rivers and streams, even affecting life at the inlets to the sea. Reservoir sedimentation often results in increased downstream sedimentation.

Figure 14.2 Hydroelectric power dam diagram.

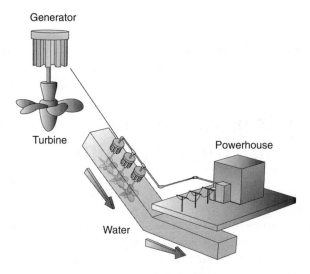

Figure 14.3 Multistage hydroelectric power generating system diagram.

Changes in water flow resulting from construction of dams also lead to changes in the transportation of sediments, which in turn reduces the water nutrient quality essential for survival of aquatic life.

Dam reservoirs of hydropower plants affect the flow of a watercourse and disturb natural groundwater levels in surrounding areas, which in turn influences the quality of the water and the sediment transport of the watercourse which often results in area runoff and ground erosion.

Entrapment of nutrients in reservoirs usually results in excessive accumulation of ground bed fertilization which often leads to an increased growth of algae, which in turn may cause anaerobic conditions and a lack of oxygen in the deep-water layers which can destroy aquatic life.

Warm weather conditions which cause water evaporation may also cause a concentration of nutrients, leading to excessive fertilization.

As a result of changes to water quality a habitat reproduction for some species could be hindered or prevented during the spawning period. Submerged grounds and water flow changes also alter the fauna and vegetation habitats and can cause animal extinctions.

Effect of Dam Construction on Local Population

Large hydropower plants with dams require large reservoir and discharge areas, which in some instances force habitants to evacuate from the grounds. Adverse social consequences for the local population usually result in relocation and transfer of indigenous groups of people which may endanger their entire cultural system. Evidence of such social disturbance has been associated with the construction of the Three Gorges

Dam in China where thousands of villagers were forcefully displaced and relocated to newer grounds.

Similarly, construction of the Aswan Dam in Egypt resulted in a wholesale displacement of hundreds of villages and damaged or destroyed historical and cultural landscapes, ancient monuments, holy places, and burial grounds of great importance.

Impact of Hydropower Dam Construction on Human Health

Hydroelectric power plants increase the incidence of water-related diseases caused by pathogens, such as typhus, cholera, and dysentery, and of infections by tapeworms and roundworms. Some of the diseases that have increased as a result of construction of the Aswan Dam include bilharzia, malaria, filariasis, sleeping sickness, and yellow fever.

Reservoirs with large, stagnant waters and slow water-level variations offer favorable growth conditions for a variety of pathogens. Excessive aquatic vegetation growth resulting from fertilizer concentrations blocks ultraviolet rays and provides a fertile ground for infection carrier bacteria which would have otherwise been destroyed by sunlight. Vegetation also promotes growth of mosquito species carrying malaria and filariasis. Seepage of contaminated water from reservoirs into groundwater wells increases the risk of infection by the spread of pathogens into drinking water.

Hydroelectric Power Technology

As referenced earlier hydropower energy generation is achieved by turbines and generators that convert the dynamic or static energy of water to mechanical and then electric energy. Turbines and generators are either located within the dam structure or in the vicinity of the dam, and are driven by pressurized water that is transported from the dam through penstocks or pipelines. Compared with conventional thermal power generating plants, hydropower technology is much more efficient and can produce twice the amount of power. The efficiency of hydroelectric plants can be attributed to the fact that the kinetic energy of water falling a vertical distance represents an energy that gets converted into mechanical rotary power without loss as opposed to the conversion of biomass calories into energy which has a significant intrinsic loss of efficiency.

Equipment such as turbines and generators associated with hydropower are based on well-established and well-developed technologies that are relatively simple to manufacture, are highly reliable, and have an extended life span of about 50 years. On the contrary, thermal combustion–based power generation technology equipment is relatively complicated, has a shorter life span, and has lower reliability.

CLASSIFICATION OF HYDROPOWER ENERGY FACILITIES

Hydropower technologies are classified into two types of operational categories, namely conventional and pumped storage types. Power plants are in turn rated for power capacity (such as big or small), head of water (low, medium, or high), type of turbine used (such as Kaplan, Francis, or Pelton), and finally the location and type of dam or reservoir.

Conventional hydropower systems   Conventional hydropower plants derive energy from rivers, streams, canal systems, and reservoirs. This category of power-generating stations is further divided into two subcategories; one is known as impoundment and the other as diversion. Impoundment-type hydropower generating stations use dam structures to store water. Water from the reservoirs is released and the flow is controlled by vanes that maintain a constant water level. In diversion-type hydropower technology, portions of the river water are diverted through a canal or penstock; however, some installations require a dam.

Pumped storage hydropower plants   A pumped storage hydroelectricity plant is constructed from two reservoirs built at different altitudes. During periods of high electric demand, water from the high reservoir is released to the lower reservoir to generate electricity. Power generation results from the release of kinetic energy which is created by the discharge through high-pressure shafts which direct the water through turbines connected to generator-motors. Upon completion of the power generation period during the daytime when the demand and cost of energy is high, water is pumped back to the upper reservoir at nighttime for storage when the cost of energy and the energy demand are low.

Even though pumped storage facilities consume more energy than they can generate, they are used by power utility companies to provide peak power production when needed. In some installations, pumped storage plants operate on a full-cycle basis.

HYDROELECTRIC PLANT EQUIPMENT

The major components and machinery used in hydroelectric plants consist of dam water flow controls, reservoir controls, turbine controls, electric generator controls, power transformation equipment needed to convert electricity from low voltages to the high voltages required for power transmission and distribution, transmission lines required to conduct electricity from the hydropower plant to the electric distribution system, and finally the penstock system which carries water to the turbines. Figure 14.4 is a graphic representation of a pumped type hydroelectric system.

Turbines used in hydroelectric power systems are classified in several ways based on the method of functional operation such as impulse or reaction turbine; another classification is based on the way the turbine is constructed, such as the shaft arrangement or the feed of water. Turbines are also designed in a manner to allow them to operate as a pump or as a combination of both.

As an example, impulse turbines use a special nozzle that converts the water under pressure into a fast-moving jet. The jet of water is then directed at the turbine wheel or the runner, which converts the kinetic energy of the water into shaft rotational power. Another example of a turbine is a Francis turbine which uses the full head of water available to generate rotational power. Most hydraulic turbines consist of a

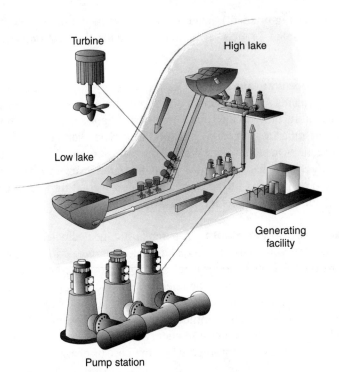

Figure 14.4 Pumped hydroelectric power generating system diagram.

shaft-mounted waterwheel or runners that are located within a water passage that conducts water from higher elevations to a lower one below the dam.

Without exception all hydraulic turbine generators are designed to turn at a constant speed. This constant speed is achieved by a device called a governor, which is a rotating ball mechanism that is balanced by rotational centrifugal force and keeps each generator unit operating at its proper speed by controlling the flow-control gates in the water passage.

Pelton turbine   The Pelton turbine is basically based on the same principles as a classic waterwheel. This type of turbine is used in applications where heads exceed 40 m or more. In some instances the turbine is used for heads as high as 2000 m. In settings where the water head is lower than 250 m, Francis turbines are given preference.

Francis turbine   The main difference of this technology compared to the Pelton turbine is that the runner is completely submerged in water, which results in a decrease of water pressure from the inlet to the outlet. Water flow into the turbine is directed radially toward the center. The guide vanes within the turbine are arranged so that the energy of the water is largely converted into rotary motion.

Kaplan turbine   This type of turbine is designed for use in the situation where the water head is low but there are high flow rates. In the Kaplan turbine the water flows through the propeller and sets it in rotation.

The design of the turbine is such that the area through which the water flows is as big as it can be to allow the entire blade areas to be swept by water currents. This makes the technology very suitable for large volume flows where the head is only a few meters. Water enters the turbine laterally and is deflected by the guide vanes and then flows axially through the propeller striking the blades when exiting. The construction of this type of turbine is relatively simple. The applications of this type of turbines are limited to heads that range from 1 to 30 m, which also requires a relatively larger flow of water.

Engineering and construction of large hydroelectric power plants require extensive capital investment and involve very sophisticated manufacturing technology, electro-mechanical equipment, significant feasibility studies, environmental impact reports, and civil construction activities. Large-scale hydropower stations also require in-depth environmental investigation and social impact considerations.

Case Studies of Hydroelectric Power Plants

HOOVER DAM

The following coverage of the Hoover Dam was obtained from the U.S. Department of the Interior, Bureau of Reclamations Web site http://www.usbr.gov/lc/hooverdam/service/index.html. This Web site provides the story of the development of the dam's construction which includes very interesting historical and engineering footnotes.

Hoover Dam is the highest and third largest concrete dam in the United States. The dam, power plant, and high-voltage switchyards are located in the Black Canyon of the Colorado River on the Arizona-Nevada state line. Lake Mead, the reservoir behind the dam, holds an average 2-year flow of the Colorado River. Hoover Dam's authorized principal objective was to regulate the river flow, improve navigation, and improve flood control. The second objective was to deliver stored water for irrigation and other domestic uses, and the third objective was to harvest electric energy. Lake Mead also provides outstanding outdoor, water-based recreation opportunities and is home to a myriad of wildlife.

Project development plan   Waters of the Colorado River are impounded by Hoover Dam. This water is released when needed to meet downstream demands for irrigation or domestic water, or when the dam is being operated under flood-control criteria.

The water is released at a time and in a way to meet the water delivery need and to maximize other benefits, including power generation. Irrigation water is provided to numerous projects in the lower Colorado River Basin, including the Imperial Irrigation

District and Coachella Valley Water District through the All-American Canal system; which includes the Gila, Yuma, and Yuma Auxiliary projects; the Palo Verde project near Blythe, California; the Colorado River Indian Reservation project; and the Central Arizona project. A dependable supply of water for domestic purposes also is provided to the semiarid southern California coastal region, to central and southern Arizona, and to southern Nevada. The water for southern California is diverted at Lake Havasu and transported through the Metropolitan Water District's Colorado River aqueducts to the district's area of use. The water for Arizona is also diverted from Lake Havasu, by the Central Arizona Project aqueduct, and transported into the state's interior. Southern Nevada withdraws its water from Lake Mead through the Robert B. Griffith Water Project, formerly called the Southern Nevada Water Project.

Hoover Dam and Lake Mead facilities   Hoover Dam is located about 35 mi from Las Vegas, Nevada. It is constructed of a very thick concrete and measures 726.4 ft high and 1244 ft long at the crest. The dam contains about 3,250,000 yd^3 of concrete; the total concrete in the dam and appurtenant works is 4,400,000 yd^3. The reservoir behind the dam, known as Lake Mead, has a total storage capacity of 32,471,000 acre-ft. Following completion of a sedimentation survey conducted in 1963–64, it was calculated that the total storage capacity had been reduced to 27,377,000 acre-ft, which includes 1.5 million acre-ft of space reserved exclusively for flood control.

To bypass and control the river during construction, four 50-ft-diameter concrete-lined tunnels were constructed through the canyon walls, two on each side of the river. The tunnels averaged about 4000 ft in length. After completion of the dam construction, the upstream tunnel entrances were closed by huge steel gates, and concrete plugs were placed near the midpoint of each tunnel. The downstream sections were

Figure 14.5   Hoover Dam hydroelectric power generating station. *Courtesy of U.S. Bureau of Reclamations.*

incorporated into the dam's spillway and outlet works features. A total of 315,000 yd^3 of concrete was used to line the diversion tunnels.

The dam complex includes two drum-gate controlled channel spillways, one on each side of the canyon. Each spillway discharges through an inclined, concrete-lined tunnel that connects with the remaining portion of the original outer diversion tunnel downstream from the tunnel plug. The crest of each spillway is surmounted by three piers that divide the crest into four 100-ft sections, each equipped with a 16- by 100-ft drum gate. The spillway capacity at reservoir elevation is 1229 ft, and the discharge capacity is 64,800 ft^3/s. At a surface elevation of 1229 ft with the gates lowered, the discharge capacity of the spillways is 400,000 ft^3/s.

The dam design consists of a combination of four penstock and outlet units, each originating at one of the four intake towers located upstream of the dam, and installed in a tunnel located at the back of the dam abutments. The penstock and outlet units originating at the upstream intake towers are installed in the inner pair of the four tunnels originally used for river diversion. The two penstock and outlet units originating at the downstream intake towers are installed in tunnels located about 170 ft above the lower units. Beyond the penstock outlets, each downstream unit branches into outlet pipes that terminate in the Arizona and Nevada canyon-wall valve houses.

The capacity of the canyon-wall and tunnel-plug outlets at the reservoir water surface at an elevation of 1225 ft has a discharge capacity of 52,600 ft^3/s. The total release capacity through the canyon-wall outlet works, tunnel plug outlet works, and the generating units, is 100,600 ft^3/s.

Hydroelectric power plant   The power plant is located at the toe of the dam and extends downstream 650 ft along each canyon wall. The turbines are designed to operate at heads ranging from 420 to 590 ft. The final generating unit, designated N-8, was installed at Hoover Dam in 1961, giving the dam a total of 17 commercial generating units. Installation of Unit N-8 brought the power plant's rated capacity to 1,850,000 hp. Two station-service units, rated at 3500 hp each, increased the plant total power output capacity to 1,857,000 hp. In terms of electric energy, the total rated capacity for the plant was 1,344,800 kW. The dam complex also includes two transformer station-service units each rated at 2400 kW. Between 1982 and 1993, the 17 commercial generating units were replaced with new turbines and new transformers, which raised the hydroelectric power capacity to its current level of 2,991,000 hp or 2,074,000 kW.

The dam, power plant, and all facilities are owned, operated, and maintained by the U.S. government. Prior to 1987, the power plant, transformer, and switching facilities were operated and maintained by the Los Angeles Department of Water and Power, and Southern California Edison Co.

Brief history   By the treaty concluding the Mexican War in 1849, and by the Gadsden Purchase of 1853, the United States acquired the territories of New Mexico, Arizona, and California. Discovery of gold in California in 1849 brought hordes of adventurers westward. They crossed the Colorado River near Yuma, Arizona, and at Needles, California. In 1857, Lieutenant J. C. Ives traveled 400 mi up the river by boat from the Gulf of California to the Black Canyon, the present site of Hoover Dam. He reported the region to be valueless.

In 1869, Major J. W. Powell of the Geological Survey succeeded in leading a river expedition down the canyon of the Colorado. The expedition traveled from the Green River in Utah to the Virgin River in Nevada—through more than a thousand miles of unknown rapids and treacherous canyons.

In 1875, a route was mapped for a canal to irrigate southern California's rich but arid land. Construction of the canal began about 20 years later, and in 1901 the first water from the Colorado River flowed through the Imperial Canal into the Imperial Valley.

The river, annually fed by melting snows in the Rocky Mountains, typically swelled to a raging flood in the spring and then dried to a trickle in the late summer and fall, so crops were frequently destroyed. Farmers built levees to keep out the river, but even when the levees held, crops withered and died when the river ran too low to be diverted into the canals.

In 1905, a disastrous flood burst the banks of the river, and it flowed for nearly 2 years into the Salton Sink in the Imperial Valley, creating what is now known as the Salton Sea. The river was eventually turned back into its original channel, but the continuing threats of floods remained.

Faced with constantly recurring cycles of flood and drought, residents of the southwest appealed to the then-Reclamation Service to solve the problem. Engineers began extensive studies of the river in search of a feasible plan for its control. In 1918, a plan was conceived for regulation of the river by building a single dam of unprecedented height in Boulder Canyon, about 8 mi upstream of the dam's eventual location. The Colorado River Compact, signed at Santa Fe, New Mexico, on November 24, 1922, cleared the way for construction of the dam by allocating most of the river's estimated flow between the upper and lower basins of the river and providing for later division of what was thought to be water excess to these allocations.

The project was authorized by the act of December 21, 1928, subject to the terms of the Colorado River Compact. The act authorized the construction of a dam and power plant in either the Boulder or Black canyon, and the All-American Canal System in southern California. The Boulder Canyon Project Adjustment Act, dated July 19, 1940, provided for certain changes in the original plan.

On October 1, 1977, in conformance with the Public Law Department of Energy Organization Act of August 4, 1977, the power marketing function which included transmission lines and attendant facilities of the Bureau of Reclamation was transferred to the Department of Energy. However, operation and maintenance of the federal hydroelectric generating plants along the Colorado River remained under the Bureau of Reclamation's jurisdiction. Effective October 9, 1977, the Boulder Canyon Project Hoover Dam and portions of the Parker-Davis Dam Project were combined for administration purposes into one operational unit, which is now called the Lower Colorado Dams Project.

On August 17, 1984, Congress passed the Hoover Power Plant Act of 1984. This act authorized an increase in the capacity of the existing generating equipment at the Hoover Dam power plant, and the improvement of parking, visitor facilities, and roadways and

Figure 14.6 Hoover dam historical power generating system sketch. *Courtesy of U.S. Bureau of Reclamation.*

other facilities to contribute to the safety and sufficiency of visitor access to the Hoover Dam and Power Plant.

Project planning and construction   The Boulder Canyon Project is characterized by the extraordinary. The height and base thickness of the dam, the size of the power units, the dimensions of the fusion-welded plate-steel pipes, the novel system of artificially cooling the concrete, the speed and coordination of construction, and other major features of the project were without precedent for their time. The magnitude of the construction introduced many new problems and intensified many usual ones, requiring investigations of an extensive and diversified character to ensure structures representing the utmost in efficiency, safety, and economy of construction and operation. Construction was begun in Black Canyon in 1931, and the dam was dedicated on September 30, 1935. The first generator of the powerhouse was in full operation on October 26, 1936. The last generator went into operation on December 1, 1961. In 1962, the construction railroad spur from Boulder City, Nevada, to the dam was sold and removed.

Little changed at the dam until construction of new visitor facilities at Hoover Dam was initiated in 1986. The initial items of work included relocation of two existing electric transmission towers and realignment of a small portion of the highway on the Nevada side of the dam. Construction was halted in 1988 because of a lack of funding, but resumed again in 1989 with the excavation of a new elevator shaft in the Nevada canyon wall. Construction of the visitor building and parking garage was initiated in 1991, and the new facilities—the visitor center, parking structure, and a new penstock viewing platform—were opened to the public on June 21, 1995. A detailed time line of the Hoover Dam is covered at the end of this section.

Benefits of the dam

IRRIGATION   The project ensures a dependable water supply for irrigating more than 1 million acres of land in southern California and southwestern Arizona, and over 400,000 acres in Mexico. These irrigated lands supply large amounts of produce and other agricultural products for the nation's markets.

MUNICIPAL AND INDUSTRIAL   Hoover Dam helps ensure a dependable water supply for municipal, industrial, and other domestic uses in southern Nevada, Arizona, and southern California. More than 16 million people and numerous industries in these three states receive Colorado River water that was stored by Hoover Dam.

RECREATION AND FISH AND WILDLIFE   Surrounded by rugged mountains and canyon walls, Hoover Dam and Lake Mead are outstanding scenic and recreation attractions. More than 1 million people a year now take guided tours of Hoover Dam, and more than 9 million people visit the Lake Mead National Recreation Area (LMNRA), America's first designated national recreation area, annually. Lakes Mead and Mohave, formed by the Davis Dam, are administered by the National Park Service. Its concessionaires provide facilities such as lodge and trailer accommodations, boats for hire, and sightseeing boat trips on the lake as well as through Black Canyon below the dam. Other popular activities are camping, picnicking, swimming, boating, water skiing, and year-round fishing for striped bass, large-mouth bass, and other game fish. A large part of the area is open to hunting.

Hoover Dam is also a major tourism site. Guided tours of the dam have been provided by the Bureau of Reclamation since 1936, with only a brief hiatus during World War II. Prior to 1995, physical limitations at the dam restricted the number of guided tours to about 750,000 people a year. New visitor facilities opened at the dam in June 1995 made it possible for more people to take a guided tour of the dam and power plant, and, in fiscal year 1999 (October 1, 1998, to September 30, 1999), nearly 1.2 million people took guided tours of the facility.

HYDROELECTRIC POWER   Hoover Dam is one of the world's largest producers of electric power, generating, on average, 4 billion kWh of firm hydroelectric energy annually. This energy played a vital role in the production of airplanes and other equipment during World War II, and it also was instrumental in the development of industrial expansion in the southwest.

Firm power generated at Hoover Dam is provided to 15 contractors in the states of California, Arizona, and Nevada under contracts that were signed in 1987 and will expire in 2017. The approximate percentage of firm power delivered to each state is as follows: Nevada, 23.4 percent; Arizona, 19 percent; and California, 57.6 percent.

FLOOD CONTROL   Hoover Dam has virtually ended the possibility of devastating floods striking the lower reaches of the river as they did prior to project construction. The benefits from controlling floods are reflected in the $25 million of the project cost that was allocated by Congress to flood control. For the period 1950 through 1998, flood control benefits provided by Hoover Dam and other structures on the mainstream Colorado River are approximated at nearly $1 billion.

Chronology of the Hoover Dam construction

1540. Alarcon discovers the Colorado River and explores its lower reaches. Cardenas discovers the Grand Canyon.

1776. Father Escalante explores the upper Colorado and its tributaries.

1857. Lt. J. C. Ives navigates the Colorado River and, with his steamboat *The Explorer*, reaches the end of Black Canyon.

1869. Major John Wesley Powell makes the first recorded trip through the Grand Canyon.

1902. President Theodore Roosevelt signs the Reclamation Act. Reclamation engineers begin their long series of investigations and reports on control and use of the Colorado River.

1905–1907. The Colorado River breaks into the Imperial Valley, causing extensive damage and creating the Salton Sea.

1916. An unprecedented flood pours down the Gila River into the Colorado, and flood waters sweep into Yuma Valley.

1918. Arthur P. Davis, reclamation director and chief engineer, proposes control of the Colorado River by a dam of unprecedented height in Boulder Canyon on the Arizona-Nevada border.

1919. All-American Canal Board recommends construction of the All-American Canal, and a bill is introduced to authorize its construction.

1920. Congress passes the Kinkaid Act authorizing the secretary of the interior to investigate problems of the Imperial Valley.

1922. The Fall-Davis report entitled "Problems of Imperial Valley and Vicinity," prepared under the Kinkaid Act and submitted to Congress on February 28, recommends construction of the All-American Canal and a high dam on the Colorado River at or near Boulder Canyon. Representatives of the seven Colorado River Basin states sign the Colorado River Compact in Santa Fe, New Mexico on November 24. The first of the Swing-Johnson bills to authorize a high dam and canal is introduced in Congress.

1924. Weymouth report expands Fall-Davis report and further recommends Boulder Canyon project construction.

1928. Colorado River Board of California reports favorably on the feasibility of the project. The Boulder Canyon Project Act, introduced by Senator Johnson and Representative Swing, passes in the Senate on December 14 and in the House on December 18, and is signed by President Calvin Coolidge on December 21.

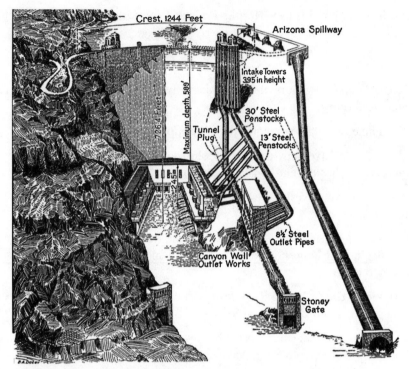

Crest, 1244 Feet

Arizona Spillway

Intake Towers
395 in height

30' Steel
Penstocks

Maximum depth, 589'

726.4 Feet

Tunnel
Plug

13' Steel
Penstocks

Canyon Wall
Outlet Works

8½' Steel
Outlet Pipes

Stoney
Gate

Figure 14.7 **Hoover dam historical power generating system sketch.** *Diagram courtesy of U.S. Bureau of Reclamations.*

1929. Six of the seven basin states approve the Colorado River Compact. Boulder Canyon Project Act declared effective on June 25.

1930. Contracts for the sale of electric energy to cover dam and power plant financing are completed.

1931. The Bureau of Reclamation opens bids for the construction of Hoover Dam and Power Plant on March 4, awards the contract to Six Companies on March 11, and gives the contractor notice to proceed on April 20.

1932. The river is diverted around the dam site on November 14. The repayment contract for the construction of the All-American Canal is completed with the Imperial Irrigation District.

1933. The first concrete is placed on June 6.

1934. All-American Canal construction begins in August. The repayment contract between the United States and the Coachella Valley Water District covering the cost of Coachella Main Canal is executed on October 15.

1935. The dam starts impounding water in Lake Mead on February 1. The last concrete is placed in the dam on May 29. President Franklin D. Roosevelt dedicates the dam on September 30.

1936. The first generator, N-2, goes into full operation on October 26. The second generator, N-4, goes into operation on November 14. The third generator, N-1, starts production on December 28.

1937. Generators N-3 and A-8 begin operation on March 22 and August 16, respectively.

1938. Lake Mead storage reaches 24 million acre-ft, and the lake extends 110 mi upstream. Generators N-5 and N-6 begin operation on June 26 and August 31, respectively.

1939. Storage in Lake Mead reaches 25 million acre-ft, more than 8 trillion gal. Generators A-7 and A-6 begin operations on June 19 and September 12th, respectively. With an installed capacity of 704,800 kW, the Hoover Power Plant is the largest hydroelectric facility in the world—a distinction held until surpassed by Grand Coulee Dam in 1949.

1940. Power generation for the year totals 3 billion kWh. All-American Canal is placed in operation. The Metropolitan Water District of Southern California successfully tests its Colorado River Aqueduct.

1941. Lake Mead elevation reaches 1220.45 ft above sea level on July 30; the lake is 580 ft deep and 120 mi long. Spillways are tested on August 6, the first time they have ever been used. Generator A-1 is placed in service on October 9. The dam closes to the public at 5:30 p.m. on December 7, and traffic moves over the dam under convoy for the duration of World War II.

1947. The Eightieth Congress passes legislation officially designating the Boulder Canyon Project's key structure "Hoover Dam" in honor of President Herbert Hoover.

1961. The power installation at Hoover Dam is complete when the final generating unit, N-8, goes on line on December 1. The installed generating capacity of the Hoover Power Plant, including station service units, reaches 1,334,800 kW.

1985. Hoover Dam celebrates its fiftieth anniversary. The majority of the cost of the Boulder Canyon project has now been repaid to the federal treasury.

ASWAN DAM

Construction history   Originally, Aswan Dam was designed by a British engineer Sir William Willcocks in 1899. The construction works were undertaken by two notable engineers, Sir Benjamin Baker and Sir John Aird. The project was completed in 1902. The gravity dam measured 1900 m long and 54 m high. Upon completion of the project, it was found that the initial design was inadequate to curb water spillage; hence the design was modified and the height of the dam was raised.

In 1946 the Aswan Dam came close to a catastrophic overflow point, at which time the Egyptian government decided to construct a new dam which would be built 6 kilometers up the Nile River. Upon completion of the initial investigation by soviet engineers, project planning began in 1952, which coincided with the Nasser revolution. Initially

Figure 14.8 View of Aswan Dam, Egypt.

the 270 million U.S. dollars was to be obtained with U.S. and British construction loans; however, the agreement due to the revolution and siding of Nasser with the USSR and arms agreement with USSR was canceled in July 1956. In the early 1800s when France and Britain subsidized the construction of the Suez Canal, it was agreed that the cost of completing the canal would be transferred to Egypt; however, soon after the revolution Nasser nationalized the Suez Canal and imposed tolls to subsidize the High Dam project. Shortly thereafter, Britain and France signed a treaty to hand over the Suez Canal to Egypt, but in concert with Israel they attacked Egypt to occupy the Suez Canal. As a result the United States and the USSR forced France and Israel to cease fire and withdraw leaving the canal to the Egyptians. At the end of the hostilities, the Egyptian government decided to accept the USSR's dam construction as a gift from the USSR people. Soviets engineers from the Zuk Hydroproject Institute provided the design, support technicians, and heavy machinery to complete the project.

Construction of the El Saad and Al Ali dams began in 1960 and was completed in 1970. The first stage of the dam was in 1964, and the reservoir began filling through 1976 when it reached to its fill capacity.

The reservoir construction became a significant concern for archaeologists since the dam construction site included 24 ancient Egyptian national treasures such as Abu Simbel and the Debod temple. With the intervention of the United Nations Educational, Scientific, and Cultural Organization (UNESCO) the sites were surveyed and excavated and 24 major monuments were moved to safer locations and sent to countries that helped in and financed the recovery and restoration works. Figure 14.9 is a photograph of displacement of historical monuments by UNESCO.

Figure 14.9 Displacement of historical monuments by UNESCO.

Dam construction and benefits   The Aswan Dam measures 3600 m in length and 980 wide at the base. It is 40 m wide at the crest and 111 m high. The dam reservoir holds 43 million m³ of water and passes 11,000 m³ of water through the dam turbines every second. Emergency spillways also have a discharge capacity of 5000 per second The reservoir is named Lake Nasser and is 480 km wide and has a surface area of 6000 square meters and holds 150,000 to 165,000 m³ of water.

The dam has twelve 175-MW generators and has an aggregate power generation capacity of 2.1 GW. Construction of the recurrence of seasonal floods and the dam has mitigated the reservoir and has created a new fishing industry.

Environmental issues   Construction of the dam has created environmental difficulties and concerns which include flooding of lower grounds in Nubia and the forced displacement of over 90,000 people. Lake Nasser has flooded valuable archeological sites and silt deposits from yearly floods that provided nutrients to the Nile floodplain. The fertile agricultural plains is held behind the dam. Ever since completion of the dam, silt deposited in the reservoir bed has lowered the water storage capacity of Lake Due to depletion of nutrient from the Nile. As a result Mediterranean fishing has been reduced considerably. Construction of the dam has also eroded the top soil in farmlands located down the Nile River.

The Nile Delta inundation has created a situation where the land is now used for rice crop plantation and brick construction industry.

The growth of plants in the reservoir basin has given rise to snails which thrive in the lake.

Aswan Dam timeline

1902. Completion of the first Aswan Dam (6400 ft. long, 176.5 ft high). Despite heightened efforts in 1912, 1929, and 1933, the dam was flooded by the construction of the High Dam.

1953. A group of military officers overthrew the Egyptian monarchy and adopted the idea of a new dam along the Nile river.

1954. Abdel Nasser become president of Egypt and begins conversing in British, German, and French about construction of the dam.

1954. The United States and Great Britain offer loans of $270 million, but back out. Nasser begins favoring the USSR and other communist states.

1958. The Soviets agree to finance construction only if Soviet equipment and engineering methods are used on the project.

1960. Construction began on the High Dam after a power station had been built upstream at the existing dam to provide needed energy for operations at the construction site.

1970. Construction is complete. The second dam is a rock- and earth-filled dam acting as a ridge across the river.

1976. The High Dam Reservoir was first filled.

THREE GORGES DAM

The initial proposal for constructing the dam on the Yangtze River was proposed in 1919; however, because of a lack of resources the idea of dam construction was set aside until 1954 when communist China adopted a major construction plan to construct a giant hydroelectric power system that would also serve flood control purposes, which in turn faced opposition. Thus the execution was postponed until the 1978 economic revolution, when the need for extended electric power became essential for China's growing industrial infrastructure.

The Chinese State in 1989 decided to suspend the plans for 5 years. In 1992 Li Peng with political persuasion convinced the National People's Congress to push the project forward.

Upon resettlement of numerous villages, physical preparations for the project started in 1994. Because of costs, project officials were forced to revise the project operations approved by the National People's Congress. At present in view of petitions by

53 Chinese academicians, President Jiang Zemin agreed to delay the filling of the reservoir and relocation of the local population until such time as scientists could determine the impact of the main reservoir sedimentation problems. To date the construction of the dam has continued without restraint.

Local culture and aesthetic values The 370-mi-long reservoir of the Three Gorges Dam is expected to inundate 1,300 archeological sites, and alter natural setting of the land referred to as The Three Gorges. At present the Chinese government is relocating all historical relics to higher grounds.

Navigation The Yangtze River project will incorporate ship locks which will increase river shipping activity. Upon completion of the project, shipping is expected to become safer, since the gorges in the past have been dangerous to navigate. Figure 14.10 is a photograph of the Three Gorges Dam lock.

Flood control The main dam reservoir which has a capacity to hold 24,452,283 U.S. tons of floodwater will mitigate the frequency of floods from once every 10 years to once every 100 years. On the downside it is estimated that the Yangtze will add 530 metric tons of silt into the reservoir per year and will render the project useless in preventing floods.

Potential construction concerns During construction of the dam in the year 2000 a crack appeared in the dam which has caused international critics to fear a

Figure 14.10 Three Gorges Dam Yangtze Lock in China.

potential catastrophe. In September 2004 the *China Times* reported that heavily armed guards have been deployed to the site to protect it from possible terrorist attack.

Construction timetable

1993–1997. The Yangtze River was diverted.

1998–2003. The first group of generators began to generate power.

2004–2009. The entire project is to be completed by 2009, when all 26 generators will be able to generate power.

15

NUCLEAR POWER

Introduction

In this chapter we will review nuclear fission and nuclear fusion reactor power technologies. Even though power generated from nuclear fission is not a sustainable source of energy, understanding the basic concepts of this technology will enable the reader to appreciate essential differences between it and fusion nuclear reactor systems, which will in the future provide the ultimate source of sustainable energy resources.

Another important consideration given to fission reactor technology is its present-day importance as a significant source of energy, providing about 17 percent of the world's electricity. In some countries, such as France, 75 percent of the electric power is generated by nuclear power. In the United States, 100 nuclear power plants provide about 15 percent of the nation's electricity. According to the International Atomic Energy Agency (IAEA), at present there are more than 400 nuclear power plant installations around the world. To understand nuclear power generation we will first review the physics of atomic fission.

Properties of Uranium

Uranium, created during the Big Bang and the planetary formation of stars, is one of the most abundant elements on Earth. When old stars explode, the scattered dust aggregate and fuse together to form new planets. Uranium-238, designated U-238 (as discussed later in this chapter), has a half-life that is greater that 4.5 billion years, which is the same age as Earth; therefore it is still present in very large quantities. U-238 makes up about 99 percent of the uranium found on our planet. However, U-235 is about 0.7 percent of

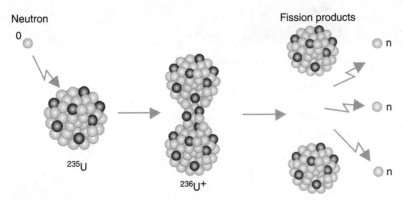

Neutron

Fission products

^{235}U

$^{236}U+$

Figure 15.1 Fission reaction process.

the remaining uranium that is found naturally. Another form of the element uranium, U-234, is even scarcer and is formed by the decay of U-238. When uranium-238 decays, it goes through several stages referred to as alpha and beta decay; it eventually becomes stable when it reaches its last chain of reaction and becomes U-234. Figure 15.1 depicts the fission reaction process.

Uranium-235 has a unique property which makes it the main candidate for bomb and nuclear power production. Similar to U-238, it decays naturally and in the process generates radiation called alpha rays and also undergoes spontaneous fission before stabilization. U-235 also has a unique property in that it can undergo induced fission. This phenomenon occurs when a free neutron is impacted into the nucleus of U-235, which readily becomes unstable and splits into two lighter subcomponent elements. In the process, newly created elements emit gamma radiation as they settle into a new stable state. In general, the ejection of neutrons depends on how the U-235 atoms are split.

In atomic reactors multiple collisions of neutrons with U-235 reach a boundary limit, referred to as the critical state, when each neutron ejected from a collision causes multiple fissions to occur. The splitting process occurs extremely quickly, on the order of picoseconds (1×10^{-12} s). The multiplicity of collisions results in enormous amounts of energy being released in the form of heat and gamma radiation. Upon the splitting of the first atom, the two atoms that result from the fission release beta radiation and gamma radiation.

Heat energy released from the fission process is a result of the fact that resulting subelements plus the neutrons weigh less than the original U-235 atom. The difference in weight is converted directly to energy at a rate governed by Einstein's famous $E = mc^2$ equation.

When U-235 decays, the energy released amounts to 200 mega-electronvolts (MeV). The equivalent energy conversion of 1 eV translates into 1.602×10^{-12} ergs, each 1×10^7 ergs being equal to 1 J or 1 watt-second. Commutation of the small energy released from one atomic collision when translated into 1 lb of uranium is equivalent to the energy generated by burning 1 million gal of gasoline. It should be noted that uranium is one of the heaviest elements and is therefore far denser than many elements; a single pound of it is the size of a small orange, whereas the holding tank for a million gallons of gasoline would be the size of a five-story building with a footing that measures 2500 ft^2.

In order to increase the efficiency of fission reactions for commercial power production, uranium is concentrated in a process referred to as enrichment, where the total fuel mass is purified to elevate the U-235 concentration to 2 to 3 percent or more.

Nuclear Fission Power Plants

Nuclear reactors are constructed from several process chambers which typically consist of a nuclear reaction chamber, a heat transfer or exchanger compartment, and finally an electric power steam turbine section. The heat generation process within the reactor takes place according to the nuclear fission processes described earlier.

The main nuclear chamber is constructed from a honeycomb of metallic tubes which resemble a Gatling gun chamber. A robotic fuel rod insertion mechanism holds numerous tubular fuel rods, each containing enriched uranium pellets that have a diameter approximately the same as that of a dime and a length of about 1 inch. Fuel rods are collected together into bundles and are kept cool by being submerged in a pressurized water vessel. In holding compartments the rods naturally reach a slight supercritical state. This process is critically timed since if uranium is left in a supercritical state, it would eventually overheat and melt. Figure 15.2 is a diagrammatic presentation of the CANDU reactor fission reactor process.

To prevent meltdown, moderator rods are constructed of materials such as carbon or heavy water encapsulated in cylinders that absorbs neutrons is inserted into the bundle using a robotic mechanism raises or lowers them into the reactor chamber. Raising and lowering the moderators allows for control of the rate of the nuclear reaction

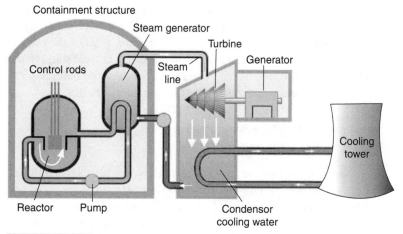

Figure 15.2 **Components of a fission reactor plant.** *Graphic courtesy of AECL.*

chamber. To lower the heat energy generation the rods are pushed into the chamber. In the event of an accident or the need for fuel replacement, moderators are completely immersed into the reaction chamber, thus shutting the reactor down.

CONSTRUCTION OF A NUCLEAR POWER PLANT

From the exterior, a nuclear power plant resembles a conventional fossil fuel fired plant with the exception that the reactor's pressure vessel is typically housed inside a thick liner dome which is intended to shield radiation. That liner in turn is housed within a much larger steel containment vessel which houses the reactor core as well as fuel-handling cranes. The containment vessel is also protected by an outer concrete building or a dome, referred to as a secondary containment, which is designed to withstand severe earthquakes or accidental collision with a large jet airliner. Secondary containment structures are also specially engineered to prevent the escape of radiation in the event of accidents or a meltdown such as the one at Three Mile Island.

Lack of proper secondary structure containment measures in Chernobyl created a major disaster in the Ukraine which caused large amounts of radioactive radiation that affected all of Europe. Plutonium-239, a by-product of the fission process, is an extremely dangerous cancer-causing agent.

The extreme temperatures generated within the reaction chamber heat liquefied carbon dioxide or a molten metal, such as sodium or potassium solution salt, which is directed to a heat exchange chamber where as a result of heat transfer, the water is converted into steam which drives steam turbines which produce electric power. The principal role of the heat exchanger is to isolate the radioactive superheated reactor liquid from passing radioactive isotopes to the turbines. Figure 15.3 is photograph of the San Onofre nuclear power reactor domes.

SUBCRITICALITY, CRITICALITY, AND SUPERCRITICALITY

As described earlier, when a U-235 atom splits, it gives off two or three neutrons. In turn free neutrons from each fission event hit another U-235 nucleus and create a chain reaction up until such time as the entire mass of uranium reaches the critical state. In order to operate, nuclear reactors must be maintained in a critical state. In the event of insufficient free neutron collisions with other U-235 atoms, the fuel mass is said to be in a subcritical state. This state can be ended with induced fission.

ADVANTAGES AND DISADVANTAGES OF FISSION NUCLEAR REACTORS

Unlike fossil fuel power plants, in general, nuclear power plants when designed properly produce electric energy without atmospheric contamination. From an environmental standpoint a coal-fired power plant, in addition to releasing contaminants such as carbon dioxide, sulfur, and countless other contaminants, also releases some radioactivity into the atmosphere far more than any nuclear power plant. However nuclear power generation is also associated with some significant problems which

Figure 15.3 San Onofre nuclear power reactor domes in California.
Photo courtesy of SCE.

include contaminations associated with uranium mining and the purification process, malfunction of nuclear power plants which can be a major source of public hazard such as the Chernobyl and Three Mile Island disasters, and containment of spent fuel which is extremely toxic radioactivity that can last for thousands of years. Another critical issue concerning the spent fuel is the storage and transport of nuclear fuel to and from plants since an accident or terrorist attack would lead to very serious consequences.

EFFECTS OF NUCLEAR RADIATION

In elementary physics we learn that atoms consist of three subatomic particles, namely, protons, neutrons, and electrons. Protons and neutrons bind together to form the nucleus of the atom, while the electrons surround and orbit the nucleus. Protons and electrons have opposite charges—electrons are negative and protons are positive—which attract and react to each other. In most instances the number of electrons and protons is equal in number and charge value. Neutrons are neutral chargeless particles, and their main purpose in the nucleus is to bind the protons together. Since protons all have the same positive charge, they tend to repel each other; however, the neutrons act as a glue that holds the protons tightly together in the nucleus.

The number of protons in the nucleus, or the atomic number, determines the characteristic behavior of each atom. For example, if an atom that consists of 13 protons is combined with an atom with 14 neutrons, the resulting element will have a combined atomic weight of 27, which in the Mendeleev periodic table is called aluminum-27.

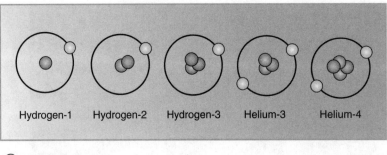

Hydrogen-1   Hydrogen-2   Hydrogen-3   Helium-3   Helium-4

○ Orbital electrons

● Protons

● Neutrons

Figure 15.4 **Isotopes of hydrogen and helium.**

Essentially, a large majority of naturally occurring elements are a combination of different atoms that have been fused together in the process of stellar creation.

In some instances an element in nature appears with different atomic structures where one element form has some extra neutrons. In this case these element forms are referred to as isotopes of each other. As an example, 70 percent of copper has an atomic weight of 63, whereas another form of copper has a weight of 65. In this example the more common copper isotope has 29 protons and 34 neutrons, and the other has 29 protons and 33 neutrons. Both isotopes of the same element have the same physical characteristics and are quite stable.

In some instances various isotopes of some elements are radioactive and others are not. As an example, hydrogen in nature appears in several isotopes, some of which are radioactive. In general, hydrogen has one proton and no neutrons and is designated as H-1 since it has only one proton in the nucleus. A second form of hydrogen isotope, which is found in very small quantities in nature, is referred to as H-2, also called deuterium, and has one proton and one neutron. Both H-1 and H-2 have an identical physical appearance; however, H-2 is toxic in large concentrations. The deuterium isotope, which is 0.15 percent of the total hydrogen found in nature, is also very stable. Another hydrogen isotope, referred to as tritium, or H-3, is composed of one proton and two neutrons; however, unlike the two previous isotopes, this one is unstable such that in time due to radioactive decay it is converted into helium-3 which has two protons and one neutron. In other instances some elements such as uranium isotopes appear in nature only in radioactive forms. Other radioactive elements that appear in nature are polonium, astatine, radon, francium, radium, actinium, thorium, and protactinium.

RADIOACTIVE DECAY

Radioactive decay is a natural process that occurs when an atom of a radioactive isotope spontaneously decays into another element by three types of processes, namely, alpha decay, beta decay, and spontaneous fission while generating different types of radioactive rays, including alpha rays, beta rays, gamma rays, and neutron rays.

Americium, which has an atomic weight of 241, is a radioactive element frequently used in smoke detectors which constantly emanates alpha particles. Alpha particles, which are composed of two protons and two neutrons, have the same equivalent weight as that of a helium-4 nucleus. In the process of emitting the alpha particles, the americium-241 atom loses weight and becomes a neptunium-237 atom. Americium-241 is known to decay at a rate where it loses half of its atoms in 458 years; as such, 458 years is the half-life of americium-241.

As a rule all radioactive elements appearing in nature, depending on the type of isotope, have a different half-life which can range from fractions of a second to millions of years. For example, another isotope of the same element americium-243 has a half-life of 7370 years.

Tritium, or H-3, is an isotope that undergoes beta decay where a neutron in the nucleus spontaneously turns into a proton, an electron, and a third particle called an antineutrino. In beta decay processes the nucleus ejects the electron and antineutrino, while the proton remains in the nucleus. The ejected electron is referred to as a beta particle and becomes a helium-3 atom.

In spontaneous fission, however, an atom instead of throwing off an electron fissions into two subparticles and throws off an alpha or beta particle.

RADIATION DANGER

Radioactive radiation, which results from the natural decay of elements, without exception is dangerous to living organisms. Emission of electrons is the principal cause of cell death and genetic mutations which cause cancer to humans and animals.

Because alpha particles are large and their rays are long in wavelength, alpha particles do not penetrate very far into matter, so they do not cause any harm. On the contrary beta particles being somewhat smaller can penetrate matter, but their harm is limited to ingestion and respiration. Beta particles are not capable of penetrating some materials such as aluminum foil or Plexiglas. Gamma rays, like x rays, can only be stopped by lead. Because of a lack of electric charge, neutrons can readily penetrate through very deep layers of matter and can only be blocked by extremely thick layers of concrete or liquids such as water.

Gamma rays and neutrons have an extreme potential for penetrating all sorts of matter and can destroy human and animal cells. In fact the principle of the neutron bomb is based upon dispersion of gamma rays which can affect living creatures without damaging material goods and buildings.

NUCLEAR RADIATION ACCIDENTS

The International Atomic Energy Agency (IAEA) was set up by the United Nations in 1957. One function was to act as an auditor of world nuclear safety. It prescribes safety procedures and the reporting of even minor incidents. Its role has been strengthened in the last decade. Every country that operates nuclear power plants has a nuclear safety inspectorate who work closely with the IAEA.

Safety is also a prime concern for those working in nuclear plants. Radiation doses are controlled by the use of remote handling equipment for many operations in the core of the reactor. Other controls include physical shielding and limiting the time workers spend in areas with significant radiation levels. These are supported by continuous monitoring of individual doses and of the work environment to ensure very low radiation exposure compared with other industries.

One mandated safety indicator is the calculated frequency of degraded core or core melt accidents. The U.S. Nuclear Regulatory Commission (NRC) specifies that reactor designs must meet a 1 in 10,000 year core damage frequency, but modern designs exceed this. U.S. utility requirements are 1 in 100,000, the best currently operating plants are about 1 in 1 million, and those likely to be built in the next decade are almost 1 in 10 million. The Three Mile Island accident in 1979 was the only one in a reactor conforming to NRC safety criteria, and this was contained as designed, without radiological harm to anyone.

Regulatory requirements today are that any core-melt accident must be confined to the plant itself, without the need to evacuate nearby residents. The main safety concern has always been the possibility of an uncontrolled release of radioactive material, leading to contamination and consequent radiation exposure off-site. At Chernobyl this tragically happened, and the results were severe, once and for all vindicating the extra expense involved in designing to high safety standards.

The use of nuclear energy for electricity generation can be considered extremely safe. Every year over 1000 people die in coal mines to provide this widely used fuel for electricity. There are also significant health and environmental effects arising from fossil fuel use.

CASE STUDY—SAN ONOFRE

The San Onofre power plant is located approximately 70 mi south of Los Angeles, California. It incorporates two nuclear reactors, each powering three generators. Southern California Edison, the plant's majority owner, has recently signed an agreement with Mitsubishi Heavy Industries to manufacture a replacement for existing 40-year-old generators. If approved, the project would be the largest capital project in the plant's history. The current power generation capacity of the power plant is 2200 mW.

San Diego Gas & Electric, which has a 20 percent stake in San Onofre, shown in Figure 15.5 has decided not to help pay for the improvements because the steam generator replacement project is too expensive and would cost its 1.2 million customers $163 million to pay its share.

CANDU REACTORS

The CANDU reactor was designed by Atomic Energy Canada Limited as an alternative to reactors that use enriched uranium at concentrations of 2 to 5% U-235. The reactor fuel is constructed from pellets of uranium dioxide with a 0.7% concentration of U-235 which appears in nature, renders power production considerably cheaper, and can theoretically extend the life of the reactor.

Figure 15.5 San Onofre nuclear generating station. *Photo courtesy of SCE.*

Most underdeveloped countries find unprocessed fuel attractive since it does not require costly enrichment facilities. The Nuclear Treaty (NPT), which safeguards regimes under the auspices of the International Atomic Energy Agency, regulates access to nuclear materials such as enriched uranium.

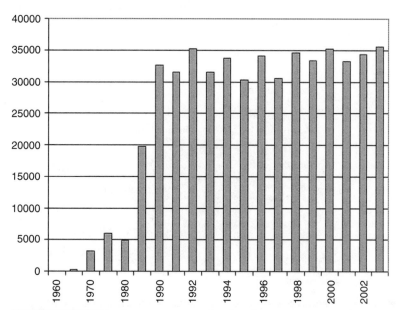

Figure 15.6 Nuclear energy production in megawatts. *Chart courtesy of Ontario Hydro. Canada.*

The moderators in the reactor (used to control the rate of fission reaction) are held in large tanks and are penetrated in the core of the reactor by several horizontal pressure tubes. Channels for the fuel are cooled by a flow of water under high pressure in the primary cooling circuit, reaching 290°C. The high pressure within the tank prevents heavy water from boiling. As in the pressurized water reactor, the primary coolant generates steam in a secondary circuit to drive the turbines. The pressure tube design means that the reactor can be refueled continuously, without down time, as the channels can be accessed individually.

CANDU reactors are designed to be constructed without large pressure vessels. The large pressure vessels commonly used in light-water reactors are expensive, and require heavy industry that is lacking in many countries. Instead, the reactor pressurizes only small tubes that actually contain the fuel. The tubes are constructed of a special alloy (zircaloy) that is relatively transparent to neutrons. Figure 15.7 is a graphic presentation of a fission reactor power plant.

A CANDU fuel assembly consists of a bundle of 37 half-meter-long fuel rods (ceramic fuel pellets in tubes) plus a support structure, with 12 bundles lying end to end in a channel. Control rods penetrate an egg crate shape holding compartment called calandria vertically, and a secondary shutdown involves injecting a gadolinium nitrate solution to the moderator. The heavy-water moderator circulating through the body of the calandria vessel also yields some waste heat.

Since the bulk of the reactor is maintained at a relatively low pressure, the equipment used to monitor and act on the core is quite a bit less complex. It only has to cope with high radiation and high neutron flux. In particular, the control rods and emergency equipment are simpler and more reliable than in other reactor types.

The reactor has the least of any known type. This is partly because so much of the reactor operates at high pressures. It is also caused by the unique fuel-handling

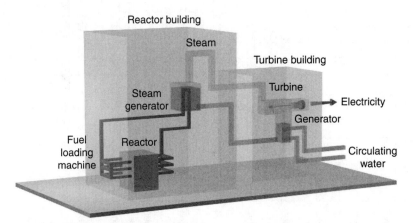

Figure 15.7 **Graphic presentation of a fission reactor power plant.**

Graphic courtesy Canadian Nuclear Association.

Figure 15.8 CANDU reactor CANFLEX fuel rod. *Photo courtesy of AECL.*

system. The pressure tubes containing the fuel rods can be individually opened, and the fuel rods changed without the reactor shut off. Use of heavy-water moderators increases the operation efficiency because re-fuelling the fuel assemblies is considered to be to the most efficient feature of the reactor core replacement while the reactor is in operation. On the other hand most other reactor designs require insertion of degradable poisons in order to lower the high reactivity when refueling. Figure 15.8 is a photograph of the CANDU reactor CANFLEX fuel rod.

Figure 15.9 CANDU reactor Deuterium moderator tower.

Another advantage of the fuel management system is that the reactors can potentially be operated as low-temperature breeder reactors, a type of nuclear reactor that uses a mixture of plutonium-uranium or thorium U-233 as fuel.

The breeder reactors essentially breed isotopes of plutonium in the core with a blanket of depleted or natural uranium. They use liquid sodium as the main heat transfer agent, and because sodium has a very high boiling point, reactor coolant is operated at high temperatures under atmospheric pressure without boiling. The high temperature in turn produces high-pressure steam which contributes toward high-efficiency power plant operation.

Unlike water, sodium in not a corrosive agent, and in its molten form can be circulated in heat transfer loops for a long time without any damage to its holding chamber or piping.

The heavy water, as described previously, slows down the neutrons that create nuclear fission. Heavy water consists of deuterium, which as discussed earlier is a nonradioactive isotope of hydrogen and a single atom. Deuterium atoms, which represent about 1.5 percent of hydrogen found in nature, are much more efficient in retarding or blocking the neutron activity resulting from fission. Deuterium, which appears in very low concentrations in lakes, is separated through a relatively complex process which is added to the initial capital cost.

In CANDU reactors the heat generated by the fission process is absorbed by water that is circulated within an isolated pressurized vessel. The primary superheated heavy water is pumped to a heat exchanger where it transfers its heat energy to light water

Figure 15.10 New Brunswick Power Plant, Canada. *Photo courtesy of AECL.*

which is vaporized to steam which in turn runs the turbines and generates electricity. The heavy water has a separate heat exchanger and circulation system for cooling the moderator.

The CANDU reactor's cooling water tubes are pressurized to 1525 lb/in^2, which is lower than that of pressurized water reactor system designs. The heavy water in the moderator system is not pressurized.

CANDU reactors incorporate considerable control system redundancy which allows equipment to operate at extended cycles. The reactor operation cycle time for CANDU reactor number 7 in 1994 reached 894 days. Fuel consumption or burn-up in a CANDU reactor is 6500 to 7500 megawatt-days (MWd) per metric ton uranium (MTU), which favorably compares to the 33,000 to 50,000 MWd/MTU obtained by many equivalent reactors.

The reactor power stations are designed to house multiple generator turbines which are enclosed in a vacuum building for containment protection. The vacuum building shown in Figure 15.10 in the New Brunswick plant is constructed as a large cylindrical building which houses eight reactors.

In general, CANDU reactor power stations have between one and eight reactors per site. The Pickering facility east of Toronto located on Lake Ontario has eight reactors per site. Nominal electric energy output from the plant is 600 MW per unit. In addition to the 15 installations in the Canadian provinces, CANDU reactors have been providing power in Argentina, India, South Korea, Pakistan, and Romania.

USED-UP FUEL

Used nuclear fuel is both hot and radioactive. It is stored under water in large cooling pools for up to 2 years after use, until it cools. Some of the used fuel will still remain radioactive for up to several thousand years. The storage of used fuel is a major concern for environmentalists. To date there are no ideal solutions for disposing of the wasted fuel. At present, storage of waste in underground salt mines is the only partially

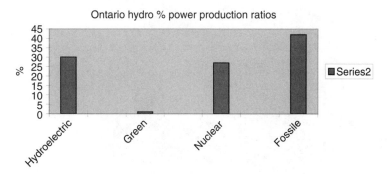

Figure 15.11 Ontario Hydro representation of proportional energy general in MW. *Courtesy of AECL.*

Figure 15.12 Canadian CANDU reactor calandria. *Photo courtesy of AECL.*

Figure 15.13 CANDU reactor spent fuel storage containers. *Photo courtesy of CAEL.*

acceptable solution. Figure 15.13 shows a photograph of specially designed spent fuel holding storage compartment.

Decommissioning of nuclear reactors, once they have completed their useful service, is another issue that raises serious concerns about the use of nuclear energy.

One of the waste materials from a nuclear reactor is plutonium, which is known to cause cancer in extremely small doses of exposure. Plutonium is also the principal fission material used in nuclear weapons. As an example, even though India's purchase of a CANDU reactor was based on the premises of its use for peaceful purposes, the reactor waste was used to develop the first atomic bomb.

Fusion Reactors

The following discussion of physics and research development of fusion reactor technology is based on articles published by the European Fusion Development Agreement FEDA and the Joint European Torus (JET) organization. For more detailed information on fusion power research and development, interested readers can find extensive information on the Web sites of these organizations http://www.jet.efda.org/. and http://www.jet.efda.org/pages/content/fusion5.html

PRINCIPLES OF FUSION REACTORS

The principle of fusion power is based on fusing light nuclei isotopes of hydrogen, which results in the release of significant amounts of energy. The process of energy release by a fusion reaction is identical to that which occurs in the sun and other stars. Essentially energy is released in a fusion reaction process when a combination of the hydrogen isotopes of deuterium and tritium are heated at high temperatures of 100,000,000 kelvin in a magnetically confined environment for only a few seconds. At present the most promising fusion power reactor configuration is based on early Russian experimental work on fusion and is referred to as a Tokomak, which implied a torus-shaped magnetic chamber.

The original large-scale experimental fusion research device, which British scientists experimented with during the 1940s and 1950s, was housed in an aviation hangar located in Harwell. The device, which was called the Zero Energy Toroidal Assembly (ZETA), was at first a highly classified secret project, but it was declassified in the late 1950s when Khrushchev and Bulganin visited England. Accompanying them was the renowned Russian fusion physicist academician Kurchatov who gave a lecture on the topic of "The Possibility of Producing Thermonuclear Reactions in a Gas Discharge" that disclosed the Soviet Union's research activity in the field and eventually lead to a joint international cooperation. When this project was declassified, the United Kingdom undertook the construction of the first fusion reactor research laboratory at Culham. At present, the international nations participating in the research include the European Union, the United States, Russia, Japan, China, Brazil, Canada, and Korea.

As it is said, a Russian scientist expressing an opinion about the development of fusion reactor technology prophetically projected, "We will not harness the potential of fusion until it becomes a necessity."

FUSION REACTION PROCESS

As mentioned before, fusion occurs only at extremely high temperatures of about 100 million kelvin. At these extreme temperatures, the deuterium (D) and tritium (T) gas mixture (covered earlier) becomes plasma, a hot, electrically charged or ionized gas. In plasma, the electrons are stripped from the atomic nuclei, and become ions. In order for the positively charged ions to fuse, their temperature or energy must be adequate to overcome their natural charge repulsion.

In order to harness fusion energy, scientists and engineers are currently performing experiments to control very high temperature plasmas. Much lower temperature plasmas are now widely used in industry, especially for semiconductor manufacture. However, the control of high-temperature fusion plasmas presents several major science and engineering challenges, some of which include heating plasma to in excess of 100 million kelvin and methods to confine plasma, by sustaining it in a magnetically confined environment so that the fusion reaction can become established.

CONDITIONS FOR A FUSION REACTION

In order to achieve sustained fusion, the appropriate parameters, such as plasma temperature, density, and confinement time, must be achieved. The simultaneous occurrence of these three conditions is referred to as fusion or triple product. In order to fuse deuterium and tritium, the product must exceed a certain weight which is called the Lawson criterion, named after British scientist John Lawson who in 1955 formulated the fundamental theory. Attaining the conditions to satisfy the Lawson criterion ensures that the plasma can exceed the breakeven point where the fusion power generated exceeds the energy input power required to heat and sustain the plasma.

Density   The density of fuel ions, which is the quantity per cubic meter, must be sufficiently large for fusion reactions to take place at the required rate. Reduced density or dilution of fuel which usually results from gas impurities, such as an accumulation of helium ions which are produced during the fusion process, can reduce the performance efficiency. In the event of dilution, fuel must be replaced and the helium products referred to as the "ash" must be removed.

Energy confinement   The energy confinement time is a measure of how long the energy in the plasma is retained before being lost. The confinement time is defined as the ratio of the thermal energy contained in the plasma and the power input required to

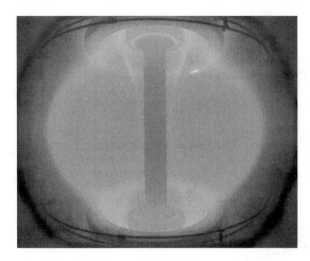

Figure 15.14 Plasma within the Tokamak reactor. *Courtesy of EFDA-JET.*

maintain these conditions. At the Joint European Torus (JET) project, in order to retain the energy for the minimal amount of time, magnetic fields are used to isolate the hot plasma from the cold vessel walls for as long as possible. Losses of magnetically confined plasma are mainly due to radiation. The confinement time increases dramatically with plasma size since large volumes retain heat much better than small volumes.

In order for sustained fusion to occur, the plasma temperature must be maintained at 100 to 200 million kelvin, the energy confinement time must be at least 1 to 2 s, and the core density of plasma must have at least 2 to 3×10^{20} particles per cubic meter, which is approximately 1/1000 g per cubic meter.

Magnetic plasma confinement   Plasma is composed of positively charged ion particles and negative electrons. In fusion reactors extremely powerful magnetic fields are used to isolate the plasma from the walls of the containment vessel which enable the plasma to be heated to temperatures in excess of 100 million kelvin. Isolation of the plasma from the magnetic field containment chamber reduces the conductive heat loss through the vessel wall and minimizes the release of impurities from the vessel walls into the plasma, which cause contamination and cooling of the plasma by radiation.

In a magnetic field the charged plasma particles are forced to spiral along the magnetic field lines. The most promising magnetic confinement system is a toroidal doughnut-shaped ring used in the most advanced experimental fusion reactor called the Tokamak currently used by the Joint European Torus research team. In the future a much larger toroidal fusion reactor sponsored by an international research effort and named the International Thermonuclear Experimental Reactor (ITER) will be constructed in Europe.

Other nonmagnetic plasma confinement such as inertial confinement and cold fusion laser induced systems are also being investigated. Figure 15.15 is a graphic presentation of principle of magnetic plasma confinement.

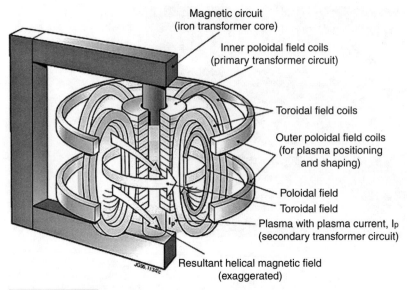

Magnetic circuit
(iron transformer core)

Inner poloidal field coils
(primary transformer circuit)

Toroidal field coils

Outer poloidal field coils
(for plasma positioning
and shaping)

Poloidal field

Toroidal field

Plasma with plasma current, I_p
(secondary transformer circuit)

Resultant helical magnetic field
(exaggerated)

Figure 15.15 **Principle of magnetic plasma confinement.** *Courtesy of*
EFDA-JET.

THE TOKAMAK

In a Tokamak toroidal fusion reactor the magnetic confinement is produced and maintained by a large number of magnetic field coils which surround the vacuum vessel shown
in Figure 15.16. The poloidal field around the plasma cross section pushes the plasma

Figure 15.16 **Graphic**
cross section of Tokamak.
Courtesy of EFDA-JET.

away from the walls of the confinement vessel and maintains the plasma's shape and stability. The poloidal field is induced by an internal current, which is driven in the plasma and is also one of the plasma heating mechanisms, and by external coils that are positioned around the perimeter of the vessel.

The main plasma current is induced in the plasma by a large transformer that changes current in the primary winding of a solenoid constructed of multiturn coil wound onto a large iron core which induces a powerful current of up to 5 million A in the plasma which acts as the transformer secondary circuit.

NUCLEAR FUSION HEATING OF THE PLASMA

As mentioned earlier, one of the main requirements for fusion is to heat the plasma particles to very high temperatures or energies. The heating process deployed in the JET fusion reactor is shown in Figure 15.17.

OHMIC HEATING AND CURRENT DRIVE

The main core transformer discussed earlier in addition to inducing currents of up to 5 million A (5 MA) also isolates the plasma from magnetic containment vessel walls. The current inherently heats the plasma by energizing electrons and ions within the plasma confined in a perpendicular direction to the toroid which results in many megawatts of heating power.

NEUTRAL BEAM HEATING

In addition to ohmic heating, beams of high-energy neutral deuterium or tritium atoms are injected into the plasma, transferring their energy to the plasma via collisions with the plasma ions. The neutral beams are first injected into the plasma by applying an

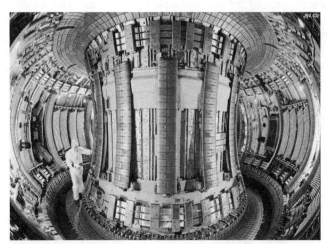

Figure 15.17 Interior of JET's Tokamak. *Courtesy of EFDA-JET.*

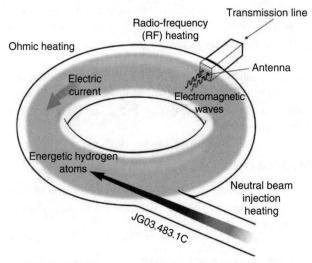

Figure 15.18 Plasma heating diagram of JET's Tokamak. *Courtesy of EFDA-J.*

accelerating voltage of up to 140,000 V; however, since the beam of charged ions could not penetrate the confining magnetic field, the beams are neutralized by turning the ions into neutral atoms before injection into the plasma. The beam neutralization process which requires additional energy is derived from special heating systems.

RADIO-FREQUENCY HEATING

By use of electromagnetic wave frequency resonance plasma confined within the vessel is rotate around the toroid. Electromagnetic wave resonance provides the means to accelerate or slow the rotation of the plasma particles which allows for the transfer of energy to the plasma at the precise location where the radio waves resonate which allows localized heating at a particular location in the plasma.

In the JET Tokamak, eight antennae in the vacuum vessel propagate waves in the frequency range of 25 to 55 MHz into the core of the plasma. These waves are tuned to resonate with particular ions in the plasma, thus heating them up.

Waves are also used to drive current in the plasma which can push electrons traveling in a specific direction. The electron direction control in the JET Tokamak is achieved by a lower hybrid microwave (at 3.7 GHz) accelerator which generates a plasma current of up to 3 MA.

SELF-HEATING OF PLASMA

Helium ions, or so-called alpha particles, are produced when deuterium and tritium fuse and remain trapped within the plasma, which are in turn pumped away through the diverter. The neutrons, which are neutral, escape the magnetic field. Future fusion reactors currently under consideration will capture neutrons and use them as a source of fusion power to produce electricity.

The fusion energy contained within helium ions heats the deuterium and tritium fuel ions by collisions and continues the fusion reaction. When this self-heating mechanism becomes sufficient to maintain the required plasma temperature for fusion, the reaction becomes self-sustaining, which means that there is no more energy requirement to heat the plasma externally. This condition is referred to as ignition.

PLASMA MONITORING AND MEASUREMENT

Measuring the key plasma properties, such as temperature, density, and radiation losses, is extremely important in the understanding of plasma behavior. However, because the plasma is contained in a vacuum vessel and its properties, which are characterized by extremely low density and high plasma temperatures, cannot be measured by conventional methods, plasma parametric diagnostics require an innovative process of physical information detection and analysis.

The measurement techniques used are categorized as active or passive. In active plasma diagnostics, the plasma is probed by use of laser beams, microwaves, or probes to verify how the plasma responds to externally induced control. For instance, by use of inteferometers, the passage of a microwave beam through the plasma slows down in the presence of the plasma compared to when the passage is through a vacuum. This allows measurement of the refractive index of the plasma from which the density of plasma ions and electrons can be interpreted. Use of such real-time diagnostics is undertaken in a manner so as to ensure that the probing mechanism does not significantly affect the behavior of the plasma.

Passive measurement of plasma diagnostics such as radiation and particle movements in the plasma allow the scientist to deduce how the plasma behaves under certain conditions. For instance, during deuterium-to-tritium fusion, neutron detectors measure the flux of neutrons emitted from the plasma. All wavelengths of radiated waves such as visible, ultraviolet and, x-ray waves are also measured, which provides a detailed knowledge of fusion. Figure 15.19 below depicts graphic presentation of measurements to verify properties of plasma within the reactor.

FUSION AS A FUTURE ENERGY SOURCE

As discussed earlier, emission of CO_2 from burning fossil fuels is producing climatic changes. As global demand for energy continues to increase year by year and as the world population grows ever so rapidly, humankind becomes more and more dependent on energy supplies. The need for mitigating environmental pollution and survival of our lifestyle as we know it depends on developing new sources of sustainable and renewable of energy.

In recent years future energy supply was discussed in a European Union (EU) green paper published in 2000 which provides the European strategy for the security of energy supply. The main concern of the EU energy council was the dependency of Europe which imports 50 percent of its energy from outside. If not mitigated, in 2030 the percentage of

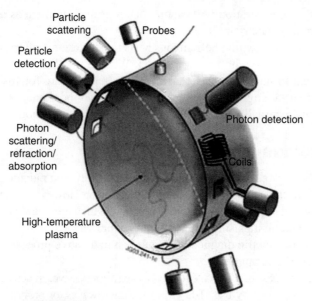

Particle scattering

Probes

Particle detection

Photon scattering/ refraction/ absorption

Photon detection

Coils

High-temperature plasma

Figure 15.19 **Techniques used for measuring the properties of plasmas.** *Courtesy of EFDA-JET.*

imported energy is projected to be 70 percent. In view of a long-term energy policy the report highlighted the importance of fusion reactor research and development. At national, European, and international levels, the future energy supply is becoming one of the key issues; as such fusion offers a valuable alternative to fossil fuel.

In the past, JET has provided valuable research data on optimizing plasma stability and confinement, which has provided the basis for the design of the next generation of fusion reactor devices, referred to as ITER, which is an international collaboration with seven partners including the European Union, Japan, the United States, South Korea, Russia, China, and India. The research project will be a more advanced and larger version of the JET Tokamak. ITER will be capable of producing 500 MW of fusion power, which is 10 times that needed to heat the plasma. In comparison, JET currently produces a fusion power that is 70 percent of the power required to heat the plasma. At present the ITER research center is located at Cadarache in France. The ITER reactor is scheduled to operate by the year 2015. During the operation of ITER, a parallel materials testing program will be undertaken to develop and evaluate materials needed for construction of commercial power plants. Power plants designed and built using the experience from both of these facilities are expected to be operational within 30 years.

Advantages of fusion energy   Fusion offers significant potential advantages as a future source of energy.

Fuels   Deuterium is abundant since it can be extracted from all forms of water. If the world's entire electricity needs were to be provided by fusion power stations, present deuterium supplies from water would last for millions of years.

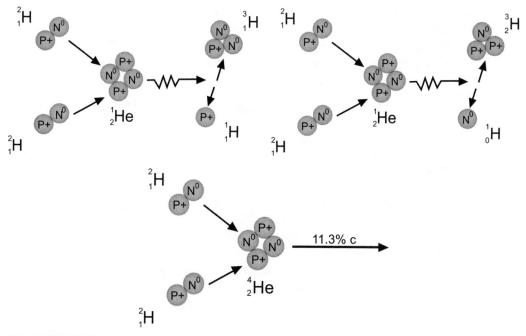

Figure 15.20 Fusion fuel production process. *Courtesy of EFDA-JET.*

Tritium does not occur naturally and will be bred from fusion of two deuterium atoms. Lithium is produced when fusion occurs between deuterium and lithium.

Lithium is the lightest metallic element and is plentiful in the earth's crust. If all the world's electricity needs were to be provided by fusion, known lithium reserves would last for at least 1000 years.

The energy gained from a fusion reaction is enormous. Only 10 g of deuterium, which can be extracted from 500 L of water, and 15 g of tritium, which can be produced from 30 g of lithium, reacting in a fusion power plant would produce enough energy for the lifetime electricity needs of an average person.

Inherent safety The fusion process is inherently safe since the amount of deuterium and tritium in the plasma at any one time is very small, just a few grams, and the conditions required for fusion to occur which requires plasma temperature and confinement are difficult to attain. Any deviation from these conditions will result in a rapid cooling of the plasma and its termination. Therefore there are no circumstances in which the plasma fusion reaction can run out of control and reach a critical condition such as a meltdown in a fission reactor.

Environmental advantages Unlike conventional fission nuclear power plants, fusion power stations will not produce any pollution or greenhouse gases and will not contribute to global warming. Since fusion is a nuclear process, power plant structure due to the energetic action of the neutrons become radioactive; however, the resulting radioactivity will decay within 50 years. In addition, unlike fission reactors, there will be no radioactive waste produced from the fusion reaction since the only by-product produced is an inert helium gas.

INTERNATIONAL RESEARCH AND TECHNOLOGY DEVELOPMENT

At present the United Kingdom has contributed significantly toward research of fusion reactors, and in the past several decades has been the focal research center for the Joint European Torus (JET) project. Key research programs in the United Kingdom are

MAST. The acronym stands for mega amp spherical Tokamak, which has been in operation since 1999. This program is a successor to START, short for small tight-aspect ratio Tokamak, which was put into operation at Culham from 1991 until 1998. This was the first high-temperature spherical Tokamak in the world.

EFDA-JET facilities. The UK Atomic Energy Authority operates the JET facilities on behalf of Europe. The experimental program is coordinated by a European unit at Culham, and involves scientists from all over Europe.

FUTURE FUSION REACTOR RESEARCH AND DEVELOPMENT

The International Thermonuclear Experimental Reactor (ITER) located in France, referenced earlier, will have the same but much larger magnetic geometry as the JET

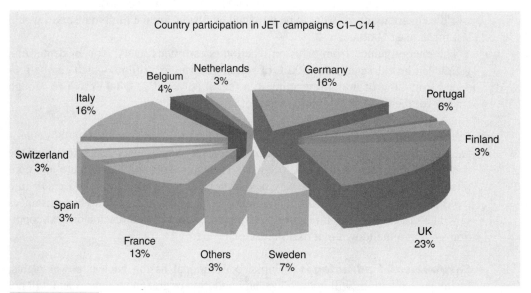

Figure 15.21 JET International research participation diagram. *Courtesy of EFDA-JET.*

Global Primary Energy Use to 2100

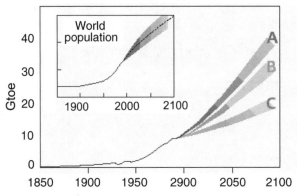

A - High growth presents a future of impressive technological improvements and high economic growth.

B - Middle course describes a future with less ambitious, though perhaps more realistic, technological improvements and more intermediate econimic growth.

C - Ecologically driven presents a "rich and green" future. That includes both substantial technological progress and unprecedented international cooperation centered explicity on environmental production and international equality.

Figure 15.22 **Projected global energy use.** *Courtesy of EFDA-JET.*

Tokamak, but will have several additional key technologies essential for development of future power stations. The ITER design will allow the fusion reactor to be able to operate for very much longer periods and will help to demonstrate the scientific and technological feasibility of fusion power. It is hoped that ITER will be the first fusion device designed to achieve sustained burns at which point the reactor becomes self-sustaining. ITER, which is currently being built at Cadarache, France, is an international project that involves collaboration of the European Union, the United States, Japan, the Russian Federation, China, South Korea, and India.

16

POLLUTION ABATEMENT

Air Pollution Abatement

Air pollution is a major global problem that has created numerous health problems that did not exist in past generations and has resulted in an unprecedented number of illnesses such as asthma and cancer which cause thousands of deaths each year. Air pollution is responsible for a wide variety of environmental problems such as damage to forests, crops, and natural earth and aquatic habitats, all of which have had a significant impact on world economics. As the environmental deterioration becomes more severe, the costs of rehabilitation and pollution abatement become increasingly important. At present, there are numerous techniques available to mitigate environmental pollution.

It should be noted that air pollution is not a recent phenomenon. Ever since the onset of the Industrial Revolution, environmental contamination and atmospheric pollution has been on the rise. The severity of air pollution and the public's awareness, however, is relatively recent. Ever since the end of World War II, the pace of industrialization has significantly increased the so-called anthropogenic, or human-made, air pollution contaminating the atmosphere with numerous toxic gases such as sulfurous oxides (SO_x), nitrous oxides (NO_x), ozone, hydrocarbons, chlorofluorocarbons (CFCs), and heavy metals. Air pollution concentrations and the composition of pollutants are directly attributable to the specific global geographic location of industrialized countries. Naturally occurring pollution, such as from forest fires and volcanic eruptions, can also cause a significant deterioration in air quality but cannot be controlled or mitigated by policy action.

Anthropomorphic atmospheric pollution released from identifiable sources is referred to as primary pollution, while pollutants that occur as a result of a chemical reaction with or in the atmosphere are referred to as secondary pollutants. In turn, categories of primary pollution sources are subdivided into various classes such as mobile or stationary, combustion or noncombustion, area or point sources, and direct or indirect. Mobile sources of pollution include automobiles, trains, and airplanes. Stationary sources include fossil fuel power generation plants whose emissions have a significant impact on air quality.

Another class of so-called nontraditional forms of air pollution includes noise, odor, heat, ionizing radiation, and electromagnetic fields which are associated with the function of various types of equipment such as internal combustion engines, sewage treatment plants, metal smelters, and jet aircraft.

Effects of Pollution on Human and Animal Life

Air pollution has been linked to countless human and environmental health problems. In 1952 an incident known as the London "black fog" pollution killed thousands of people and ever since has been an ongoing general health concern. Pollution has also been identified as the main factor in building deterioration, global warming, and widespread damage to the rain forests, and if unchecked, will have serious economic consequences and alter the human way of life as we know it.

The costs of not solving global pollution regardless of the burden of capital investment required are simply unthinkable, and the resulting consequences difficult to predict or calculate.

The following are some of the effects of air pollution episodes which in recent memory caused a significant loss of life. In 1930 severe air pollution in Belgium's Meuse Valley caused the deaths of 60 citizens. In 1948 as a result of air pollution nearly 7000 individuals lost their lives in Donora, Pennsylvania. In 1952 the infamous London black fog claimed the lives of approximately 4000 people. Similarly air pollution in 1873 referred to as the "fogs" killed thousand of Londoners. At present, high levels of air pollution generated in England is attributed to the residential and industrial use of coal-fired furnaces.

Athens has recently experienced a sixfold increase in the number of deaths resulting from air pollution. Similarly, Hungary and India have recently reported that air pollution in metropolitan areas is equivalent to smoking 10 cigarettes a day. Some researchers are convinced that a combination of sulfur dioxide and water in the air has created a situation where people breathe toxic metals and gases deep into their lungs causing as many as 50,000 deaths in the United States. Some of the diseases directly related to air pollution include:

- Chronic respiratory and cardiovascular diseases
- Alteration of bodily functions such as lung ventilation and oxygen transport
- Reduced work and athletic performance
- Sensory irritation of the eyes, nose, and throat
- Aggravation of existing diseases such as asthma
- Storage of harmful substances in the body

In a recent study in the United States it was determined that every ton of sulfur diox-ide emitted into the atmosphere causes over 3000 health-related illnesses which affect communities. This translates into $25 billion of medical expenses resulting from the emissions from midwestern coal-fired power plants alone. Atmospheric pollutants also produce indirect effects on human health since aside from acid rain, emission of toxic and carcinogenic metals such as aluminum, cadmium, lead, and mercury are soluble in water and the environment when they are leached from soils and lake sediments into aquatic environments where they contaminate water supplies. Acidic water also dissolves toxic metals from municipal and home water systems, thereby poisoning and contaminating drinking water.

Pollution Abatement Equipment

A group of air pollution abatement technologies referred to as selective catalytic reduc-tion are equipment that are designed to reduce emitted gases such as NO_x, N_2O, CO, and VOC through use of the catalysis process. One such technology by Ducon utilizes ceramic honeycomb-shaped or plate-type catalysts constructed from titanium dioxide as a base mate-rial with active coatings of vanadium pentoxide and tungsten trioxide that have been arranged at different levels. The working temperature of the catalyst ranges between 600 and 800°F. For NO_x systems Ducon uses ammonia injection systems kept in storage tanks; the ammonia is vaporized and injected into the ceramic grids. These units are capa-ble of trapping about 90 percent of NO_x, N_2O, CO, and high-VOC gases.

BAGHOUSE FILTER

A filter referred to as a baghouse supplies pulse-air and reverse-air filters for industrial and utility applications. These filters trap industrial and mining residues and particulates such as coal, cement, steel, and mining chemicals. Baghouses are state-of-the-art, high-efficiency, modular filters designed to handle gas volumes from several hundred to sev-eral thousand cubic feet per minute. Baghouse filters are custom designed for efficient high particulate removal. In large applications such as large power plants the baghouse filters can measure up to 35 ft long and are constructed from a wide variety of materi-als, such as polyester, Nomax, acrylic, Teflon, glass, Ryton, and fiberglass, that provide 90 percent particulate removal including mercury which is released from coal-fired boil-er exhaust gases.

DRY FLUE GAS DESULFURIZATION SYSTEMS

Flue gas removal is accomplished by use of several filtration technologies such as semi-dry removal (SDR) process systems, dry flue gas desulfurization (FGD) systems, as well as dry injection–type systems. The SDR filters are custom designed to absorb gaseous pollutants utilizing a spray dryer and a baghouse filter. Specially designed air atomizing nozzles provide the liquid reagent slurry to the spray dryer for high-efficiency removal of

Figure 16.1   Wet flue gas filtration on coal-fired boiler. *Photo courtesy of Ducon Industries.*

Figure 16.1   (*Continued*)

Figure 16.1 (*Continued*)

sulfur dioxide (SO_2), hydrochloric acid (HCl), mercury, and other toxic components from flue gases.

Dry injection systems use lime, trona, activated carbon, and other dry reagents to remove sulfur dioxide, dioxins, mercury, hydrogen chloride, and fumes. For higher absorption the reagent is added in a recirculating-type reactor and is injected into the filter. The particulates are then recovered downstream using a baghouse filter. Dry injection systems remove over 90 percent of SO_2.

Dry-type scrubbers used with this desulfurization process generate a dry waste product that is disposed of by conventional flyash handling equipment.

MULTITUBE CYCLONES

Another product manufactured for trapping pollution particulates is referred to as a multitube cyclone. These filters are constructed from long tubes that trap particulates by creating an air cyclone by an air suction process. This type of equipment have no moving parts, has a sturdy construction, and is designed to trap a wide range of gases at various inlet loadings. This technology has successfully served the industry for over 50 years. Cyclones are capable of removing up to 99.9 percent of particles of 7 microns and larger in size.

Figure 16.2 Dry gas filtration on glass furnace. *Photo courtesy of Ducon Industries.*

Figure 16.2 (*Continued*)

Figure 16.2 (*Continued*)

WET AND DRY PRECIPITATORS

Wet electrostatic precipitators This technology uses a wet electrostatic precipitator process that traps particulates and gases by bombarding the media with electrons. Plates polarized by electronic control achieve over 99.9 percent precipitation and removal of particles measuring 0.01 to 0.05 micron in size. A condensing-type precipitator is used to trap metal oxides.

The notable advantage of these precipitators is that they are nonclogging due to widely spaced electrodes, have a simple construction, are heavy duty and maintenance-free, and are also resistant to corrosion.

Dry electrostatic precipitators This type of precipitator uses dry electrostatic precipitation to trap particulates. The precipitators use high-voltage power for particulate polarization and sometimes use 100-kVA electric energy in a single precipitator. The units, which can be used in a modular fashion, each may measure from 10 to 30 ft in height. Standard modules are installed in parallel or series arrangements to allow for a full range of collection surface requirements.

Figure 16.3   Wet flue gas filtration on limestone-fired utility boiler.

Photo courtesy of Ducon Industries.

ACTIVATED CARBON FILTER

Carbon absorbers are custom filters designed to remove odor-causing components such as hydrogen sulfide, mercaptans, organic acids, aldehydes, and ketones in various industrial and municipal applications. These units are constructed in single-bed forms and are available in capacities of up to 10,000 ft^3/m. Dual-bed units are also utilized for higher capacities.

W-3 DYNAMIC SCRUBBER

These are compact, modular scrubbers that have a self-cleaning wet fan feature that provides high dust collection efficiency in minimum space with low water requirements.

MULTIVANE CENTRIFUGAL SCRUBBER

Centrifugal scrubbers use a multivane centrifugal design and are particularly well suited for applications involving heavy dust loads and large particulates. The operation is based on the principles of centrifugal action which takes place between gas stream and liquid and are designed with either spin vanes or spray manifolds for liquid distribution. These types of scrubbers are capable of removing particles in the 1- to 8-micron range.

WET APPROACH VENTURI SCRUBBER

These classes of scrubbers feature a wetted wall inlet which eliminates wet-dry line buildup and permits direct recycling of high-solids liquid. The scrubbing liquid is distributed by spray nozzles in a Venturi tube which creates a mist that results in efficient scrubbing of particulates. These scrubbers are capable of removing 99.9 percent of particulates in the submicron range. Chemical absorption towers (CATs) are designed to absorb and remove SO_2, HCl, H_2S, TRS, organic material, odors, fumes, and other toxic components. A proprietary technology by Ducon provides spray towers using special spray nozzles that have up to a 99.9 percent particulate entrapment efficiency.

CHEMICAL ABSORPTION TOWERS—FLUE GAS DESULFURIZATION

A class of filtration systems classified as gas desulfurization systems are designed to remove sulfur dioxide from flue gases using a variety of reagents, such as caustic, lime, limestone, ammonia, flyash, magnesium oxide, soda ash, seawater, or double alkali. The efficiency of the removal of these technologies can exceed 99 percent.

CHEMICAL STRIPPERS

Chemical strippers are engineered units that remove benzene, volatile organic compounds, and other inorganic chemicals and gases from groundwater, wastewater streams, and water supplies. Strippers are capable of removing benzene and chlorinated hydrocarbons at flow rates ranging from 1 to 100,000 gal/min.

Figure 16.4 Wet flue gas scrubber on wood-fired boiler. *Photo courtesy of Ducon Industries.*

Sewage Treatment

Sewage treatment is a process that removes the majority of contaminants from waste-water or sewage. The process produces a liquid effluent sludge suitable for disposal to the environment. In general, sewage produced from residential and industrial buildings is channeled to a treatment plant by pipes and underground canals. The most primitive treatment of sewage and most wastewaters is realized by the separation of solids and liquids, by use of a settlement process.

As mentioned earlier, sewage is the liquid waste from toilets, bathtubs, showers, and kitchens, which may also contain some liquid waste from industry and commerce. Sometimes sewage may also include surface and storm water resulting from rain drained from roofs.

In the United States and Canada, sewage has a separate conduit from surface and storm water systems; however, European countries transport liquid waste discharges and storm water together to a common treatment facility, called an integrated sewer system.

The site where the water is processed is called a sewage treatment plant. Sewage treatment plants use mechanical, biological, and chemical processes to purify the water.

- Mechanical treatment processes consist of the removal of large objects, sand, and precipitants from the influx or influent.
- Biological treatment processes consist of oxidation or an oxidizing bed also referred to as an aerated system and post-precipitation which involves the removal of solids by filtration from the sewage.
- Chemical treatment processes are usually combined with settlement processes that remove solids during filtration. The combined process is referred to as a physical-chemical treatment.

PRIMARY TREATMENT PROCESS

The main purpose of the primary treatment is to reduce oils, grease, fats, sand, grit, and coarse solids. This step is accomplished by the use of special machinery; hence, it is a mechanical treatment.

Removal of large objects from the effluent is achieved by straining and removing all large objects such as rags, sticks, condoms, sanitary napkins, tampons, cans, and fruit, that are deposited in the sewer system. This process involves a manual or auto-mated mechanically raked screen. In some instances manual intervention is required to avoid damage to sensitive equipment that can be caused by large objects.

Sand and sedimentation removal   Sand and grit removal is accomplished by use of a grit channel where the velocity of the incoming wastewater is carefully con-trolled to allow grit and stones to settle but still maintain the majority of the organic material within the flow. Such machinery is referred to as a detritor or sand catcher. Sand and stones are removed at the primary stages of the process to avoid damage to the pump and other equipment in subsequent treatment stages. The treatment process includes a sand washer, also called a grit classifier, followed by a conveyor that transports the sand to a container for disposal. The contents from the sand catcher are also fed into an incinerator along with the sludge; however, in some instances the sand and grit are sent to a landfill.

Effluent screening   Screening or maceration is a process that removes floating large objects such as rugs and cardboard from the effluents by means of fixed or rotating screens. Collected materials through this screening process are returned to the sludge or sent to a landfill or for incineration. Solid objects collected by maceration are also shredded into pieces by rotating knifes.

Primary sedimentation   Upon completion of the large object removal, the sewage is passed through large circular or rectangular tanks called sedimentation basins. The tanks are large enough that human waste and solids settle and floating material such as grease and plastics can rise to the surface where they are skimmed off. The main purpose of the primary stage is to produce a homogeneous liquid capable of being treated biologically and a sludge or scum that can be processed or treated separately. Primary settlement tanks are usually equipped with mechanically driven sludge and scum scrapers and pumps that automatically dump the material into hoppers which are transported to sludge treatment equipment.

Figure 16.5   Secondary treatments. *Courtesy of Hyperion Sewage Treatment Plant, Los Angeles, CA.*

SECONDARY TREATMENT

The secondary treatment is a process designed to alter the biological content of the sewage derived from human waste, food waste, soaps, and detergent. The biological breakdown is accomplished by the use of oxygen which destroys bacteria living within the effluent. In general, bacteria and protozoa consume biodegradable soluble organic material such as sugars, fats, and carbon . Secondary treatment systems are classified into two categories, namely, fixed film or suspended. In fixed film systems the biomass grows on media and the sewage passes over its surface. In suspended systems such as in activated sludge the biomass is well mixed with the sewage. Essentially film process systems require smaller space than an equivalent suspended growth system; however, suspended growth systems are more efficient in removing suspended solids.

Roughing filters   Roughing filters are constructed from synthetic media which treat and trap industrial organic waste. These filters have a tall, cylindrical shape and are designed to allow large amounts of hydraulic liquid and the flow of air. In some installations the air is forced through the media by means of blowers.

Activated sludge   Activated sludge plants use a variety of mechanisms and processes that dissolve oxygen within the clarified effluent to generate a biological floc that substantially removes organic material. It also traps particulate material and converts ammonia to nitrite which is eventually transformed into nitrogen gas.

Oxidizing filter beds   In this process stage a trickling filter bed, constructed of plastic media, spreads sewage liquor onto a large deep basin made up of coke or carbonized coal, chips, or specially perforated synthetic or plastic media which have

Figure 16.6 **View of filtration system equipment.** *Courtesy of Hyperion Sewage Treatment Plant, Los Angeles, CA.*

large surface areas that provide support for a thin biofilm. The sewage liquor is distributed by means of large paddle arms that rotate at a central pivot. The drained water collected within the basin is then exposed to the forced source of air which percolates up through the bottom of the bed, keeping it aerated or aerobic. The biological films of bacteria such as protozoa and fungi collected on the surface of the filter media digest and reduce the organic content. In some plants slowly rotating paddles submerged in the liquor are used to create aeration.

Secondary sedimentation   This last stage of the secondary treatment filters low-level suspended biological and organic materials.

TERTIARY TREATMENT

The tertiary treatment is the final stage of the filtration process where the quality of the effluent is raised to the required local standards before it is discharged into the sea, river, lake, ground, and natural environment. Depending upon local sewage treatment regulation, tertiary treatment is also followed by disinfection.

Effluent polishing   Effluent polishing is a process where the processed sewage water is passed through sand filters that remove much of the residual suspended matter. An activated carbon filter is also used to remove residual toxins.

Lagooning   Lagooning is a filtration stage which provides additional settlement and improved biological property of the treated water. The process involves storage of the treated sewage water in large human-made ponds or lagoons. These lagoons have a very large surface area which provides natural aerobic conditions that encourages natural macrophytes reeds and feeding invertebrates such as Daphnia and species of Rotifera to filter and digest fine particulates. In some instances human-made engineered wetlands are designed to provide natural settings provided by the lagoons.

Nutrient removal   In general, residential and industrial sewage contains high levels of nutrients such as nitrogen, ammonia, and phosphorus that in certain forms may be toxic to aquatic fish and invertebrates. Nutrients also encourage the growth of unwanted weeds or algae that produce toxins and promote bacteria growth which eventually depletes oxygen in the water and suffocates desirable fish. The transport of nutrients by streams and rivers to lakes or shallow seas can also cause severe oxygen starvation or eutrophication which kills sweet water fish. The removal of nitrogen and phosphorus from wastewater is done either by a biological process or by chemical precipitation.

Nitrogen removal is accomplished by biological conversion or reduction of nitrogen from the ammonia to nitrate, called nitrification, where nitrate is then converted to nitrogen gas in a process called denitrification, which is released to the atmosphere.

Phosphorus removal is achieved by a biological process in which specific bacteria are introduced into the treated sewage water. Phosphorus removal can also be achieved by chemical precipitation by use of ferric chloride or aluminum salt called alum.

Disinfection   The purpose of disinfection is to destroy and reduce the number of living organisms such as pathogens in the water. The effectiveness of disinfection is subject to properties of the water such as turbidity and pH level, the type of disinfection used, its dosage water temperature, and other environmental variables. Elevated water turbidity diminishes disinfection since solid matter can shield organisms from exposure. Longer periods of exposure to disinfection increases bacterial reduction. Common methods of disinfection include ozone, chlorine, or ultraviolet (UV) light. Chloramines used generally to disinfect potable water are not used in European wastewater treatment; however because of the low cost of the process, it is the most common type of wastewater disinfection in North America.

The major disadvantage of disinfection by the chlorination process is that residual organic material can generate chlorinated-organics referred to as chloramines are considered carcinogenic and harmful to the environment and are toxic to aquatic species.

Ultraviolet light disinfection is considered to be the safest since it destroys without impacting the quality of the wastewater. UV radiation damages the genetic structure of viruses and other pathogens, making them incapable of reproduction. The main disadvantage of UV disinfection is that the process requires frequent lamp replacement and maintenance and its effectiveness diminishes with the presence of floating particulates that block the UV rays.

Ozone O_3 exposure is also another way of destroying bacteria. Ozone is produced by passing regular oxygen O_2 through a voltage arc which creates a third oxygen atom forming O_3. Ozone is an exceedingly powerful oxidizer, which reacts and oxidizes most organic material, thereby destroying any microorganisms that it comes in to contact with. Ozone use in disinfection is much safer than chlorine since it destroys bacteria without introducing additional chemicals into the water. The disadvantage of ozone is its production requires sophisticated process equipment and maintenance personnel.

BATCH REACTORS

Batch reactors allow for compact simplified plants that occupy less space to treat wastewater and meet required environmental standards. Batch reactors essentially combine several processes of treatment into single stages and provide an alternative to constructing large treatment facilities.

An example of a combined process is secondary treatment and settlement where activated sludge is mixed with raw incoming sewage and aerated. The resultant is then allowed to settle which produces a high-quality effluent. This type of treatment technology is deployed in many parts of the world.

The main disadvantage of batch reactors is that the filtration process requires precise control and timing which can only be achieved by means of computerized automation which necessitates a sophisticated level of operation and maintenance.

SLUDGE TREATMENT

The coarse solids and secondary biosolids accumulated in a wastewater treatment process consist of toxic organic and inorganic compounds and metals such as lead and chromium. A process referred to as digestion is implemented to reduce the amount of organic matter and the disease-causing microorganisms present in the solids. The digestion process consists of anaerobic digestion, aerobic digestion, and composting.

Anaerobic digestion   Anaerobic digestion is either a thermophilic (heat loving) process where sludge is fermented in special tanks that are heated to about 38°C or mesophilic digestion where sludge is kept in large tanks for weeks to allow natural mineralization of the sludge. Thermophilic digestion generates biogas which contains a large proportion of methane which is filtered, dehumidified compressed, and used to heat the tanks and run microturbines as described in earlier chapters. In large treatment plants where large quantities of methane are harvested, sufficient energy can be generated which produces sufficient electricity to run the digestion facility machinery. Methane digestion by anaerobic processes requires about 30 days to produce the gas.

Aerobic digestion   Aerobic digestion is a bacterial process where digestion occurs in the presence of oxygen referred to as aerobic condition where special bacteria consume organic matter and convert it into carbon dioxide. Upon depletion of organic matter bacteria die automatically and are in turn consumed by other bacteria. This process is called endogenous respiration where solids in the water are reduced. The aerobic digestion process comparatively is significantly faster than the anaerobic digestion. Moreover, the capital costs required for the process are much lower; however, because of higher energy requirements, the operating costs are greater.

Composting   Composting is essentially an aerobic process which involves mixing the wastewater solids with other sources of hydrocarbon by-products such as sawdust, straw, or wood chips. In the presence of oxygen, bacteria digest the wastewater solids and the mixed carbon source and in the process produce a large amount of heat which destroys bacterial pathogens and microorganisms and produces solid fertilizers that are used as soil builders and fertilizers in agriculture.

SLUDGE DISPOSAL

Remnant sludge left over from treatment is a thickened cake that is produced by dewatering of the settlement tank and lagoon sludge by centrifugal spinning machinery. The thickened cake is disposed of by injection into land, a dangerous practice which can produce leachets that can be mixed in deposited landfills or incinerated. In some instances liquid sludge is disposed of by injection into land, a dangerous practice which can produce leachets that can contaminate ground waters.

Emerging Future Technologies—Bioreactors

A new technology which is expected to emerge within the next few years is based on the use of natural photosynthesis by plants which absorb carbon to grow while producing oxygen. Experiments conducted at laboratories indicate that a special species of algae in addition to absorbing carbon dioxide can also absorb oxide and dioxide, a contributor to acid rain. An experimental vessel called a bioreactor intakes flue gas emissions at 55°C and mixes with water which embeds a recently discovered cyanobacteria algae called *Chroogloeocystis siderophila* found in the hot springs of Yellowstone National Park. By spreading the algae inside the bioreactor by means of parabolic mirror reflectors and plastic fiber-optic cable, it was found that the algae devoured the toxic gases and thrived in the process. At present an issue that concerns scientists is that if these algae are released outside the bioreactor, they might constitute an invasive species with unknown consequences.

UNIT CONVERSION AND DESIGN REFERENCE TABLES

Renewable Energy Tables and Important Solar Power Facts

1 Recent analysis by the Department of Energy (DOE) shows that by year 2025, one-half of the new U.S. electricity generation could come from the sun.

2 The United States generated only 4 GW (1 GW is 1000 MW) of solar power. By the year 2030, it is estimated to be 200 GW.

3 A typical nuclear power plant generates about 1 GW of electric power, which is equal to 5 GW of solar power (daily power generation is limited to an average of 5 to 6 hours per day).

4 Global sales of solar power systems have been growing at a rate of 35 percent in the past few years.

5 It is projected that by the year 2020, the United States will be producing about 7.2 GW of solar power.

6 Shipment of U.S. solar power systems has fallen by 10 percent annually, but has increased by 45 percent throughout Europe.

7 Annual sales growth globally has been 35 percent.

8 Present cost of solar power modules on the average is $2.33/W. By 2030 it should be about $0.38/W.

9 World production of solar power is 1 GW/year.

10 Germany has a $0.50/W grid feed incentive that will be valid for the next 20 years. The incentive is to be decreased by 5 percent per year.

11 In the past few years, Germany installed 130 MW of solar power per year.

12 Japan has a 50 percent subsidy for solar power installations of 3- to 4-kW systems and has about 800 MW of grid-connected solar power systems. Solar power in Japan has been in effect since 1994.

13 California, in 1996, set aside $540 million for renewable energy, which has provided a $4.50/W to $3.00/W buyback as a rebate.

14 In the years 2015 through 2024, it is estimated that California could produce an estimated $40 billion of solar power sales.

15 In the United States, 20 states have a solar rebate program. Nevada and Arizona have set aside a state budget for solar programs.

16 Projected U.S. solar power statistics are shown in the following table:

	2004	2005
Base installed cost per watt	$6.50–$9.00	$1.93
Annual power production (MW)	120	31,000
Employment	20,000	350,000
Cell efficiency (%)	20	22–40
Module performance (%)	8–15	20–30
System performance (%)	6–12	18–25

17 Total U.S. production has been just about 18 percent of global production.

18 For each megawatt of solar power produced, we employ 32 people.

19 A solar power collector, sized 100×100 mi, in the southwest United States could produce sufficient electric power to satisfy the country's yearly energy needs.

20 For every kilowatt of power produced by nuclear or fossil fuel plants, $1/2$ gal of water is used for scrubbing, cleaning, and cooling. Solar power practically does not require any water usage.

21 Significant impact of solar power cogeneration:
- Boosts economic development.
- Lowers cost of peak power.
- Provides greater grid stability.
- Lowers air pollution.
- Lowers greenhouse gas emissions.
- Lowers water consumption and contamination.

22 A mere 6.7 mi/gal efficiency increase in cars driven in the United States could offset our share of imported Saudi oil.

23 State of solar power technology at present:
- Crystalline
- Polycrystalline
- Amorphous
- Thin- and thick-film technologies

24 State of solar power technology in future:
- Plastic solar cells
- Nano-structured materials
- Dye-synthesized cells

Energy Conversion Table

UNITS

Energy units

1 J (joule) = 1 Ws = 4.1868 cal

1 GJ (gigajoule) = 10 E9 J

1 TJ (terajoule) = 10 E12 J

1 PJ (petajoule) = 10 E15 J

1 kWh (kilowatt-hour) = 3,600,000 J

1 toe (tonne oil equivalent) = 7.4 barrels of crude oil in primary energy = 7.8 barrels in total final consumption = 1270 m^3 of natural gas = 2.3 metric tonnes of coal

Mtoe (million tonne oil equivalent) = 41.868 PJ

Power Electric power is usually measured in watts (W), kilowatts (kW), megawatts (MW), and so forth. Power is energy transfer per unit of time.

1 kW = 1000 W

1 MW = 1,000,000

1 GW = 1000 MW

1 TW = 1,000,000 MW

Power (e.g., in W) may be measured at any point in time, whereas energy (e.g., in kWh) has to be measured during a certain period, for example, a second, an hour, or a year.

Unit Abbreviations

m = meter = 3.28 feet (ft)

s = second

h = hour

W = Watt

hp = horsepower

J = joule

cal = calorie

toe = tonnes of oil equivalent

Hz = hertz (cycles per second)

10 E–12 = pico (p) = 1/1000,000,000,000

10 E–9 = nano (n) = 1/1,000,000,000

10 E–6 = micro (µ) = 1/1000,000

10 E–3 = milli (m) = 1/1000

10 E–3 = kilo (k) = 1000 = thousands

10 E–6 = mega (M) = 1,000,000 = millions

10 E–9 = giga (G) = 1,000,000,000

10 E–12 = tera (T) = 1,000,000,000,000

10 E–15 = peta (P) = 1,000,000,000,000,000

Wind Speeds

1 m/s = 3.6 km/h = 2.187 mi/h = 1.944 knots

1 knot = 1 nautical mile per hour = 0.5144 m/s = 1.852 km/h = 1.125 mi/h

Voltage Drop Formulas and dc Cable Charts

VOLTAGE DROP CALCULATION FOR COPPER WIRES

	WIRE	THHN AMPACITY	THWN AMPACITY	MCM	CONDUIT (in)
	2,000			2,016,252	
	1,750			1,738,503	
	1,500			1,490,944	
	1,250			1,245,699	
	1,000	615	545	999,424	
	900	595	520	907,924	
Single phase VD = A × Ft × 2K/C.M	800	565	490	792,756	
	750	535	475	751,581	
	700	520	460	698,389	4
Three phase VD = A × Ft × 2KX.866/C.M.	600	475	420	597,861	4
	500	430	380	497,872	4
	400	380	335	400,192	4
Three phase VD = A × Ft × 2KX.866X1.5/C.M.2 pole	350	350	310	348,133	3
	300	320	285	299,700	3
	250	290	255	248,788	3
	4/0	260	230	211,600	2½
A-Ampers	3/0	225	200	167,000	2
L-Distance from source of supply to load	2/0	195	175	133,100	2
C.M. = Cross-sectional area of conductor in circular mills	1/0	170	150	105,600	2
	1	150	130	83,690	1½
12 for copper more than 50% loading	2	130	115	66,360	1¼
11 for copper less than 50% loading	3	110	100	52,620	1½
18 for aluminum	4	95	85	41,740	1½
	6	75	65	26,240	1
	8	55	50	15,510	¾
	10	30	30	10,380	½
	12	20	20	6,530	½

240 VOLT AC OR DC CABLE CHART = VOLTAGE DROP OF 2% NEC CODE ALLOWED CABLE DISTANCES

AMPS	WATTS	AWG #14	AWG #12	AWG #10	AWG #8	AWG #6	AWG #4	AWG #2	AWG #1/0	AWG #2/0	AWG #3/0
2	480	338	525								
4	960	150	262	413							
6	1,440	113	180	262	450						
8	1,920	82	180	218	338	266					
10	2,400	67	105	173	270	427					
15	3,600	45	67	105	180	285	450				
20	4,800		52	82	144	218	338	540			
25	6,000			67	105	173	270	434			
30	7,200			53	90	142	225	360	578		
40	9,600				67	250	173	270	434	540	
50	12,000				54	82	137	218	345	434	547

120 VOLT AC OR DC CABLE CHART = VOLTAGE DROP OF 2% NEC CODE ALLOWED CABLE DISTANCES

AMPS	WATTS	AWG #14	AWG #12	AWG #10	AWG #8	AWG #6	AWG #4	AWG #2	AWG #1/0	AWG #2/0	AWG #3/0
2	240	169	262								
4	480	75	131	206							
6	720	56	90	131	225						
8	960	41	90	109	169	266					
10	1,200	34	52	86	135	214					
15	1,800	22	34	52	90	142	225				
20	2,400		26	41	72	109	169	270			
25	3,000			34	52	86	135	217			
30	3,600			26	45	71	112	180	289		
40	4,800				34	125	86	135	217	270	
50	6,000				27	41	68	109	172	217	274

48 VOLT DC CABLE CHART = VOLTAGE DROP OF 2% NEC CODE ALLOWED CABLE DISTANCES

AMPS	WATTS	AWG #14	AWG #12	AWG #10	AWG #8	AWG #6	AWG #4	AWG #2	AWG #1/0	AWG #2/0	AWG #3/0
1	48	135	210	330	540						
2	96	67	105	166	270	426					
4	192	30	53	82	135	214					
6	288	22	36	53	90	142	226				
8	384	17	26	43	67	106	173				
10	480	14	21	34	54	86	135	216			
15	720	9	14	21	36	57	90	144	231		
20	960		10	17	30	43	67	108	174	216	274
35	1,200			14	21	34	214	86	138	174	219
30	1,440			10	18	29	45	72	115	138	182
40	1,920				14	21	34	54	86	115	137
50	2,400				9	17	27	43	69	86	138

24 VOLT DC CABLE CHART = VOLTAGE DROP OF 2% NEC CODE ALLOWED CABLE DISTANCES

AMPS	WATTS	AWG # 14	AWG # 12	AWG # 10	AWG # 8	AWG # 6	AWG # 4	AWG # 2	AWG # 1/0	AWG # 2/0	AWG # 3/0
1	24	68	105	165	270						
2	48	34	52	83	135	213					
4	96	15	26	41	68	107					
6	144	11	18	26	45	71	113				
8	192	8	13	22	34	53	86				
10	240	7	10	17	27	43	68	108			
15	360	4	7	10	18	28	45	72	116		
20	480		5	8	15	22	34	54	87	108	137
25	600			7	10	17	27	43	69	87	110
30	720			5	9	14	22	36	58	69	91
40	960				7	10	17	27	43	58	68
50	1,200				4	8	14	22	34	43	89

12 VOLT DC CABLE CHART = VOLTAGE DROP OF 2% NEC CODE ALLOWED CABLE DISTANCES

AMPS	WATTS	AWG # 14	AWG # 12	AWG # 10	AWG # 8	AWG # 6	AWG # 4	AWG # 2	AWG # 1/0	AWG # 2/0	AWG # 3/0
1	12	84	131	206	337	532					
2	24	42	66	103	168	266	432	675			
4	48	18	33	52	84	133	216	337	543	672	
6	72	14	22	33	56	89	141	225	360	450	570
8	96	10	16	27	42	66	108	168	272	338	427
10	120	9	13	22	33	53	84	135	218	270	342
15	180	6	9	13	22	35	56	90	144	180	228
20	240		7	10	16	27	42	67	108	135	171
25	300			8	13	22	33	54	86	108	137
30	360			7	11	18	28	45	72	90	114
50	480				8	13	21	33	54	67	85

CROSS REFERENCE OF AMERICAN WIRE GAUGE (AWG) AND METERIC (mm) SYSTEM			
AMG	mm²	AMG	mm²
30	0.05	6	16
28	0.08	4	25
26	0.14	2	35
24	0.25	1	50
22	0.34	1/0	55
21	0.38	2/0	70
20	0.5	3/0	95
18	0.75	4/0	120
17	1	300 MCM	150
16	1.5	350 MCM	185
14	2.5	500 MCM	240
12	4	600 MCM	300
10	6	750 MCM	400
8	10	1,000 MCM	500

Solar Photovoltaic Module Tilt Angle Correction Table

SOLAR PANEL ORIENTATION TILT CORRECTION FACTOR SOLAR TILT ANGLE FROM HORIZONTAL						
TILT ANGLE/FACING	0	15	30	45	60	90
South	0.89	0.97	1.00	0.97	0.88	0.56
SSE or SSW	0.89	0.97	0.99	0.96	0.87	0.57
SE or SW	0.89	0.95	0.96	0.93	0.85	0.59
ESE or WSW	0.89	0.92	0.91	0.87	0.79	0.57
East or West	0.89	0.88	0.84	0.78	0.7	0.51

TILT ANGLE EFFICIENCY MULTIPLIER TABLE

	COLLECTOR TILT ANGLE FROM HORIZONTAL (DEGREES)					
	0	15	30	45	60	90
FRESNO						
South	0.90	0.98	1.00	0.96	0.87	0.55
SSE, SSW	0.90	0.97	0.99	0.96	0.87	0.56
SE, SW	0.90	0.95	0.96	0.92	0.84	0.68
ESE, WSW	0.90	0.92	0.91	0.87	0.79	0.57
E, W	0.90	0.88	0.86	0.78	0.70	0.51
DAGGETT						
South	0.88	0.97	1.00	0.97	0.88	0.56
SSE, SSW	0.88	0.96	0.99	0.96	0.87	0.58
SE, SW	0.88	0.94	0.96	0.93	0.85	0.59
ESE, WSW	0.88	0.91	0.91	0.86	0.78	0.57
E, W	0.88	0.87	0.83	0.77	0.69	0.51
SANTA MARIA						
South	0.89	0.97	1.00	0.97	0.88	0.57
SSE, SSW	0.89	0.97	0.99	0.96	0.87	0.58
SE, SW	0.89	0.95	0.96	0.93	0.86	0.59
ESE, WSW	0.89	0.92	0.91	0.87	0.79	0.67
E, W	0.89	0.88	0.84	0.78	0.70	0.52
LOS ANGELES						
South	0.89	0.97	1.00	0.97	0.88	0.57
SSE, SSW	0.89	0.97	0.99	0.96	0.87	0.58
SE, SW	0.89	0.95	0.96	0.93	0.85	0.69
ESE, WSW	0.89	0.92	0.91	0.87	0.79	0.57
E, W	0.89	0.88	0.85	0.78	0.70	0.51
SAN DIEGO						
South	0.89	0.98	1.00	0.97	0.88	0.57
SSE, SSW	0.89	0.97	0.99	0.96	0.87	0.58
SE, SW	0.89	0.95	0.96	0.92	0.54	0.59
ESE, WSW	0.89	0.92	0.91	0.87	0.79	0.57
E, W	0.89	0.88	0.85	0.78	0.70	0.51

Solar Insolation Table for Major Cities in the United States*

STATE	CITY	HIGH	LOW	AVG.	STATE	CITY	HIGH	LOW	AVG.
AK	Fairbanks	5.87	2.12	3.99	GA	Griffin	5.41	4.26	4.99
AK	Matanuska	5.24	1.74	3.55	III	Honolulu	6.71	5.59	6.02
AL	Montgomery	4.69	3.37	4.23	IA	Ames	4.80	3.73	4.40
AR	Bethel	6.29	2.37	3.81	ill	Boise	5.83	3.33	4.92
AR	Little Rock	5.29	3.88	4.69	ill	Twin Falls	5.42	3.42	4.70
AZ	Tucson	7.42	6.01	6.57	IL	Chicago	4.08	1.47	3.14
AZ	Page	7.30	5.65	6.36	IN	Indianapolis	5.02	2.55	4.21
AZ	Phoenix	7.13	5.78	6.58	KS	Manhattan	5.08	3.62	4.57
CA	Santa Maria	6.52	5.42	5.94	KS	Dodge City	4.14	5.28	5.79
CA	Riverside	6.35	5.35	5.87	KY	Lexington	5.97	3.60	4.94
CA	Davis	6.09	3.31	5.10	LA	Lake Charles	5.73	4.29	4.93
CA	Fresno	6.19	3.42	5.38	LA	New Orleans	5.71	3.63	4.92
CA	Los Angeles	6.14	5.03	5.62	LA	Shreveport	4.99	3.87	4.63
CA	Soda Springs	6.47	4.40	5.60	MA	E. Wareham	4.48	3.06	3.99
CA	La Jolla	5.24	4.29	4.77	MA	Boston	4.27	2.99	3.84
CA	Inyokern	8.70	6.87	7.66	MA	Blue Hill	4.38	3.33	4.05
CO	Grandbaby	7.47	5.15	5.69	MA	Natick	4.62	3.09	4.10
CO	Grand Lake	5.86	3.56	5.08	MA	Lynn	4.60	2.33	3.79
CO	Grand Junction	6.34	5.23	5.85	MD	Silver Hill	4.71	3.84	4.47
CO	Boulder	5.72	4.44	4.87	ME	Caribou	5.62	2.57	4.19
DC	Washington	4.69	3.37	4.23	ME	Portland	5.23	3.56	4.51
FL	Apalachicola	5.98	4.92	5.49	MI	Sault Ste. Marie	4.83	2.33	4.20
FL	Belie Is.	5.31	4.58	4.99	MI	E. Lansing	4.71	2.70	4.00
FL	Miami	6.26	5.05	5.62	MN	St. Cloud	5.43	3.53	4.53
FL	Gainesville	5.81	4.71	5.27	MO	Columbia	5.50	3.97	4.73
FL	Tampa	6.16	5.26	5.67	MO	St. Louis	4.87	3.24	4.38
GA	Atlanta	5.16	4.09	4.74	MS	Meridian	4.86	3.64	4.43

(Continued)

(*Continued*)

STATE	CITY	HIGH	LOW	AVG.	STATE	CITY	HIGH	LOW	AVG.
MT	Glasgow	5.97	4.09	5.15	PA	Pittsburg	4.19	1.45	3.28
MT	Great Falls	5.70	3.66	4.93	PA	State College	4.44	2.79	3.91
MT	Summit	5.17	2.36	3.99	RI	Newport	4.69	3.58	4.23
NM	Albuquerque	7.16	6.21	6.77	SC	Charleston	5.72	4.23	5.06
NB	Lincoln	5.40	4.38	4.79	SD	Rapid City	5.91	4.56	5.23
NB	N. Omaha	5.28	4.26	4.90	1N	Nashville	5.2	3.14	4.45
NC	Cape Hatteras	5.81	4.69	5.31	1N	Oak Ridge	5.06	3.22	4.37
NC	Greensboro	5.05	4.00	4.71	TX	San Antonio	5.88	4.65	5.3
ND	Bismarck	5.48	3.97	5.01	TX	Brownsville	5.49	4.42	4.92
NJ	Sea Brook	4.76	3.20	4.21	TX	El Paso	7.42	5.87	6.72
NV	Las Vegas	7.13	5.84	6.41	TX	Midland	6.33	5.23	5.83
NV	Ely	6.48	5.49	5.98	TX	Fort Worth	6.00	4.80	5.43
NY	Binghamton	3.93	1.62	3.16	UT	Salt Lake City	6.09	3.78	5.26
NY	Ithaca	4.57	2.29	3.79	UT	Flaming Gorge	6.63	5.48	5.83
NY	Schenectady	3.92	2.53	3.55	VA	Richmond	4.50	3.37	4.13
NY	Rochester	4.22	1.58	3.31	WA	Seattle	4.83	1.60	3.57
NY	New York City	4.97	3.03	4.08	WA	Richland	6.13	2.01	4.44
OH	Columbus	5.26	2.66	4.15	WA	Pullman	6.07	2.90	4.73
OH	Cleveland	4.79	2.69	3.94	WA	Spokane	5.53	1.16	4.48
OK	Stillwater	5.52	4.22	4.99	WA	Prosser	6.21	3.06	5.03
OK	Oklahoma City	6.26	4.98	5.59	WI	Madison	4.85	3.28	4.29
OR	Astoria	4.76	1.99	3.72	WV	Charleston	4.12	2.47	3.65
OR	Corvallis	5.71	1.90	4.03	WY	Lander	6.81	5.50	6.06
OR	Medford	5.84	2.02	4.51					

*Values are given in kilowatt-hours per square meter per day.

Longitude and Latitude Tables

	LONGITUDE	LATITUDE		LONGITUDE	LATITUDE
ALABAMA			Barstow AP	34° 51~ N	116° 47~ W
Alexander City	32° 57~ N	85° 57~ W	Blythe AP	33° 37~ N	114° 43~ W
Anniston AP	33° 35~ N	85° 51~ W	Burbank AP	34° 12~ N	118° 21~ W
Auburn	32° 36~ N	85° 30~ W	Chico	39° 48~ N	121° 51~ W
Birmingham AP	33° 34~ N	86° 45~ W	Concord	37° 58~ N	121° 59~ W
Decatur	34° 37~ N	86° 59~ W	Covina	34° 5~ N	117° 52~ W
Dothan AP	31° 19~ N	85° 27~ W	Crescent City AP	41° 46~ N	124° 12~ W
Florence AP	34° 48~ N	87° 40~ W	Downey	33° 56~ N	118° 8~ W
Gadsden	34° 1~ N	86° 0~ W	El Cajon	32° 49~ N	116° 58~ W
Huntsville AP	34° 42~ N	86° 35~ W	El Cerrito AP (S)	32° 49~ N	115° 40~ W
Mobile AP	30° 41~ N	88° 15~ W	Escondido	33° 7~ N	117° 5~ W
Mobile Co	30° 40~ N	88° 15~ W	Eureka/Arcata AP	40° 59~ N	124° 6~ W
Montgomery AP	32° 23~ N	86° 22~ W	Fairfield-Trafis AFB	38° 16~ N	121° 56~ W
Selma-Craig AFB	32° 20~ N	87° 59~ W	Fresno AP (S)	36° 46~ N	119° 43~ W
Talladega	33° 27~ N	86° 6~ W	Hamilton AFB	38° 4~ N	122° 30~ W
Tuscaloosa AP	33° 13~ N	87° 37~ W	Laguna Beach	33° 33~ N	117° 47~ W
ALASKA			Livermore	37° 42~ N	121° 57~ W
Anchorage AP	61° 10~ N	150° 1~ W	Lompoc, Vandenberg AFB	34° 43~ N	120° 34~ W
Barrow (S)	71° 18~ N	156° 47~ W	Long Beach AP	33° 49~ N	118° 9~ W
Fairbanks AP (S)	64° 49~ N	147° 52~ W	Los Angeles AP (S)	33° 56~ N	118° 24~ W
Juneau AP	58° 22~ N	134° 35~ W	Los Angeles CO (S)	34° 3~ N	118° 14~ W
Kodiak	57° 45~ N	152° 29~ W	Merced-Castle AFB	37° 23~ N	120° 34~ W
Nome AP	64° 30~ N	165° 26~ W	Modesto	37° 39~ N	121° 0~ W

	Lat.	Long.		Lat.	Long.
ARIZONA			Monterey	36° 36~ N	121° 54~ W
Douglas AP	31° 27~ N	109° 36~ W	Napa	38° 13~ N	122° 17~ W
Flagstaff AP	35° 8~ N	111° 40~ W	Needles AP	34° 36~ N	114° 37~ W
Fort Huachuca AP (S)	31° 35~ N	110° 20~ W	Oakland AP	37° 49~ N	122° 19~ W
Kingman AP	35° 12~ N	114° 1~ W	Oceanside	33° 14~ N	117° 25~ W
Nogales	31° 21~ N	110° 55~ W	Ontario	34° 3~ N	117° 36~ W
Phoenix AP (S)	33° 26~ N	112° 1~ W	Oxnard	34° 12~ N	119° 11~ W
Prescott AP	34° 39~ N	112° 26~ W	Palmdale AP	34° 38~ N	118° 6~ W
Tucson AP (S)	32° 7~ N	110° 56~ W	Palm Springs	33° 49~ N	116° 32~ W
Winslow AP	35° 1~ N	110° 44~ W	Pasadena	34° 9~ N	118° 9~ W
Yuma AP	32° 39~ N	114° 37~ W	Petaluma	38° 14~ N	122° 38~ W
ARKANSAS			Pomona Co	34° 3~ N	117° 45~ W
Blytheville AFB	35° 57~ N	89° 57~ W	Redding AP	40° 31~ N	122° 18~ W
Camden	33° 36~ N	92° 49~ W	Redlands	34° 3~ N	117° 11~ W
El Dorado AP	33° 13~ N	92° 49~ W	Richmond	37° 56~ N	122° 21~ W
Fayetteville AP	36° 0~ N	94° 10~ W	Riverside-March AFB (S)	33° 54~ N	117° 15~ W
Fort Smith AP	35° 20~ N	94° 22~ W	Sacramento AP	38° 31~ N	121° 30~ W
Hot Springs	34° 29~ N	93° 6~ W	Salinas AP	36° 40~ N	121° 36~ W
Jonesboro	35° 50~ N	90° 42~ W	San Bernadino, Norton AFB	34°8~ N	117° 16~ W
Little Rock AP (S)	34° 44~ N	92° 14~ W	San Diego AP	32° 44~ N	117° 10~ W
Pine Bluff AP	34° 18~ N	92° 5~ W	San Fernando	34° 17~ N	118° 28~ W
Texarkana AP	33° 27~ N	93° 59~ W	San Francisco AP	37° 37~ N	122° 23~ W
CALIFORNIA			San Francisco Co	37° 46~ N	122° 26~ W
Bakersfield AP	35° 25~ N	119° 3~ W	San Jose AP	37° 22~ N	121° 56~ W

(Continued)

	LONGITUDE	LATITUDE		LONGITUDE	LATITUDE
CALIFORNIA (Continued)			Fort Myers AP	26° 35~ N	81° 52~ W
San Louis Obispo	35° 20~ N	120° 43~ W	Fort Pierce	27° 28~ N	80° 21~ W
Santa Ana AP	33° 45~ N	117° 52~ W	Gainsville AP (S)	29° 41~ N	82° 16~ W
Santa Barbara MAP	34° 26~ N	119° 50~ W	Jacksonville AP	30° 30~ N	81° 42~ W
Santa Cruz	36° 59~ N	122° 1~ W	Key West AP	24° 33~ N	81° 45~ W
Santa Maria AP (S)	34° 54~ N	120° 27~ W	Lakeland Co (S)	28° 2~ N	81° 57~ W
Santa Monica CIC	34° 1~ N	118° 29~ W	Miami AP (S)	25° 48~ N	80° 16~ W
Santa Paula	34° 21~ N	119° 5~ W	Miami Beach Co	25° 47~ N	80° 17~ W
Santa Rosa	38° 31~ N	122° 49~ W	Ocala	29° 11~ N	82° 8~ W
Stockton AP	37° 54~ N	121° 15~ W	Orlando AP	28° 33~ N	81° 23~ W
Ukiah	39° 9~ N	123° 12~ W	Panama City, Tyndall AFB	30° 4~ N	85° 35~ W
Visalia	36° 20~ N	119° 18~ W	Pensacola Co	30° 25~ N	87° 13~ W
Yreka	41° 43~ N	122° 38~ W	St. Augustine	29° 58~ N	81° 20~ W
Yuba City	39° 8~ N	121° 36~ W	St. Petersburg	27° 46~ N	82° 80~ W
COLORADO			Stanford	28° 46~ N	81° 17~ W
Alamosa AP	37° 27~ N	105° 52~ W	Sarasota	27° 23~ N	82° 33~ W
Boulder	40° 0~ N	105° 16~ W	Tallahassee AP (S)	30° 23~ N	84° 22~ W
Colorado Springs AP	38° 49~ N	104° 43~ W	Tampa AP (S)	27° 58~ N	82° 32~ W
Denver AP	39° 45~ N	104° 52~ W	West Palm Beach AP	26° 41~ N	80° 6~ W
Durango	37° 17~ N	107° 53~ W	**GEORGIA**		
Fort Collins	40° 45~ N	105° 5~ W	Albany, Turner AFB	31° 36~ N	84° 5~ W
Grand Junction AP (S)	39° 7~ N	108° 32~ W	Americus	32° 3~ N	84° 14~ W
Greeley	40° 26~ N	104° 38~ W	Athens	33° 57~ N	83° 19~ W
Lajunta AP	38° 3~ N	103° 30~ W	Atlanta AP (S)	33° 39~ N	84° 26~ W

Location	Longitude	Latitude	Location	Longitude	Latitude
Leadville	106° 18~ W	39° 15~ N	Augusta AP	81° 58~ W	33° 22~ N
Pueblo AP	104° 29~ W	38° 18~ N	Brunswick	81° 29~ W	31° 15~ N
Sterling	103° 12~ W	40° 37~ N	Columbus, Lawson AFB	84° 56~ W	32° 31~ N
Trinidad	104° 20~ W	37° 15~ N	Dalton	84° 57~ W	34° 34~ N
CONNECTICUT			Dublin	82° 54~ W	32° 20~ N
Bridgeport AP	73° 11~ W	41° 11~ N	Gainsville	83° 41~ W	34° 11~ N
Hartford, Brainard Field	72° 39~ W	41° 44~ N	Griffin	84° 16~ W	33° 13~ N
New Haven AP	73° 55~ W	41° 19~ N	LaGrange	85° 4~ W	33° 1~ N
New London	72° 6~ W	41° 21~ N	Macon AP	83° 39~ W	32° 42~ N
Norwalk	73° 25~ W	41° 7~ N	Marietta, Dobbins AFB	84° 31~ W	33° 55~ N
Norwick	72° 4~ W	41° 32~ N	Savannah	81° 12~ W	32° 8~ N
Waterbury	73° 4~ W	41° 35~ N	Valdosta-Moody AFB	83° 12~ W	30° 58~ N
Widsor Locks, Bradley Fld	72° 41~ W	41° 56~ N	Waycross	82° 24~ W	31° 15~ N
DELAWARE			**HAWAII**		
Dover AFB	75° 28~ W	39° 8~ N	Hilo AP (S)	155° 5~ W	19° 43~ N
Wilmington AP	75° 36~ W	39° 40~ N	Honolulu AP	157° 55~ W	21° 20~ N
DISTRICT OF COLUMBIA			Kaneohe Bay MCAS	157° 46~ W	21° 27~ N
Andrews AFB	76° 5~ W	38° 5~ N	Wahiawa	158° 2~ W	21° 3~ N
Washington, National AP	77° 2~ W	38° 51~ N	**IDAHO**		
FLORIDA			Boise AP (S)	116° 13~ W	43° 34~ N
Belle Glade	80° 39~ W	26° 39~ N	Burley	113° 46~ W	42° 32~ N
Cape Kennedy AP	80° 34~ W	28° 29~ N	Coeur D'Alene AP	116° 49~ W	47° 46~ N
Daytona Beach AP	81° 3~ W	29° 11~ N	Idaho Falls AP	112° 4~ W	43° 31~ N
E Fort Lauderdale	80° 9~ W	26° 4~ N	Lewiston AP	117° 1~ W	46° 23~ N

(Continued)

IDAHO (Continued)

Location	LONGITUDE	LATITUDE
Moscow	46° 44~ N	116° 58~ W
Mountain Home AFB	43° 2~ N	115° 54~ W
Pocatello AP	42° 55~ N	112° 36~ W
Twin Falls AP (S)	42° 29~ N	114° 29~ W

ILLINOIS

Location	LONGITUDE	LATITUDE
Aurora	41° 45~ N	88° 20~ W
Belleville, Scott AFB	38° 33~ N	89° 51~ W
Bloomington	40° 29~ N	88° 57~ W
Carbondale	37° 47~ N	89° 15~ W
Champaign/Urbana	40° 2~ N	88° 17~ W
Chicago, Midway AP	41° 47~ N	87° 45~ W
Chicago, O'Hare AP	41° 59~ N	87° 54~ W
Chicago Co	41° 53~ N	87° 38~ W
Danville	40° 12~ N	87° 36~ W
Decatur	39° 50~ N	88° 52~ W
Dixon	41° 50~ N	89° 29~ W
Elgin	42° 2~ N	88° 16~ W
Freeport	42° 18~ N	89° 37~ W
Galesburg	40° 56~ N	90° 26~ W
Greenville	38° 53~ N	89° 24~ W
Joliet	41° 31~ N	88° 10~ W
Kankakee	41° 5~ N	87° 55~ W
La Salle/Peru	41° 19~ N	89° 6~ W

Location	LONGITUDE	LATITUDE
La Porte	41° 36~ N	86° 43~ W
Marion	40° 29~ N	85° 41~ W
Muncie	40° 11~ N	85° 21~ W
Peru, Grissom AFB	40° 39~ N	86° 9~ W
Richmond AP	39° 46~ N	84° 50~ W
Shelbyville	39° 31~ N	85° 47~ W
South Bend AP	41° 42~ N	86° 19~ W
Terre Haute AP	39° 27~ N	87° 18~ W
Valparaiso	41° 31~ N	87° 2~ W
Vincennes	38° 41~ N	87° 32~ W

IOWA

Location	LONGITUDE	LATITUDE
Ames (S)	42° 2~ N	93° 48~ W
Burlington AP	40° 47~ N	91° 7~ W
Cedar Rapids AP	41° 53~ N	91° 42~ W
Clinton	41°50~ N	90° 13~ W
Council Bluffs	41° 20~ N	95° 49~ W
Des Moines AP	41° 32~ N	93° 39~ W
Dubuque	42° 24~ N	90° 42~ W
Fort Dodge	42° 33~ N	94° 11~ W
Iowa City	41° 38~ N	91° 33~ W
Keokuk	40° 24~ N	91° 24~ W
Marshalltown	42° 4~ N	92° 56~ W
Mason City AP	43° 9~ N	93° 20~ W
Newton	41° 41~ N	93° 2~ W

City	Latitude	Longitude		City	Latitude	Longitude
Macomb	40° 28~ N	90° 40~ W		Ottumwa AP	41° 6~ N	92° 27~ W
Moline AP	41° 27~ N	90° 31~ W		Sioux City AP	42° 24~ N	96° 23~ W
Mt Vernon	38° 19~ N	88° 52~ W		Waterloo	42° 33~ N	92° 24~ W
Peoria AP	40° 40~ N	89° 41~ W		**KANSAS**		
Quincy AP	39° 57~ N	91° 12~ W		Atchison	39° 34~ N	95° 7~ W
Rantoul, Chanute AFB	40° 18~ N	88° 8~ W		Chanute AP	37° 40~ N	95° 29~ W
Rockford	42° 21~ N	89° 3~ W		Dodge City AP (S)	37° 46~ N	99° 58~ W
Springfield AP	39° 50~ N	89° 40~ W		El Dorado	37° 49~ N	96° 50~ W
Waukegan	42° 21~ N	87° 53~ W		Emporia	38° 20~ N	96° 12~ W
INDIANA				Garden City AP	37° 56~ N	100° 44~ W
Anderson	40° 6~ N	85° 37~ W		Goodland AP	39° 22~ N	101° 42~ W
Bedford	38° 51~ N	86° 30~ W		Great Bend	38° 21~ N	98° 52~ W
Bloomington	39° 8~ N	86° 37~ W		Hutchinson AP	38° 4~ N	97° 52~ W
Columbus, Bakalar AFB	39° 16~ N	85° 54~ W		Liberal	37° 3~ N	100° 58~ W
Crawfordsville	40° 3~ N	86° 54~ W		Manhattan, Ft Riley (S)	39° 3~ N	96° 46~ W
Evansville AP	38° 3~ N	87° 32~ W		Parsons	37° 20~ N	95° 31~ W
Fort Wayne AP	41° 0~ N	85° 12~ W		Russell AP	38° 52~ N	98° 49~ W
Goshen AP	41° 32~ N	85° 48~ W		Salina	38° 48~ N	97° 39~ W
Hobart	41° 32~ N	87° 15~ W		Topeka AP	39° 4~ N	95° 38~ W
Huntington	40° 53~ N	85° 30~ W		Wichita AP	37° 39~ N	97° 25~ W
Indianapolis AP	39° 44~ N	86° 17~ W		**KENTUCKY**		
Jeffersonville	38° 17~ N	85° 45~ W		Ashland	38° 33~ N	82° 44~ W
Kokomo	40° 25~ N	86° 3~ W		Bowling Green AP	35° 58~ N	86° 28~ W
Lafayette	40° 2~ N	86° 5~ W		Corbin AP	36° 57~ N	84° 6~ W

(Continued)

399

KENTUCKY (Continued)	LONGITUDE	LATITUDE
Covington AP	39° 3~ N	84° 40~ W
Hopkinsville, Ft Campbell	36° 40~ N	87° 29~ W
Lexington AP (S)	38° 2~ N	84° 36~ W
Louisville AP	38° 11~ N	85° 44~ W
Madisonville	37° 19~ N	87° 29~ W
Owensboro	37° 45~ N	87° 10~ W
Paducah AP	37° 4~ N	88° 46~ W
LOUISIANA		
Alexandria AP	31° 24~ N	92° 18~ W
Baton Rouge AP	30° 32~ N	91° 9~ W
Bogalusa	30° 47~ N	89° 52~ W
Houma	29° 31~ N	90° 40~ W
Lafayette AP	30° 12~ N	92° 0~ W
Lake Charles AP (S)	30° 7~ N	93° 13~ W
Minden	32° 36~ N	93° 18~ W
Monroe AP	32° 31~ N	92° 2~ W
Natchitoches	31° 46~ N	93° 5~ W
New Orleans AP	29° 59~ N	90° 15~ W
Shreveport AP (S)	32° 28~ N	93° 49~ W
MAINE		
Augusta AP	44° 19~ N	69° 48~ W
Bangor, Dow AFB	44° 48~ N	68° 50~ W
Caribou AP (S)	46° 52~ N	68° 1~ W
Lewiston	44° 2~ N	70° 15~ W

MICHIGAN	LONGITUDE	LATITUDE
Adrian	41° 55~ N	84° 1~ W
Alpena AP	45° 4~ N	83° 26~ W
Battle Creek AP	42° 19~ N	85° 15~ W
Benton Harbor AP	42° 8~ N	86° 26~ W
Detroit	42° 25~ N	83° 1~ W
Escanaba	45° 44~ N	87° 5~ W
Flint AP	42° 58~ N	83° 44~ W
Grand Rapids AP	42° 53~ N	85° 31~ W
Holland	42° 42~ N	86° 6~ W
Jackson AP	42° 16~ N	84° 28~ W
Kalamazoo	42° 17~ N	85° 36~ W
Lansing AP	42° 47~ N	84° 36~ W
Marquette Co	46° 34~ N	87° 24~ W
Mt Pleasant	43° 35~ N	84° 46~ W
Muskegon AP	43° 10~ N	86° 14~ W
Pontiac	42° 40~ N	83° 25~ W
Port Huron	42° 59~ N	82° 25~ W
Saginaw AP	43° 32~ N	84° 5~ W
Sault Ste. Marie AP (S)	46° 28~ N	84° 22~ W
Traverse City AP	44° 45~ N	85° 35~ W
Ypsilanti	42° 14~ N	83° 32~ W
MINNESOTA		
Albert Lea	43° 39~ N	93° 21~ W
Alexandria AP	45° 52~ N	95° 23~ W

Location	Latitude	Longitude
Millinocket AP	45° 39~ N	68° 42~ W
Portland (S)	43° 39~ N	70° 19~ W
Waterville	44° 32~ N	69° 40~ W
MARYLAND		
Baltimore AP	39° 11~ N	76° 40~ W
Baltimore Co	39° 20~ N	76° 25~ W
Cumberland	39° 37~ N	78° 46~ W
Frederick AP	39° 27~ N	77° 25~ W
Hagerstown	39° 42~ N	77° 44~ W
Salisbury (S)	38° 20~ N	75° 30~ W
MASSACHUSETTS		
Boston AP	42° 22~ N	71° 2~ W
Clinton	42° 24~ N	71° 41~ W
Fall River	41° 43~ N	71° 8~ W
Framingham	42° 17~ N	71° 25~ W
Gloucester	42° 35~ N	70° 41~ W
Greenfield	42° 3~ N	72° 4~ W
Lawrence	42° 42~ N	71° 10~ W
Lowell	42° 39~ N	71° 19~ W
New Bedford	41° 41~ N	70° 58~ W
Pittsfield AP	42° 26~ N	73° 18~ W
Springfield, Westover AFB	42° 12~ N	72° 32~ W
Taunton	41° 54~ N	71° 4~ W
Worcester AP	42° 16~ N	71° 52~ W
Bemidji AP	47° 31~ N	94° 56~ W
Brainerd	46° 24~ N	94° 8~ W
Duluth AP	46° 50~ N	92° 11~ W
Fairbault	44° 18~ N	93° 16~ W
Fergus Falls	46° 16~ N	96° 4~ W
International Falls AP	48° 34~ N	93° 23~ W
Mankato	44° 9~ N	93° 59~ W
Minneapolis/St. Paul AP	44° 53~ N	93° 13~ W
Rochester AP	43° 55~ N	92° 30~ W
St. Cloud AP (S)	45° 35~ N	94° 11~ W
Virginia	47° 30~ N	92° 33~ W
Willmar	45° 7~ N	95° 5~ W
Winona	44° 3~ N	91° 38~ W
MISSISSIPPI		
Biloxi—Keesler AFB	30° 25~ N	88° 55~ W
Clarksdale	34° 12~ N	90° 34~ W
Columbus AFB	33° 39~ N	88° 27~ W
Greenville AFB	33° 29~ N	90° 59~ W
Greenwood	33° 30~ N	90° 5~ W
Hattiesburg	31° 16~ N	89°15~ W
Jackson AP	32° 19~ N	90° 5~ W
Laurel	31° 40~ N	89° 10~ W
Mccomb AP	31° 15~ N	90° 28~ W
Meridian AP	32° 20~ N	88° 45~ W

(Continued)

	LONGITUDE	LATITUDE
MISSISSIPPI (Continued)		
Natchez	31° 33~ N	91° 23~ W
Tupelo	34° 16~ N	88° 46~ W
Vicksburg Co	32° 24~ N	90° 47~ W
MISSOURI		
Cape Girardeau	37° 14~ N	89° 35~ W
Columbia AP (S)	38° 58~ N	92° 22~ W
Farmington AP	37° 46~ N	90° 24~ W
Hannibal	39° 42~ N	91° 21~ W
Jefferson City	38° 34~ N	92° 11~ W
Joplin AP	37° 9~ N	94° 30~ W
Kansas City AP	39° 7~ N	94° 35~ W
Kirksville AP	40° 6~ N	92° 33~ W
Mexico	39° 11~ N	91° 54~ W
Moberly	39° 24~ N	92° 26~ W
Poplar Bluff	36° 46~ N	90° 25~ W
Rolla	37° 59~ N	91° 43~ W
St. Joseph AP	39° 46~ N	94° 55~ W
St. Louis AP	38° 45~ N	90° 23~ W
St. Louis CO	38° 39~ N	90° 38~ W
Sikeston	36° 53~ N	89° 36~ W
Sedalia—Whiteman AFB	38° 43~ N	93° 33~ W
Sikeston	36° 53~ N	89° 36~ W
Springfield AP	37° 14~ N	93° 23~ W

	LONGITUDE	LATITUDE
McCook	40° 12~ N	100° 38~ W
Norfolk	41° 59~ N	97° 26~ W
North Platte AP (S)	41° 8~ N	100° 41~ W
Omaha AP	41° 18~ N	95° 54~ W
Scottsbluff AP	41° 52~ N	103° 36~ W
Sidney AP	41° 13~ N	103° 6~ W
NEVADA		
Carson City	39° 10~ N	119° 46~ W
Elko AP	40° 50~ N	115° 47~ W
Ely AP (S)	39° 17~ N	114° 51~ W
Las Vegas AP (S)	36° 5~ N	115° 10~ W
Lovelock AP	40° 4~ N	118° 33~ W
Reno AP (S)	39° 30~ N	119° 47~ W
Reno Co	39° 30~ N	119° 47~ W
Tonopah AP	38° 4~ N	117° 5~ W
Winnemucca AP	40° 54~ N	117° 48~ W
NEW HAMPSHIRE		
Berlin	44° 3~ N	71° 1~ W
Claremont	43° 2~ N	72° 2~ W
Concord AP	43° 12~ N	71° 30~ W
Keene	42° 55~ N	72° 17~ W
Laconia	43° 3~ N	71° 3~ W
Manchester, Grenier AFB	42° 56~ N	71° 26~ W
Portsmouth, Pease AFB	43° 4~ N	70° 49~ W

MONTANA

Billings AP	45° 48~ N	108° 32~ W
Bozeman	45° 47~ N	111° 9~ W
Butte AP	45° 57~ N	112° 30~ W
Cut Bank AP	48° 37~ N	112° 22~ W
Glasgow AP (S)	48° 25~ N	106° 32~ W
Glendive	47° 8~ N	104° 48~ W
Great Falls AP (S)	47° 29~ N	111° 22~ W
Havre	48° 34~ N	109° 40~ W
Helena AP	46° 36~ N	112° 0~ W
Kalispell AP	48° 18~ N	114°16~ W
Lewiston AP	47° 4~ N	109° 27~ W
Livingstown AP	45° 42~ N	110° 26~ W
Miles City AP	46° 26~ N	105° 52~ W
Missoula AP	46° 55~ N	114° 5~ W

NEBRASKA

Beatrice	40° 16~ N	96° 45~ W
Chadron AP	42° 50~ N	103° 5~ W
Columbus	41° 28~ N	97° 20~ W
Fremont	41° 26~ N	96° 29~ W
Grand Island AP	40° 59~ N	98° 19~ W
Hastings	40° 36~ N	98° 26~ W
Kearney	40° 44~ N	99° 1~ W
Lincoln Co (S)	40° 51~ N	96° 45~ W

NEW JERSEY

Atlantic City CO	39° 23~ N	74° 26~ W
Long Branch	40° 19~ N	74° 1~ W
Newark AP	40° 42~ N	74° 10~ W
New Brunswick	40° 29~ N	74° 26~ W
Paterson	40° 54~ N	74° 9~ W
Phillipsburg	40° 41~ N	75° 11~ W
Trenton Co	40° 13~ N	74° 46~ W
Vineland	39° 29~ N	75° 0~ W

NEW MEXICO

Alamagordo° Holloman AFB	32° 51~ N	106° 6~ W
Albuquerque AP (S)	35° 3~ N	106° 37~ W
Artesia	32° 46~ N	104° 23~ W
Carlsbad AP	32° 20~ N	104° 16~ W
Clovis AP	34° 23~ N	103° 19~ W
Farmington AP	36° 44~ N	108° 14~ W
Gallup	35° 31~ N	108° 47~ W
Grants	35° 10~ N	107° 54~ W
Hobbs AP	32° 45~ N	103° 13~ W
Las Cruces	32° 18~ N	106° 55~ W
Los Alamos	35° 52~ N	106° 19~ W
Raton AP	36° 45~ N	104° 30~ W
Roswell, Walker AFB	33° 18~ N	104° 32~ W
Santa Fe CO	35° 37~ N	106° 5~ W

(Continued)

	LONGITUDE	LATITUDE		LONGITUDE	LATITUDE
NEW MEXICO (Continued)			Greenville	35° 37~ N	77° 25~ W
Silver City AP	32° 38~ N	108° 10~ W	Henderson	36° 22~ N	78° 25~ W
Socorro AP	34° 3~ N	106° 53~ W	Hickory	35° 45~ N	81° 23~ W
Tucumcari AP	35° 11~ N	103° 36~ W	Jacksonville	34° 50~ N	77° 37~ W
NEW YORK			Lumberton	34° 37~ N	79° 4~ W
Albany AP (S)	42° 45~ N	73° 48~ W	New Bern AP	35° 5~ N	77° 3~ W
Albany Co	42° 39~ N	73° 45~ W	Raleigh/Durham AP (S)	35° 52~ N	78° 47~ W
Auburn	42° 54~ N	76° 32~ W	Rocky Mount	35° 58~ N	77° 48~ W
Batavia	43° 0~ N	78° 11~ W	Wilmington AP	34° 16~ N	77° 55~ W
Binghamton AP	42° 13~ N	75° 59~ W	Winston-Salem AP	36° 8~ N	80° 13~ W
Buffalo AP	42° 56~ N	78° 44~ W	**NORTH DAKOTA**		
Cortland	42° 36~ N	76° 11~ W	Bismarck AP (S)	46° 46~ N	100° 45~ W
Dunkirk	42° 29~ N	79° 16~ W	Devils Lake	48° 7~ N	98° 54~ W
Elmira AP	42° 10~ N	76° 54~ W	Dickinson AP	46° 48~ N	102° 48~ W
Geneva (S)	42° 45~ N	76° 54~ W	Fargo AP	46° 54~ N	96° 48~ W
Glens Falls	43° 20~ N	73° 37~ W	Grand Forks AP	47° 57~ N	97° 24~ W
Gloversville	43° 2~ N	74° 21~ W	Jamestown AP	46° 55~ N	98° 41~ W
Hornell	42° 21~ N	77° 42~ W	Minot AP	48° 25~ N	101° 21~ W
Ithaca (S)	42° 27~ N	76° 29~ W	Williston	48° 9~ N	103° 35~ W
Jamestown	42° 7~ N	79° 14~ W	**OHIO**		
Kingston	41° 56~ N	74° 0~ W	Akron-Canton AP	40° 55~ N	81° 26~ W
Lockport	43° 9~ N	79° 15~ W	Ashtabula	41° 51~ N	80° 48~ W
Massena AP	44° 56~ N	74° 51~ W	Athens	39° 20~ N	82° 6~ W
Newburgh, Stewart AFB	41° 30~ N	74° 6~ W	Bowling Green	41° 23~ N	83° 38~ W

NYC-Central Park (S)	40° 47~ N	73° 58~ W	Cambridge	40° 4~ N	81° 35'W
NYC-Kennedy AP	40° 39~ N	73° 47~ W	Chillicothe	39° 21~ N	83° 0~ W
NYC-La Guardia AP	40° 46~ N	73° 54~ W	Cincinnati Co	39° 9~ N	84° 31~ W
Niagara Falls AP	43° 6~ N	79° 57~ W	Cleveland AP (S)	41° 24~ N	81° 51~ W
Olean	42° 14~ N	78° 22~ W	Columbus AP (S)	40° 0~ N	82° 53~ W
Oneonta	42° 31~ N	75° 4~ W	Dayton AP	39° 54~ N	84° 13~ W
Oswego Co	43° 28~ N	76° 33~ W	Defiance	41° 17~ N	84° 23~ W
Plattsburg AFB	44° 39~ N	73° 28~ W	Findlay AP	41° 1~ N	83° 40~ W
Poughkeepsie	41° 38~ N	73° 55~ W	Fremont	41° 20~ N	83° 7~ W
Rochester AP	43° 7~ N	77° 40~ W	Hamilton	39° 24~ N	84° 35~ W
Rome, Griffiss AFB	43° 14~ N	75° 25~ W	Lancaster	39° 44~ N	82° 38~ W
Schenectady (S)	42° 51~ N	73° 57~ W	Lima	40° 42~ N	84° 2~ W
Suffolk County AFB	40° 51~ N	72° 38~ W	Mansfield AP	40° 49~ N	82° 31~ W
Syracuse AP	43° 7~ N	76° 7~ W	Marion	40° 36~ N	83° 10~ W
Utica	43° 9~ N	75° 23~ W	Middletown	39° 31~ N	84° 25~ W
Watertown	43° 59~ N	76° 1~ W	Newark	40° 1~ N	82° 28~ W
			Norwalk	41° 16~ N	82° 37~ W
NORTH CAROLINA			Portsmouth	38° 45~ N	82° 55~ W
Asheville AP	35° 26~ N	82° 32~ W	Sandusky Co	41° 27~ N	82° 43~ W
Charlotte AP	35° 13~ N	80° 56~ W	Springfield	39° 50~ N	83° 50~ W
Durham	35° 52~ N	78° 47~ W	Steubenville	40° 23~ N	80° 38~ W
Elizabeth City AP	36° 16~ N	76° 11~ W	Toledo AP	41° 36~ N	83° 48~ W
Fayetteville, Pope AFB	35° 10~ N	79° 1~ W	Warren	41° 20~ N	80° 51~ W
Goldsboro, Seymour-Johnson	35° 20~ N	77° 58~ W	Wooster	40° 47~ N	81° 55~ W
Greensboro AP (S)	36° 5~ N	79° 57~ W			

(Continued)

	LONGITUDE	LATITUDE		LONGITUDE	LATITUDE
OHIO (*Continued*)			Pittsburgh AP	40° 30~ N	80° 13~ W
Youngstown AP	41° 16~ N	80° 40~ W	Pittsburgh Co	40° 27~ N	80° 0~ W
Zanesville AP	39° 57~ N	81° 54~ W	Reading Co	40° 20~ N	75° 38~ W
OKLAHOMA			Scranton/Wilkes-Barre	41° 20~ N	75° 44~ W
Ada	34° 47~ N	96° 41~ W	State College (S)	40° 48~ N	77° 52~ W
Altus AFB	34° 39~ N	99° 16~ W	Sunbury	40° 53~ N	76° 46~ W
Ardmore	34° 18~ N	97° 1~ W	Uniontown	39° 55~ N	79° 43~ W
Bartlesville	36° 45~ N	96° 0~ W	Warren	41° 51~ N	79° 8~ W
Chickasha	35° 3~ N	97° 55~ W	West Chester	39° 58~ N	75° 38~ W
Enid, Vance AFB	36° 21~ N	97° 55~ W	Williamsport AP	41° 15~ N	76° 55~ W
Lawton AP	34° 34~ N	98° 25~ W	York	39° 55~ N	76° 45~ W
McAlester	34° 50~ N	95° 55~ W	**RHODE ISLAND**		
Muskogee AP	35° 40~ N	95° 22~ W	Newport (S)	41° 30~ N	71° 20~ W
Norman	35° 15~ N	97° 29~ W	Providence AP	41° 44~ N	71° 26~ W
Oklahoma City AP (S)	35° 24~ N	97° 36~ W	**SOUTH CAROLINA**		
Ponca City	36° 44~ N	97° 6~ W	Anderson	34° 30~ N	82° 43~ W
Seminole	35° 14~ N	96° 40~ W	Charleston AFB (S)	32° 54~ N	80° 2~ W
Stillwater (S)	36° 10~ N	97° 5~ W	Charleston Co	32° 54~ N	79° 58~ W
Tulsa AP	36° 12~ N	95° 54~ W	Columbia AP	33° 57~ N	81° 7~ W
Woodward	36° 36~ N	99° 31~ W	Florence AP	34° 11~ N	79° 43~ W
OREGON			Georgetown	33° 23~ N	79° 17~ W
Albany	44° 38~ N	123° 7~ W	Greenville AP	34° 54~ N	82° 13~ W
Astoria AP (S)	46° 9~ N	123° 53~ W	Greenwood	34° 10~ N	82° 7~ W
Baker AP	44° 50~ N	117° 49~ W	Orangeburg	33° 30~ N	80° 52~ W

Location	Latitude	Longitude
Bend	44° 4~ N	121° 19~ W
Corvallis (S)	44° 30~ N	123° 17~ W
Eugene AP	44° 7~ N	123° 13~ W
Grants Pass	42° 26~ N	123° 19~ W
Klamath Falls AP	42° 9~ N	121° 44~ W
Medford AP (S)	42° 22~ N	122° 52~ W
Pendleton AP	45° 41~ N	118° 51~ W
Portland AP	45° 36~ N	122° 36~ W
Portland Co	45° 32~ N	122° 40~ W
Roseburg AP	43° 14~ N	123° 22~ W
Salem AP	44° 55~ N	123° 1~ W
The Dalles	45° 36~ N	121° 12~ W
PENNSYLVANIA		
Allentown AP	40° 39~ N	75° 26~ W
Altoona Co	40° 18~ N	78° 19~ W
Butler	40° 52~ N	79° 54~ W
Chambersburg	39° 56~ N	77° 38~ W
Erie AP	42° 5~ N	80° 11'W
Harrisburg AP	40° 12~ N	76° 46~ W
Johnstown	40° 19~ N	78° 50~ W
Lancaster	40° 7~ N	76° 18'W
Meadville	41° 38~ N	80° 10~ W
New Castle	41° 1~ N	80° 22'W
Philadelphia AP	39° 53~ N	75° 15~ W
Rock Hill	34° 59~ N	80° 58~ W
Spartanburg AP	34° 58~ N	82° 0~ W
Sumter, Shaw AFB	33° 54~ N	80° 22~ W
SOUTH DAKOTA		
Aberdeen AP	45° 27~ N	98° 26~ W
Brookings	44° 18~ N	96° 48~ W
Huron AP	44° 23~ N	98° 13~ W
Mitchell	43° 41~ N	98° 1~ W
Pierre AP	44° 23~ N	100° 17~ W
Rapid City AP (S)	44° 3~ N	103° 4~ W
Sioux Falls AP	43° 34~ N	96° 44~ W
Watertown AP	44° 55~ N	97° 9~ W
Yankton	42° 55~ N	97° 23~ W
TENNESSEE		
Athens	35° 26~ N	84° 35~ W
Bristol-Tri City AP	36° 29~ N	82° 24~ W
Chattanooga AP	35° 2~ N	85° 12'W
Clarksville	36° 33~ N	87° 22~ W
Columbia	35° 38~ N	87° 2~ W
Dyersburg	36° 1~ N	89° 24'W
Greenville	36° 4~ N	82° 50'W
Jackson AP	35° 36~ N	88° 55~ W
Knoxville AP	35° 49~ N	83° 59~ W
Memphis AP	35° 3~ N	90° 0~ W

(Continued)

	LONGITUDE	LATITUDE		LONGITUDE	LATITUDE
TENNESSEE *(Continued)*			Snyder	32° 43~ N	100° 55~ W
Murfreesboro	34° 55~ N	86° 28~ W	Temple	31° 6~ N	97° 21~ W
Nashville AP (S)	36° 7~ N	86° 41'W	Tyler AP	32° 21~ N	95° 16~ W
Tullahoma	35° 23~ N	86° 5~ W	Vernon	34° 10~ N	99° 18~ W
TEXAS			Victoria AP	28° 51~ N	96° 55~ W
Abilene AP	32° 25~ N	99° 41~ W	Waco AP	31° 37~ N	97° 13~ W
Alice AP	27° 44~ N	98° 2~ W	Wichita Falls AP	33° 58~ N	98° 29~ W
Amarillo AP	35° 14~ N	100° 42~ W	**UTAH**		
Austin AP	30° 18~ N	97° 42~ W	Cedar City AP	37° 42~ N	113° 6~ W
Bay City	29° 0~ N	95° 58~ W	Logan	41° 45~ N	111° 49~ W
Beaumont	29° 57~ N	94° 1~ W	Moab	38° 36~ N	109° 36~ W
Beeville	28° 22~ N	97° 40~ W	Ogden AP	41° 12~ N	112° 1~ W
Big Spring AP (S)	32° 18~ N	101° 27~ W	Price	39° 37~ N	110° 50~ W
Brownsville AP (S)	25° 54~ N	97° 26~ W	Provo	40° 13~ N	111° 43~ W
Brownwood	31° 48~ N	98° 57~ W	Richfield	38° 46~ N	112° 5~ W
Bryan AP	30° 40~ N	96° 33~ W	St George Co	37° 2~ N	113° 31~ W
Corpus Christi AP	27° 46~ N	97° 30~ W	Salt Lake City AP (S)	40° 46~ N	111° 58~ W
Corsicana	32° 5~ N	96° 28~ W	Vernal AP	40° 27~ N	109° 31~ W
Dallas AP	32° 51~ N	96° 51~ W	**VERMONT**		
Del Rio, Laughlin AFB	29° 22~ N	100° 47~ W	Barre	44° 12~ N	72° 31~ W
Denton	33° 12~ N	97° 6~ W	Burlington AP (S)	44° 28~ N	73° 9~ W
Eagle Pass	28° 52~ N	100° 32~ W	Rutland	43° 36~ N	72° 58~ W
El Paso AP (S)	31° 48~ N	106° 24~ W	**VIRGINIA**		
Fort Worth AP (S)	32° 50~ N	97° 3~ W	Charlottesville	38° 2~ N	78° 31~ W

City	Latitude	Longitude	City	Latitude	Longitude
Galveston AP	29° 18~ N	94° 48~ W	Danville AP	36° 34~ N	79° 20~ W
Greenville	33° 4~ N	96° 3~ W	Fredericksburg	38° 18~ N	77° 28~ W
Harlingen	26° 14~ N	97° 39~ W	Lynchburg AP	37° 20~ N	79° 12~ W
Harrisonburg	38° 27~ N	78° 54~ W	Norfolk AP	36° 54~ N	76° 12~ W
Houston AP	29° 58~ N	95° 21~ W	Petersburg	37° 11~ N	77° 31~ W
Houston Co	29° 59~ N	95° 22~ W	Richmond AP	37° 30~ N	77° 20~ W
Huntsville	30° 43~ N	95° 33~ W	Roanoke AP	37° 19~ N	79° 58~ W
Killeen, Robert Gray AAF	31° 5~ N	97° 41~ W	Staunton	38° 16~ N	78° 54~ W
Lamesa	32° 42~ N	101° 56~ W	Winchester	39° 12~ N	78° 10~ W
Laredo AFB	27° 32~ N	99° 27~ W	**WASHINGTON**		
Longview	32° 28~ N	94° 44~ W	Aberdeen	46° 59~ N	123° 49~ W
Lubbock AP	33° 39~ N	101° 49~ W	Bellingham AP	48° 48~ N	122° 32~ W
Lufkin AP	31° 25~ N	94° 48~ W	Bremerton	47° 34~ N	122° 40~ W
Mcallen	26° 12~ N	98° 13~ W	Ellensburg AP	47° 2~ N	120° 31~W
Midland AP (S)	31° 57~ N	102° 11~ W	Everett, Paine AFB	47° 55~ N	122° 17~ W
Mineral Wells AP	32° 47~ N	98° 4~ W	Kennewick	46° 13~ N	119° 8~ W
Palestine Co	31° 47~ N	95° 38~ W	Longview	46° 10~ N	122° 56~ W
Pampa	35° 32~ N	100° 59~ W	Moses Lake, Larson AFB	47° 12~ N	119° 19~ W
Pecos	31° 25~ N	103° 30~ W	Olympia AP	46° 58~ N	122° 54~ W
Plainview	34° 11~ N	101° 42~ W	Port Angeles	48° 7~ N	123° 26~ W
Port Arthur AP	29° 57~ N	94° 1~ W	Seattle-Boeing Field	47° 32~ N	122° 18~ W
Goodfellow AFB	31° 26~ N	100° 24~ W	Seattle Co (S)	47° 39~ N	122° 18~ W
San Antonio AP (S)	29° 32~ N	98° 28~ W			
Sherman, Perrin AFB	33° 43~ N	96° 40~ W			

(Continued)

	LONGITUDE	LATITUDE		LONGITUDE	LATITUDE
WASHINGTON (Continued)			Green Bay AP	44° 29~ N	88° 8~ W
Seattle-Tacoma AP (S)	47° 27~ N	122° 18~ W	La Crosse AP	43° 52~ N	91° 15~ W
Spokane AP (S)	47° 38~ N	117° 31~ W	Madison AP (S)	43° 8~ N	89° 20~ W
Tacoma, McChord AFB	47° 15~ N	122° 30~ W	Manitowoc	44° 6~ N	87° 41~ W
Walla Walla AP	46° 6~ N	118° 17~ W	Marinette	45° 6~ N	87° 38~ W
Wenatchee	47° 25~ N	120° 19~ W	Milwaukee AP	42° 57~ N	87° 54~ W
Yakima AP	46° 34~ N	120° 32~ W	Racine	42° 43~ N	87° 51~ W
WEST VIRGINIA			Sheboygan	43° 45~ N	87° 43~ W
Beckley	37° 47~ N	81°7~ W	Stevens Point	44° 30~ N	89° 34~ W
Bluefield AP	37° 18~ N	81° 13~ W	Waukesha	43° 1~ N	88° 14~ W
Charleston AP	38° 22~ N	81° 36~ W	Wausau AP	44° 55~ N	89° 37~ W
Clarksburg	39° 16~ N	80° 21~ W	**WYOMING**		
Elkins AP	38° 53~ N	79° 51~ W	Casper AP	42° 55~ N	106° 28~ W
Huntington Co	38° 25~ N	82° 30~ W	Cheyenne	41° 9~ N	104° 49~ W
Martinsburg AP	39° 24~ N	77° 59~ W	Cody AP	44° 33~ N	109° 4~ W
Morgantown AP	39° 39~ N	79° 55~ W	Evanston	41° 16~ N	110° 57~ W
Parkersburg Co	39° 16~ N	81° 34~ W	Lander AP (S)	42° 49~ N	108° 44~ W
Wheeling	40° 7~ N	80° 42~ W	Laramie AP (S)	41° 19~ N	105° 41~ W
WISCONSIN			Newcastle	43° 51~ N	104° 13~ W
Appleton	44° 15~ N	88° 23~ W	Rawlins	41° 48~ N	107° 12~ W
Ashland	46° 34~ N	90° 58~ W	Rock Springs AP	41° 36~ N	109° 0~ W
Beloit	42° 30~ N	89° 2~ W	Sheridan AP	44° 46~ N	106° 58~ W
Eau Claire AP	44° 52~ N	91° 29~ W	Torrington	42° 5~ N	104° 13~ W
Fond Du Lac	43° 48~ N	88° 27~ W			

CANADA LONGITUDES AND LATITUDES

	LONGITUDE	LATITUDE		LONGITUDE	LATITUDE
ALBERTA			**MANITOBA**		
Calgary AP	51° 6~ N	114° 1~ W	Brandon	49° 52~ N	99° 59~ W
Edmonton AP	53° 34~ N	113° 31~ W	Churchill AP (S)	58° 45~ N	94° 4~ W
Grande Prairie AP	55° 11~ N	118° 53~ W	Dauphin AP	51° 6~ N	100° 3~ W
Jasper	52° 53~ N	118° 4~ W	Flin Flon	54° 46~ N	101° 51~ W
Lethbridge AP (S)	49° 38~ N	112° 48~ W	Portage La Prairie AP	49° 54~ N	98° 16~ W
McMurray AP	56° 39~ N	111° 13~ W	The Pas AP (S)	53° 58~ N	101° 6~ W
Medicine Hat AP	50° 1~ N	110° 43~ W	Winnipeg AP (S)	49° 54~ N	97° 14~ W
Red Deer AP	52° 11~ N	113° 54~ W	**NEW BRUNSWICK**		
BRITISH COLUMBIA			Campbellton Co	48° 0~ N	66° 40~ W
Dawson Creek	55° 44~ N	120° 11~ W	Chatham AP	47° 1~ N	65° 27~ W
Fort Nelson AP (S)	58° 50~ N	122° 35~ W	Edmundston Co	47° 22~ N	68° 20~ W
Kamloops Co	50° 43~ N	120° 25~ W	Fredericton AP (S)	45° 52~ N	66° 32~ W
Nanaimo (S)	49° 11~ N	123° 58~ W	Moncton AP (S)	46° 7~ N	64° 41~ W
New Westminster	49° 13~ N	122° 54~ W	Saint John AP	45° 19~ N	65° 53~ W
Penticton AP	49° 28~ N	119° 36~ W	**NEWFOUNDLAND**		
Prince George AP (S)	53° 53~ N	122° 41~ W	Corner Brook	48° 58~ N	57° 57~ W
Prince Rupert Co	54° 17~ N	130° 23~ W	Gander AP	48° 57~ N	54° 34~ W
Trail	49° 8~ N	117° 44~ W	Goose Bay AP (S)	53° 19~ N	60° 25~ W
Vancouver AP (S)	49° 11~ N	123° 10~ W	St John's AP (S)	47° 37~ N	52° 45~ W
Victoria Co	48° 25~ N	123° 19~ W	Stephenville AP	48° 32~ N	58° 33~ W

(Continued)

	LONGITUDE	LATITUDE		LONGITUDE	LATITUDE
NORTHWEST TERRITORIES			Toronto AP (S)	43° 41~ N	79° 38~ W
Fort Smith AP(S)	60° 1~ N	111° 58~ W	Windsor AP	42° 16~ N	82° 58~ W
Frobisher AP (S)	63° 45~ N	68° 33~ W	**PRINCE EDWARD ISLAND**		
Inuvik (S)	68° 18~ N	133° 29~ W	Charlottetown AP (S)	46° 17~ N	63° 8~ W
Resolute AP (S)	74° 43~ N	94° 59~ W	Summerside AP	46° 26~ N	63° 50~ W
Yellowknife AP	62° 28~ N	114° 27~ W	**QUEBEC**		
NOVA SCOTIA			Bagotville AP	48° 20~ N	71° 0~ W
Amherst	45° 49~ N	64°13~ W	Chicoutimi	48° 25~ N	71° 5~ W
Halifax AP (S)	44° 39~ N	63° 34~ W	Drummondville	45° 53~ N	72° 29~ W
Kentville (S)	45° 3~ N	64° 36~ W	Granby	45° 23~ N	72° 42~ W
New Glasgow	45° 37~ N	62° 37~ W	Hull	45° 26~ N	75° 44~ W
Sydney AP	46° 10~ N	60° 3~ W	Megantic AP	45° 35~ N	70° 52~ W
Truro Co	45° 22~ N	63° 16~ W	Montreal AP (S)	45° 28~ N	73° 45~ W
Yarmouth AP	43° 50~ N	66° 5~ W	Quebec AP	46° 48~ N	71° 23~ W
ONTARIO			Rimouski	48° 27~ N	68° 32~ W
Belleville	44° 9~ N	77° 24~ W	St Jean	45° 18~ N	73° 16~ W
Chatham	42° 24~ N	82° 12~ W	St Jerome	45° 48~ N	74° 1~ W
Cornwall	45° 1~ N	74° 45~ W	Sept. Iles AP (S)	50° 13~ N	66° 16~ W
Hamilton	43° 16~ N	79° 54~ W	Shawinigan	46° 34~ N	72° 43~ W
Kapuskasing AP (S)	49° 25~ N	82° 28~ W	Sherbrooke Co	45° 24~ N	71° 54~ W
Kenora AP	49° 48~ N	94° 22~ W	Thetford Mines	46° 4~ N	71° 19~ W
Kingston	44° 16~ N	76° 30~ W	Trois Rivieres	46° 21~ N	72° 35~ W
Kitchener	43° 26~ N	80° 30~ W	Val D'or AP	48° 3~ N	77° 47~ W
London AP	43° 2~ N	81° 9~ W	Valleyfield	45° 16~ N	74° 6~ W

North Bay AP	46° 22~ N	79° 25~ W
Oshawa	43° 54~ N	78° 52~ W
Ottawa AP (S)	45° 19~ N	75° 40~ W
Owen Sound	44° 34~ N	80° 55~ W
Peterborough	44° 17~ N	78° 19~ W
St Catharines	43° 11~ N	79° 14~ W
Sarnia	42° 58~ N	82° 22~ W
Sault Ste Marie AP	46° 32~ N	84° 30~ W
Sudbury AP	46° 37~ N	80° 48~ W
Thunder Bay AP	48° 22~ N	89° 19~ W
Timmins AP	48° 34~ N	81° 22~ W

SASKATCHEWAN

Estevan AP	49° 4~ N	103° 0~ W
Moose Jaw AP	50° 20~ N	105° 33~ W
North Battleford AP	52° 46~ N	108° 15~ W
Prince Albert AP	53° 13~ N	105° 41~ W
Regina AP	50° 26~ N	104° 40~ W
Saskatoon AP (S)	52° 10~ N	106° 41~ W
Swift Current AP (S)	50° 17~ N	107° 41~ W
Yorkton AP	51° 16~ N	102° 28~ W

YUKON TERRITORY

Whitehorse AP (S)	60° 43~ N	135° 4~ W

	LONGITUDE	LATITUDE		LONGITUDE	LATITUDE
AFGHANISTAN			**BURMA**		
Kabul	34° 35~ N	69° 12~ E	Mandalay	21° 59~ N	96° 6~ E
ALGERIA			Rangoon	16° 47~ N	96° 9~ E
Algiers	36° 46~ N	30° 3~ E	**CAMBODIA**		
ARGENTINA			Phnom Penh	11° 33~ N	104° 51~ E
Buenos Aires	34° 35~ S	58° 29~ W	**CHILE**		
Cordoba	31° 22~ S	64° 15~ W	Punta Arenas	53° 10~ S	70° 54~ W
Tucuman	26° 50~ S	65° 10~ W	Santiago	33° 27~ S	70° 42~ W
AUSTRALIA			Valparaiso	33° 1~ S	71° 38~ W
Adelaide	34° 56~ S	138° 35~ E	**CHINA**		
Alice Springs	23° 48~ S	133° 53~ E	Chongquing	29° 33~ N	106° 33~ E
Brisbane	27° 28~ S	153° 2~ E	Shanghai	31° 12~ N	121° 26~ E
Darwin	12° 28~ S	130° 51~ E	**COLOMBIA**		
Melbourne	37° 49~ S	144° 58~ E	Baranquilla	10° 59~ N	74° 48~ W
Perth	31°57~ S	115°51~ E	Bogota	4° 36~ N	74° 5~ W
Sydney	33° 52~ S	151° 12~ E	Cali	3° 25~ N	76° 30~ W
AUSTRIA			Medellin	6° 13~ N	75° 36~ W
Vienna	48° 15~ N	16° 22~ E	**CONGO**		
AZORES			Brazzaville	4°15~ S	15° 15~ E
Lajes (Terceira)	38° 45~ N	27° 5~ W	**CUBA**		
BAHAMAS			Guantanamo Bay	19° 54~ N	75° 9~ W
Nassau	25° 5~ N	77° 21~ W	Havana	23° 8~ N	82° 21~ W

BANGLADESH			
Chittagong	22° 21~ N	91° 50~ E	
BELGIUM			
Brussels	50° 48~ N	4° 21~ E	
BERMUDA			
Kindley AFB	33° 22~ N	64° 41~ W	
BOLIVIA			
La Paz	16° 30~ S	68° 9~ W	
BRAZIL			
Belem	1° 27~ S	48° 29~ W	
Belo Horizonte	19° 56~ S	43° 57~ W	
Brasilia	15° 52~ S	47° 55~ W	
Curitiba	25° 25~ S	49° 17~ W	
Fortaleza	3° 46~ S	38° 33~ W	
Porto Alegre	30° 2~ S	51° 13~ W	
Recife	8° 4~ S	34° 53~ W	
Rio De Janeiro	22° 55~ S	43° 12~ W	
Salvador	13° 0~ S	38° 30~ W	
Sao Paulo	23° 33~ S	46° 38~ W	
BELIZE			
Belize	17° 31~ N	88° 11~ W	
BULGARIA			
Sofia	42° 42~ N	23° 20~ E	
Strasbourg	48° 35~ N	7° 46~ E	
CZECHOSLOVAKIA			
Prague	50° 5~ N	14° 25~ E	
DENMARK			
Copenhagen	55° 41~ N	12° 33~ E	
DOMINICAN REPUBLIC			
Santo Domingo	18° 29~ N	69° 54~ W	
EGYPT			
Cairo	29° 52~ N	31° 20~ E	
EL SALVADOR			
San Salvador	13° 42~ N	89° 13~ W	
EQUADOR			
Guayaquil	2° 0~ S	79° 53~ W	
Quito	0° 13~ S	78° 32~ W	
ETHIOPIA			
Addis Ababa	90° 2~ N	38° 45~ E	
Asmara	15° 17~ N	38° 55~ E	
FINLAND			
Helsinki	60° 10~ N	24° 57~ E	
FRANCE			
Lyon	45° 42~ N	4° 47~ E	
Marseilles	43° 18~ N	5° 23~ E	
Nantes	47° 15~ N	1° 34~ W	
Nice	43° 42~ N	7° 16~ E	
Paris	48° 49~ N	2° 29~ E	

(Continued)

	LONGITUDE	LATITUDE		LONGITUDE	LATITUDE
FRENCH GUIANA			**IRAN**		
Cayenne	4° 56~ N	52° 27~ W	Abadan	30° 21~ N	48° 16~ E
GERMANY			Meshed	36° 17~ N	59° 36~ E
Berlin (West)	52° 27~ N	13° 18~ E	Tehran	35° 41~ N	51° 25~ E
Hamburg	53° 33~ N	9° 58~ E	**IRAQ**		
Hannover	52° 24~ N	9° 40~ E	Baghdad	33° 20~ N	44° 24~ E
Mannheim	49° 34~ N	8° 28~ E	Mosul	36° 19~ N	43° 9~ E
Munich	48° 9~ N	11° 34~ E	**IRELAND**		
GHANA			Dublin	53° 22~ N	6° 21~ W
Accra	5° 33~ N	0° 12~ W	Shannon	52° 41~ N	8° 55~ W
GIBRALTAR			**IRIAN BARAT**		
Gibraltar	36° 9~ N	5° 22~ W	Manokwari	0° 52~ S	134° 5~ E
GREECE			**ISRAEL**		
Athens	37° 58~ N	23° 43~ E	Jerusalem	31° 47~ N	35° 13~ E
Thessaloniki	40° 37~ N	22° 57~ E	Tel Aviv	32° 6~ N	34° 47~ E
GREENLAND			**ITALY**		
Narsarssuaq	61° 11~ N	45° 25~ W	Milan	45° 27~ N	9° 17~ E
GUATEMALA			Naples	40° 53~ N	14° 18~ E
Guatemala City	14° 37~ N	90° 31~ W	Rome	41° 48~ N	12° 36~ E
GUYANA			**IVORY COAST**		
Georgetown	6° 50~ N	58° 12~ W	Abidjan	5° 19~ N	4° 1~ W
HAITI			**JAPAN**		
Port Au Prince	18° 33~ N	72° 20~ W	Fukuoka	33° 35~ N	130° 27~ E

	Latitude	Longitude
HONDURAS		
Tegucigalpa	14° 6~ N	87° 13~ W
HONG KONG		
Hong Kong	22° 18~ N	114° 10~ E
HUNGARY		
Budapest	47° 31~ N	19° 2~ E
ICELAND		
Reykjavik	64°8~ N	21°56~ E
INDIA		
Ahmenabad	23° 2~ N	72° 35~ E
Bangalore	12° 57~ N	77° 37~ E
Bombay	18° 54~ N	72° 49~ E
Calcutta	22° 32~ N	88° 20~ E
Madras	13° 4~ N	80° 15~ E
Nagpur	21°9~ N	79°7~ E
New Delhi	28° 35~ N	77° 12~ E
INDONESIA		
Djakarta	6° 11~ S	106° 50~ E
Kupang	10° 10~ S	123° 34~ E
Makassar	5° 8~ S	119° 28~ E
Medan	3° 35~ N	98° 41~ E
Palembang	3° 0~ S	104° 46~ E
Surabaya	7°13~ S	112° 43~ E

	Latitude	Longitude
Sapporo	43° 4~ N	141° 21~ E
Tokyo	35° 41~ N	139° 46~ E
JORDAN		
Amman	31° 57~ N	35° 57~ E
KENYA		
Nairobi	1° 16~ S	36° 48~ E
KOREA		
Pyongyang	39° 2~ N	125° 41~ E
Seoul	37° 34~ N	126° 58~ E
LEBANON		
Beirut	33° 54~ N	35° 28~ E
LIBERIA		
Monrovia	6° 18~ N	10° 48~ W
LIBYA		
Benghazi	32° 6~ N	20° 4~ E
Tananarive	18° 55~ S	47° 33~ E
MALAYSIA		
Kuala Lumpur	3° 7~ N	101° 42~ E
Penang	5° 25~ N	100° 19~ E
MARTINIQUE		
Fort De France	14° 37~ N	61° 5~ W
MEXICO		
Guadalajara	20° 41~ N	103° 20~ W
Merida	20° 58~ N	89° 38~ W

(Continued)

	LONGITUDE	LATITUDE		LONGITUDE	LATITUDE
MEXICO (Continued)			Kiev	50° 27~ N	30° 30~ E
Mexico City	19° 24~ N	99° 12~ W	Kharkov	50° 0~ N	36° 14~ E
Monterrey	25° 40~ N	100° 18~ W	Kuibyshev	53° 11~ N	50° 6~ E
Vera Cruz	19° 12~ N	96° 8~ W	Leningrad	59° 56~ N	30° 16~ E
MOROCCO			Minsk	53° 54~ N	27° 33~ E
Casablanca	33° 35~ N	7° 39~ W	Moscow	55° 46~ N	37° 40~ E
NEPAL			Odessa	46° 29~ N	30° 44~ E
Katmandu	27° 42~ N	85° 12~ E	Petropavlovsk	52° 53~ N	158° 42~ E
NETHERLANDS			Rostov on Don	47° 13~ N	39° 43~ E
Amsterdam	52° 23~ N	4° 55~ E	Sverdlovsk	56° 49~ N	60° 38~ E
NEW ZEALAND			Tashkent	41° 20~ N	69° 18~ E
Auckland	36° 51~ S	174° 46~ E	Tbilisi	41° 43~ N	44° 48~ E
Christchurch	43° 32~ S	172° 37~ E	Vladivostok	43° 7~ N	131° 55~ E
Wellington	41° 17~ S	174° 46~ E	Volgograd	48° 42~ N	44° 31~ E
NICARAGUA			**SAUDI ARABIA**		
Managua	12° 10~ N	86° 15~ W	Dhahran	26° 17~ N	50° 9~ E
NIGERIA			Jedda	21° 28~ N	39° 10~ E
Lagos	6° 27~ N	3° 24~ E	Riyadh	24° 39~ N	46° 42~ E
NORWAY			**SENEGAL**		
Bergen	60° 24~ N	5° 19~ E	Dakar	14° 42~ N	17° 29~ W
Oslo	59° 56~ N	10° 44~ E	**SINGAPORE**		
PAKISTAN			Singapore	1° 18~ N	103° 50~ E
Karachi	24° 48~ N	66° 59~ E	**SOMALIA**		
Lahore	31° 35~ N	74° 20~ E	Mogadiscio	2° 2~ N	49° 19~ E
Peshwar	34° 1~ N	71° 35~ E			

PANAMA		
Panama City	8° 58~ N	79° 33~ W
PAPUA NEW GUINEA		
Port Moresby	9° 29~ S	147° 9~ E
PARAGUAY		
Ascuncion	25° 17~ S	57° 30~ W
PERU		
Lima	12° 5~ S	77° 3~ W
PHILIPPINES		
Manila	14° 35~ N	120° 59~ E
POLAND		
Krakow	50° 4~ N	19° 57~ E
Warsaw	52° 13~ N	21° 2~ E
PORTUGAL		
Lisbon	38° 43~ N	9° 8~ W
PUERTO RICO		
San Juan	18° 29~ N	66° 7~ W
RUMANIA		
Bucharest	44° 25~ N	26° 6~ E
RUSSIA		
Alma Ata	43° 14~ N	76° 53~ E
Archangel	64° 33~ N	40° 32~ E
Kaliningrad	54° 43~ N	20° 30~ E
Krasnoyarsk	56° 1~ N	92° 57~ E
SOUTH AFRICA		
Cape Town	33° 56~ S	18° 29~ E
Johannesburg	26° 11~ S	28° 3~ E
Pretoria	25° 45~ S	28° 14~ E
SOUTH YEMEN		
Aden	12° 50~ N	45° 2~ E
SPAIN		
Barcelona	41° 24~ N	2° 9~ E
Madrid	40° 25~ N	3° 41~ W
Valencia	39° 28~ N	0° 23~ W
SRI LANKA		
Colombo	6° 54~ N	79° 52~ E
SUDAN		
Khartoum	15° 37~ N	32° 33~ E
SURINAM		
Paramaribo	5° 49~ N	55° 9~ W
SWEDEN		
Stockholm	59° 21~ N	18° 4~ E
SWITZERLAND		
Zurich	47° 23~ N	8° 33~ E
SYRIA		
Damascus	33° 30~ N	36° 20~ E
TAIWAN		
Tainan	22° 57~ N	120° 12~ E

(Continued)

	LONGITUDE	LATITUDE		LONGITUDE	LATITUDE
TAIWAN (Continued)			Cardiff	51° 28~ N	3°10~ W
Taipei	25°2~ N	121°31~ E	Edinburgh	55° 55~ N	3° 11~ W
TANZANIA			Glasgow	55° 52~ N	4° 17~ W
Dar es Salaam	6° 50~ S	39° 18~ E	London	51° 29~ N	0° 0~ W
THAILAND			**URUGUAY**		
Bangkok	13° 44~ N	100° 30~ E	Montevideo	34° 51~ S	56° 13~ W
TRINIDAD			**VENEZUELA**		
Port of Spain	10° 40~ N	61° 31~ W	Caracas	10° 30~ N	66° 56~ W
TUNISIA			Maracaibo	10° 39~ N	71° 36~ W
Tunis	36° 47~ N	10° 12~ E	**VIETNAM**		
TURKEY			Da Nang	16° 4~ N	108° 13~ E
Adana	36° 59~ N	35° 18~ E	Hanoi	21° 2~ N	105° 52~ E
Ankara	39° 57~ N	32° 53~ E	Ho Chi Minh City (Saigon)	10° 47~ N	106° 42~ E
Istanbul	40° 58~ N	28° 50~ E	**YUGOSLAVIA**		
Izmir	38° 26~ N	27° 10~ E	Belgrade	44° 48~ N	20° 28~ E
UNITED KINGDOM			**ZAIRE**		
Belfast	54° 36~ N	5° 55~ W	Kinshasa (Leopoldville)	4° 20~ S	15° 18~ E
Birmingham	52° 29~ N	1° 56~ W	Kisangani (Stanleyville)	0° 26~ S	15° 14~ E

PHOTOVOLTAIC SYSTEM SUPPORT

HARDWARE AND STRUCTURES

The following photographs and graphics are courtesy of UNIRAC Corporation.

Figure B.1 Roof-mount semi-adjustable tilt PV support structure.

Figure B.2 Ground-mount semi-adjustable tilt PV support structure.

Figure B.3 Field-mount fixed tilt PV support structure.

Figure B.4 Field-mount semi-adjustable tilt PV support structure.

Figure B.5 Pipe-mounted semi-adjustable tilt PV support structure.

Figure B.6 Pipe-mounted fixed tilt PV support structure.

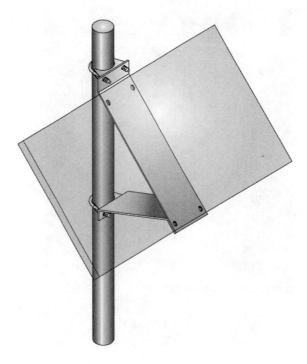

Figure B.7 Pipe-mounted fixed tilt PV support hardware.

Figure B.8 Pipe-mounted semi-adjustable tilt PV support structure.

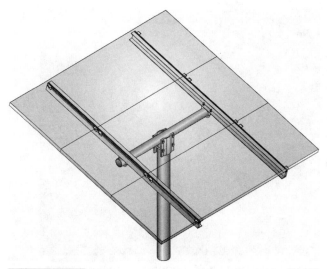

Figure B.9 Pipe-mounted manually adjustable tilt PV support structure.

Figure B.10 Ground-mount fixed tilt PV support system.

Figure B.11   Ground-mount fixed tilt PV support system hardware detail.

Figure B.12 Ground-mount fixed tilt PV support system graphics.

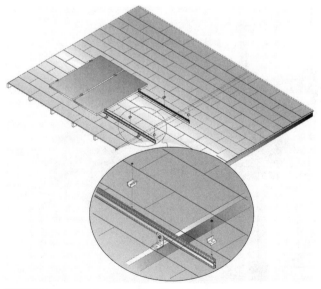

Figure B.13 Roof-mount fixed tilt PV support system using simple channel hardware.

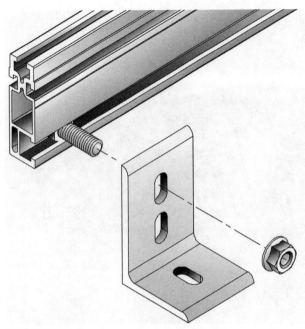

Figure B.14 Roof-mount fixed tilt PV support railing and support system hardware.

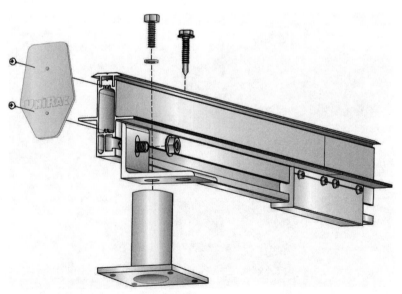

Figure B.15 Roof-mount fixed tilt PV support railing and raised support system hardware.

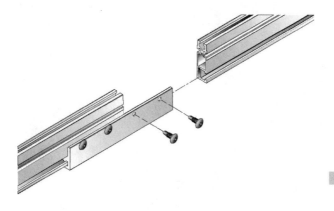

Figure B.16 Railing tie strap system hardware.

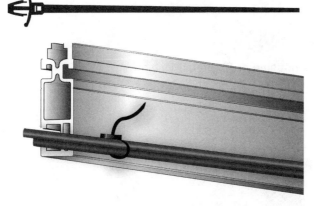

Figure B.17 Cable tie for railing hardware.

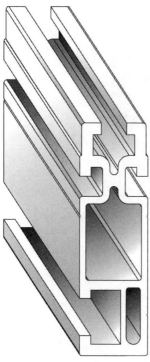

Figure B.18 Cross section of a reinforced PV support railing.

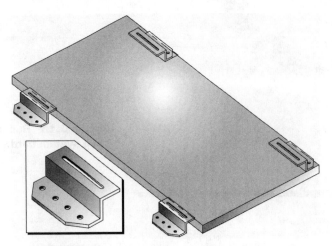

Figure B.19 Mobile housing or RV prefabricated PV support hardware.

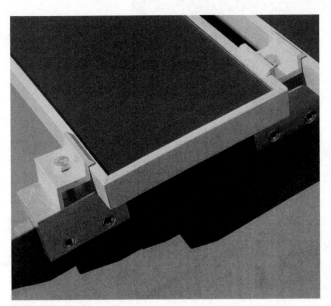

Figure B.20 V module railing attachment detail.

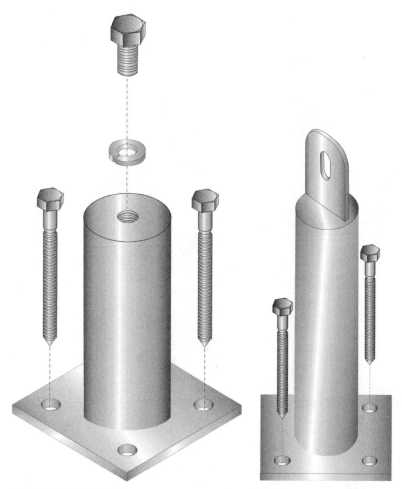

Figure B.21 Railing support stand-offs for PV support railing system hardware.

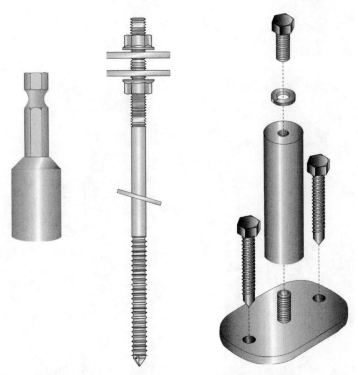

Figure B.22 PV support rail anchoring hardware.

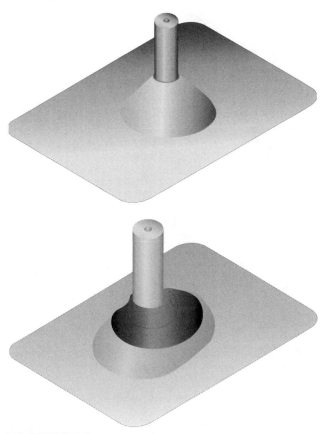

Figure B.23 Waterproof boots for PV support rail
stand-offs.

C

Certified Photovoltaic Modules

TABLE C.1 LIST OF ELIGIBLE PHOTOVOLTAIC MODULES CALIFORNIA ENERGY COMMISSION EMERGING RENEWABLES PROGRAM (FEBRUARY 2005)

MANUFACTURER NAME	MODULE MODEL NUMBER	DESCRIPTION	CEC PTC* RATING	NOTES
ASE Americas, Inc.	ASE-100-ATF/17-34	100W/17V EFG Module, framed	89.7	NA
ASE Americas, Inc.	ASE-300-DGF/17-285	285W/17V EFG Module, framed	255.4	NA
ASE Americas, Inc.	ASE-300-DGF/17-300	300W/17V EFG Module, framed	269.1	NA
ASE Americas, Inc.	ASE-300-DGF/17-315	315W/17V EFG Module, framed	283	NA
ASE Americas, Inc.	ASE-300-DGF/25-145	145W/25V EFG Module, framed	128.5	NA
ASE Americas, Inc.	ASE-300-DGF/34-195	195W/34V EFG Module, framed	173.5	NA
ASE Americas, Inc.	ASE-300-DGF/42-240	240W/42V EFG Module, framed	214	NA
ASE Americas, Inc.	ASE-300-DGF/50-260	260W/50V EFG Module, framed	232.6	NA
ASE Americas, Inc.	ASE-300-DGF/50-265	265W/50V EFG Module, framed	237	NA
ASE Americas, Inc.	ASE-300-DGF/50-285	285W/50V EFG Module, framed	255.4	NA
ASE Americas, Inc.	ASE-300-DGF/50-300	300W/50V EFG Module, framed	269.1	NA
ASE Americas, Inc.	ASE-300-DGF/50-315	315W/50V EFG Module, framed	282.8	NA
AstroPower, Inc.	AP-100	100W Single Crystal Module (was AP-1006)	88.9	NA
AstroPower, Inc.	AP-1006	100W Single Crystal Module	88.9	NA
AstroPower, Inc.	AP-110	110W Single Crystal Module (was AP-1106)	97.9	NA
AstroPower, Inc.	AP-1106	110W Single Crystal Module	97.9	NA
AstroPower, Inc.	AP-120	120W Single Crystal Module (was AP-1206)	107	NA
AstroPower, Inc.	AP-1206	120W Single Crystal Module	107	NA
AstroPower, Inc.	AP-50-GA	50W Single Crystal Module	44	NA

AstroPower, Inc.	AP-50-GT	50W Single Crystal Module	44	NA
AstroPower, Inc.	AP-55-GA	55W Single Crystal Module	48.5	NA
AstroPower, Inc.	AP-55-GT	55W Single Crystal Module	48.5	NA
AstroPower, Inc.	AP-6105	65W Single Crystal Module	58.8	NA
AstroPower, Inc.	AP-65	65W Single Crystal Module (was AP-6105)	58.8	NA
AstroPower, Inc.	AP-7105	75W Single Crystal Module	68	NA
AstroPower, Inc.	AP-75	75W Single Crystal Module (was AP-7105)	68	NA
AstroPower, Inc.	AP6-160	160W Single Crystal Module	141.1	NA
AstroPower, Inc.	AP6-170	170W Single Crystal Module	150.1	NA
AstroPower, Inc.	APi-030-MNA	30W Single Crystal Module w/o connectors (was AP-30)	26.5	NA
AstroPower, Inc.	APi-030-MNB	30W Single Crystal Module w/o connectors (was AP-30) B	26.5	NA
AstroPower, Inc.	APi-045-MNA	45W Single Crystal Module w/o connectors (was AP-45)	39.7	NA
AstroPower, Inc.	APi-045-MNB	45W Single Crystal Module w/o connectors (was AP-45) B	39.7	NA
AstroPower, Inc.	APi-050-MNA	50W Single Crystal Module w/o connectors (was AP-50)	44.2	NA
AstroPower, Inc.	APi-050-MNB	50W Single Crystal Module w/o connectors (was AP-50) B	44.2	NA
AstroPower, Inc.	APi-055-GCA	55W Single Crystal Module w/MC connectors (was AP-50)	48.9	NA
AstroPower, Inc.	APi-055-GCB	55W Single Crystal Module w/MC connectors (was AP-55-GA)	48.9	NA
AstroPower, Inc.	APi-065-MNA	65W Single Crystal Module w/o connectors (was AP-65)	57.7	NA

(Continued)

TABLE C.1   LIST OF ELIGIBLE PHOTOVOLTAIC MODULES CALIFORNIA ENERGY COMMISSION EMERGING RENEWABLES PROGRAM (FEBRUARY 2005) (Continued)

MANUFACTURER NAME	MODULE MODEL NUMBER	DESCRIPTION	CEC PTC° RATING	NOTES
AstroPower, Inc.	APi-065-MNB	65W Single Crystal Module w/o connectors (was AP-65) B	57.7	NA
AstroPower, Inc.	APi-070-MNA	70W Single Crystal Module w/o connectors (was AP-70)	62.2	NA
AstroPower, Inc.	APi-070-MNB	70W Single Crystal Module w/o connectors (was AP-70) B	62.2	NA
AstroPower, Inc.	APi-100-MCA	100W Single Crystal Module w/MC connectors (was AP-100)	88.7	NA
AstroPower, Inc.	APi-100-MCB	100W Single Crystal Module w/MC connectors (was AP-100) B	88.7	NA
AstroPower, Inc.	APi-100-MNA	100W Single Crystal Module w/o connectors (was AP-100)	88.7	NA
AstroPower, Inc.	APi-100-MNB	100 W Single Crystal Module w/o connectors (was AP-100) B	88.7	NA
AstroPower, Inc.	APi-110-MNB	110W Single Crystal Module w/o connectors (was AP-110) B	97.8	NA
AstroPower, Inc.	APi-110-MCA	110W Single Crystal Module w/MC connectors (was AP-110)	97.8	NA
AstroPower, Inc.	APi-110-MCB	110W Single Crystal Module w/MC connectors (was AP-110) B	97.8	NA
AstroPower, Inc.	APi-110-MNA	110W Single Crystal Module w/o connectors (was AP-110)	97.8	NA
AstroPower, Inc.	APi-165-MCA	165W Single Crystal Module w/MC connectors (was AP-165)	146.7	NA

Manufacturer	Model	Description		
AstroPower, Inc.	APi-165-MCB	165W Single Crystal Module w/MC connectors (was AP-165) B	146.7	NA
AstroPower, Inc.	APi-173-MCA	173W Single Crystal Module w/MC connectors (was AP-173)	154	NA
AstroPower, Inc.	APi-173-MCB	173W Single Crystal Module w/MC connectors (was AP-173) B	154	NA
AstroPower, Inc.	APx-045-MNA	45W Apex Module w/o connectors (was APx-45)	38.6	NA
AstroPower, Inc.	APx-045-MNB	45W Apex Module w/o connectors (was APx-45) B	38.6	NA
AstroPower, Inc.	APx-050-MNA	50W Apex Module w/o connectors (was APx-50)	42.9	NA
AstroPower, Inc.	APx-065-MNA	65W Apex Module w/o connectors (was APx-65)	55.7	NA
AstroPower, Inc.	APx-065-MNB	65W Apex Module w/o connectors (was APx-65) B	55.7	NA
AstroPower, Inc.	APx-070-MNA	70W Apex Module w/o connectors (was APx-70)	60.1	NA
AstroPower, Inc.	APx-070-MNB	70W Apex Module w/o connectors (was APx-70) B	60.1	NA
AstroPower, Inc.	APx-075-MNA	75W Apex Module w/o connectors (was APx-75)	64.4	NA
AstroPower, Inc.	APx-075-MNB	75W Apex Module w/o connectors (was APx-75) B	64.4	NA
AstroPower, Inc.	APx-130	130W Apex Silicon Film Module	112	NA
AstroPower, Inc.	APx-130-MCA	130W Apex Silicon Film Module w/MC connectors (was APx-130)	111.6	NA
AstroPower, Inc.	APx-130-MCB	130W Apex Silicon Film Module w/MC connectors (was APx-130) B	111.6	NA

(Continued)

TABLE C.1 LIST OF ELIGIBLE PHOTOVOLTAIC MODULES CALIFORNIA ENERGY COMMISSION EMERGING RENEWABLES PROGRAM (FEBRUARY 2005) (Continued)

MANUFACTURER NAME	MODULE MODEL NUMBER	DESCRIPTION	CEC PTC* RATING	NOTES
AstroPower, Inc.	APx-130-MNA	130W Apex Silicon Film Module w/o connectors (was APx-130)	111.6	NA
AstroPower, Inc.	APx-130-MNB	130W Apex Silicon Film Module w/o connectors (was APx-130) B	111.6	NA
AstroPower, Inc.	APx-140	140W Apex Silicon Film Module	121	NA
AstroPower, Inc.	APx-140-MCA	140W Apex Silicon Film Module w/MC connectors (was APx-140)	120.4	NA
AstroPower, Inc.	APx-140-MCB	140W Apex Silicon Film Module w/MC connectors (was APx-140) B	120.4	NA
AstroPower, Inc.	APx-140-MNA	140W Apex Silicon Film Module w/o connectors (was APx-140)	120.4	NA
AstroPower, Inc.	APx-140-MNB	140W Apex Silicon Film Module w/o connectors (was APx-140) B	120.4	NA
AstroPower, Inc.	APx-45	45W Apex Module	38.8	NA
AstroPower, Inc.	APx-50	50W Apex Module	43.2	NA
AstroPower, Inc.	APx-65	65W Apex Module	56	NA
AstroPower, Inc.	APX-75	75W Apex Module	64.8	NA
AstroPower, Inc.	APx-75	75W Apex Module	64.8	NA
AstroPower, Inc.	APX-80	80W Apex Module	69.2	NA
AstroPower, Inc.	APX-90	90W Apex Module	77.8	NA
AstroPower, Inc.	APx050-MNB	50W Apex Module w/o connectors (was APx-50) B	42.9	NA

AstroPower, Inc.	LAP-425	425W Single Crystal Large Area Panel, frameless	378	NA
AstroPower, Inc.	LAP-440	440W Single Crystal Large Area Panel, frameless	391.8	NA
AstroPower, Inc.	LAP-460	460W Single Crystal Large Area Panel, frameless	409.9	NA
AstroPower, Inc.	LAP-480	480W Single Crystal Large Area Panel, frameless	428.2	NA
AstroPower, Inc.	LAPX-300	300W Apex Large Area Panel, Frameless	259.2	NA
Atlantis Energy, Inc.	AP-F	11.8W Shingle Module (AstroPower cells)	10.7	NA
Atlantis Energy, Inc.	AP-G	12.0W Shingle Module (AstroPower cells)	10.8	NA
Atlantis Energy, Inc.	AP-H	12.2W Shingle Module (AstroPower cells)	11	NA
Atlantis Energy, Inc.	SM-II	12.2W Shingle Module (Siemens cells)	11	NA
Atlantis Energy, Inc.	SP-A	13.3W Shingle Module (Sharp cells)	11.4	NA
Atlantis Energy, Inc.	SP-B	12.7W Shingle Module (Sharp cells)	10.9	NA
Atlantis Energy, Inc.	SP-C	11.9W Shingle Module (Sharp cells)	10.2	NA
Atlantis Energy, Inc.	SX-D	11.6W Shingle Module (Solarex cells)	10.5	NA
Atlantis Energy, Inc.	SX-E	11.0W Shingle Module (Solarex cells)	9.9	NA
Baoding Yingli New Energy Resources Co. Ltd.	110(17)P1447X663	110W Crystalline Silicon Solar Cells	96.6	NA
Baoding Yingli New Energy Resources Co. Ltd.	120(17)P1447X663	120W Crystalline Silicon Solar Cells	105.6	NA
Baoding Yingli New Energy Resources Co. Ltd.	30(17)P754X350	30W Crystalline Silicon Solar Cells	26.3	NA
Baoding Yingli New Energy Resources Co. Ltd.	40(17)P516X663	40W Crystalline Silicon Solar Cells	35.1	NA
Baoding Yingli New Energy Resources Co. Ltd.	50(17)P974X453	50W Crystalline Silicon Solar Cells	43.9	NA

(Continued)

TABLE C.1 LIST OF ELIGIBLE PHOTOVOLTAIC MODULES CALIFORNIA ENERGY COMMISSION EMERGING RENEWABLES PROGRAM (FEBRUARY 2005) (Continued)

MANUFACTURER NAME	MODULE MODEL NUMBER	DESCRIPTION	CEC PTC* RATING	NOTES
Baoding Yingli New Energy Resources Co. Ltd.	75(17)P1172X541	75W Crystalline Silicon Solar Cells	65.9	NA
Baoding Yingli New Energy Resources Co. Ltd.	85(17)P1172X541	85W Crystalline Silicon Solar Cells	74.9	NA
BP Solar	BP SX 140S	140W 24V Polycrystalline Module w/multicontact conn.	122.1	NA
BP Solar	BP SX 150S	150W 24V Polycrystalline Module w/multicontact conn.	131.1	NA
BP Solar	BP2140S	140W 24V Single Crystal Module w/multicontact conn.	122.1	NA
BP Solar	BP2150S	150W 24V Single Crystal Module w/multicontact conn.	131.1	NA
BP Solar	BP270U	70W BP Solar Single Crystal Module (universal frame)	61	NA
BP Solar	BP270UL	70W Single Crystal Module	61	NA
BP Solar	BP275U	75W Single Crystal Module (universal frame)	65.5	NA
BP Solar	BP275UL	75W Single Crystal Module	65.5	NA
BP Solar	BP3115S	115W 12V Polycrystalline Module w/multicontact conn.	101.7	NA
BP Solar	BP3115U	115W 12V Polycrystalline Module, universal frame	101.7	NA
BP Solar	BP3123XR	123W Polycrystalline Module w/multicontact conn.	108.2	NA

BP Solar	BP3125S	125W 12V Polycrystalline Module w/multicontact conn.	110.8	NA
BP Solar	BP3125U	125W 12V Polycrystalline Module, universal frame	110.8	NA
BP Solar	BP3126XR	126W Polycrystalline Module w/multicontact conn.	110.8	NA
BP Solar	BP3140B	140W 24V Polycrystalline Module, New AR, multicontact; bronze frame	123.8	NA
BP Solar	BP3140S	140W 24V Polycrystalline Module, New AR w/multicontact conn.	123.8	NA
BP Solar	BP3150B	150 W 24V Polycrystalline Module, New AR, multicontact; bronze frame	132.9	NA
BP Solar	BP3150S (2003+)	150W (2003 Rating) 24V Polycrystalline Module, New AR w/multicontact conn.	133	NA
BP Solar	BP3160B (2003+)	160W (2003 Rating) 24V Polycrystalline Module, New AR, multicontact; bronze frame	142.1	NA
BP Solar	BP3160QS	160W 16V Polycrystalline Module w/multicontact conn.	142	NA
BP Solar	BP3160S (2003+)	160W (2003 Rating) 24V Polycrystalline Module, New AR w/multicontact conn.	142.1	NA
BP Solar	BP360U	60W 12V Polycrystalline Module, universal frame	53	NA
BP Solar	BP365U	65W 12V Polycrystalline Module, universal frame	57.6	NA
BP Solar	BP375S (2003+)	75W (2003 Rating) Polycrystalline Module (universal frame), new AR w/multicontact conn.	66.4	NA
BP Solar	BP375U (2003+)	75W (2003 Rating) Polycrystalline Module (universal frame), new AR	66.4	NA

(Continued)

TABLE C.1 LIST OF ELIGIBLE PHOTOVOLTAIC MODULES CALIFORNIA ENERGY COMMISSION EMERGING RENEWABLES PROGRAM (FEBRUARY 2005) (Continued)

MANUFACTURER NAME	MODULE MODEL NUMBER	DESCRIPTION	CEC PTC* RATING	NOTES
BP Solar	BP380S (2003+)	80W (2003 Rating) Polycrystalline Module (universal frame), new AR w/multicontact conn.	71	NA
BP Solar	BP380U (2003+)	80W (2003 Rating) Polycrystalline Module (universal frame), new AR	71	NA
BP Solar	BP4150H	150W 24V Single Crystal Module (universal frame), new AR	132.5	NA
BP Solar	BP4150S	150W 24V Single Crystal Module, New AR w/multicontact conn.	132.6	NA
BP Solar	BP4160H	160W 24V Single Crystal Module (universal frame), new AR	141.6	NA
BP Solar	BP4160S	160W 24V Single Crystal Module, New AR w/multicontact conn.	141.7	NA
BP Solar	BP4165B	165W 24V Monocrystalline Module, multicontact; bronze frame	146.1	NA
BP Solar	BP4165S	165W 24V Monocrystalline Module w/multicontact conn.	146.1	NA
BP Solar	BP4170H	170W 24V Single Crystal Module (universal frame), new AR	150.7	NA
BP Solar	BP4170S	170W 24V Single Crystal Module (universal frame), new AR, Multicontact	150.7	NA
BP Solar	BP4175B	175W 24V Monocrystalline Module, multicontact; bronze frame	155.2	NA
BP Solar	BP4175I	175W 24V Monocrystalline Module w/multicontact conn.; integral frame	155.2	NA

BP Solar	BP4175S	175W 24V Monocrystalline Module w/multicontact conn.	155.2	NA
BP Solar	BP475S (2003+)	75W (2003 Rating) Single Crystal Module (universal frame), new AR, multicontact conn.	66.3	NA
BP Solar	BP475U (2003+)	75W (2003 Rating) Single Crystal Module (universal frame), new AR	66.3	NA
BP Solar	BP480S (2003+)	80W (2003 Rating) Single Crystal Module (universal frame), new AR w/multicontact conn.	70.8	NA
BP Solar	BP480U (2003+)	80W (2003 Rating) Single Crystal Module (universal frame), new AR	70.8	NA
BP Solar	BP485H	85W Single Crystal Module (universal frame), new AR	75.3	NA
BP Solar	BP485S	85W Single Crystal Module (universal frame), new AR, multicontact	75.3	NA
BP Solar	BP485U	85W Single Crystal Module (universal frame), new AR	75.3	NA
BP Solar	BP5160S (2003+)	160W (2003 Rating) 24V Buried Grid Single Crystal Module w/multicontact conn.	141.3	NA
BP Solar	BP5170S (2003+)	170W (2003 Rating) 24V Buried Grid Single Crystal Module w/multicontact conn.	150.4	NA
BP Solar	BP580U (2003+)	80W (2003 Rating) Buried Grid Single Crystal Module (universal frame)	70.6	NA
BP Solar	BP585DB	85W Buried Grid Single Crystal Module, multicontact; bronze frame	75.1	NA
BP Solar	BP585KD (2003+)	85W (2003 Rating) Buried Grid Single Crystal Module (frameless) with special fasteners	75.1	NA

(Continued)

445

TABLE C.1   LIST OF ELIGIBLE PHOTOVOLTAIC MODULES CALIFORNIA ENERGY COMMISSION EMERGING RENEWABLES PROGRAM (FEBRUARY 2005) (Continued)

MANUFACTURER NAME	MODULE MODEL NUMBER	DESCRIPTION	CEC PTC* RATING	NOTES
BP Solar	BP585S	85W 12V Buried Grid Single Crystal Module w/multicontact conn.	75.1	NA
BP Solar	BP585U (2003+)	85W (2003 Rating) Buried Grid Single Crystal Module (universal frame)	75.1	NA
BP Solar	BP585UL (2003+)	85W (2003 Rating) Buried Grid Single Crystal Module	75.1	NA
BP Solar	BP590UL (2003+)	90W (2003 Rating) Buried Grid Single Crystal Module	79.7	NA
BP Solar	BP7170S	170W 24V Saturn Single Crystal Module w/multicontact conn.	151.1	NA
BP Solar	BP7175S	175W 24V Saturn Single Crystal Module w/multicontact conn.	154.9	NA
BP Solar	BP7180S	180W 24V Saturn Single Crystal Module w/multicontact conn.	160.2	NA
BP Solar	BP7185S	185W 24V Saturn Single Crystal Module w/multicontact conn.	164	NA
BP Solar	BP785S	85W 12V Saturn Single Crystal Module w/multicontact conn.	75.5	NA
BP Solar	BP790DB	90W 12V Saturn Single Crystal Module, dark frame	80	NA
BP Solar	BP790S	90W 12V Saturn Single Crystal Module w/multicontact conn.	80	NA
BP Solar	BP790U	90W 12V Saturn Single Crystal Module, universal frame	80	NA

BP Solar	BP845I	45W Millennia 2J a-Si Module (medium voltage, integral frame)	42.4	NA
BP Solar	BP850I	50W Millennia 2J a-Si Module (medium voltage, integral frame)	47.1	NA
BP Solar	BP855I	55W Millennia 2J a-Si Module (medium voltage, integral frame)	51.9	NA
BP Solar	BP970B	70W Thin-film CdTe Laminate w/mounting brackets	62.3	NA
BP Solar	BP970I	70W Thin-film CdTe Module (Integra frame)	62.2	NA
BP Solar	BP980B	80W Thin-film CdTe Laminate w/mounting brackets	71.3	NA
BP Solar	BP980I	80W Thin-film CdTe Module (Integra frame)	71.2	NA
BP Solar	BP990B	90W Thin-film CdTe Laminate w/mounting brackets	80.4	NA
BP Solar	BP990I	90W Thin-film CdTe Module (Integra frame)	80.3	NA
BP Solar	MST-43I	43W Millennia 2J a-Si Module (med. voltage, Integra frame)	40.5	NA
BP Solar	MST-43LV	43W Millennia 2J a-Si Module (low voltage, universal frame)	40.5	NA
BP Solar	MST-43MV	43W Millennia 2J a-Si Module (med. voltage, universal frame)	40.5	NA
BP Solar	MST-45LV	45W Millennia 2J a-Si Module (low voltage, universal frame)	42.4	NA
BP Solar	MST-45MV	45W Millennia 2J a-Si Module (medium voltage, universal frame)	42.4	NA
BP Solar	MST-50I	50W Millennia 2J a-Si Module (med. voltage, Integra frame)	47.1	NA

(Continued)

TABLE C.1 LIST OF ELIGIBLE PHOTOVOLTAIC MODULES CALIFORNIA ENERGY COMMISSION EMERGING RENEWABLES PROGRAM (FEBRUARY 2005) (Continued)

MANUFACTURER NAME	MODULE MODEL NUMBER	DESCRIPTION	CEC PTC° RATING	NOTES
BP Solar	MST-50LV	50W Millennia 2J a-Si Module (low voltage, universal frame)	47.1	NA
BP Solar	MST-50MV	50W Millennia 2J a-Si Module (med. voltage, universal frame)	47.1	NA
BP Solar	MST-55MV	55W Millennia 2J a-Si Module (med. Voltage, universal frame)	51.9	NA
BP Solar	MSX-110	110W Solarex Polycrystalline Module	95.6	NA
BP Solar	MSX-120	120W Solarex Polycrystalline Module	104.5	NA
BP Solar	MSX-240	240W Solarex Polycrystalline Module	209.1	NA
BP Solar	MSX-50	50W Solarex Polycrystalline Module	43.5	NA
BP Solar	MSX-56	56W Solarex Polycrystalline Module	48.7	NA
BP Solar	MSX-60	60W Solarex Polycrystalline Module	52.2	NA
BP Solar	MSX-64	64W Solarex Polycrystalline Module	55.8	NA
BP Solar	MSX-77	77W Solarex Polycrystalline Module	67	NA
BP Solar	MSX-80U	80W Solarex Polycrystalline Module (universal frame)	69.7	NA
BP Solar	MSX-83	83W Solarex Polycrystalline Module	72.3	NA
BP Solar	SX-110S	110W Solarex poly-Si Module (univ. frame, multicontact conn.)	95.6	NA
BP Solar	SX-110U	110W Solarex poly-Si Module (univ. frame)	95.6	NA
BP Solar	SX-120S	120W Solarex poly-Si Module (univ. frame, multicontact conn.)	104.6	NA

Manufacturer	Model	Description		
BP Solar	SX-120U	120W Solarex poly-Si Module (univ. frame)	104.6	NA
BP Solar	SX-160S	160W 24V Polycrystalline Module w/multicontact connection	142	NA
BP Solar	SX-40D	40W Solarex poly-Si Module (direct-mount frame)	34.8	NA
BP Solar	SX-40M	40W Solarex poly-Si Module (multimount frame)	34.8	NA
BP Solar	SX-40U	40W Solarex poly-Si Module (univ. frame)	34.8	NA
BP Solar	SX-50D	50W Solarex poly-Si Module (direct-mount frame)	43.5	NA
BP Solar	SX-50M	50W Solarex poly-Si Module (multimount frame)	43.5	NA
BP Solar	SX-50U	50W Solarex poly-Si Module (univ. frame)	43.5	NA
BP Solar	SX-55D	55W Solarex poly-Si Module (direct-mount frame)	47.8	NA
BP Solar	SX-55U	55W Solarex poly-Si Module (univ. frame)	47.8	NA
BP Solar	SX-60D	60W Solarex poly-Si Module (direct-mount frame)	52.2	NA
BP Solar	SX-60U	60W Solarex poly-Si Module (univ. frame)	52.2	NA
BP Solar	SX-65D	65W Solarex poly-Si Module (Direct-mount frame)	56.7	NA
BP Solar	SX-65U	65W Solarex poly-Si Module (univ. frame)	56.7	NA
BP Solar	SX-75	75W Solarex poly-Si Module	65.2	NA
BP Solar	SX-75TS	75W Solarex poly-Si Module (125mm cells, Low-Profile/MC)	65.4	NA
BP Solar	SX-75TU	75W Solarex poly-Si Module (125mm cells, SPJB)	65.4	NA
BP Solar	SX-80	80W Solarex poly-Si Module	69.7	NA
BP Solar	SX-85	85W Solarex poly-Si Module	74.1	NA
BP Solar	SX140B	140W 24V Polycrystalline Module w/multicontact, bronze frame	123.8	NA
BP Solar	SX150B	150W 24V Polycrystalline Module w/multicontact, bronze frame	132.9	NA
BP Solar	SX160B	160W 24V Polycrystalline Module w/multicontact, bronze frame	142	NA

(Continued)

TABLE C.1 LIST OF ELIGIBLE PHOTOVOLTAIC MODULES CALIFORNIA ENERGY COMMISSION EMERGING RENEWABLES PROGRAM (FEBRUARY 2005) (Continued)

MANUFACTURER NAME	MODULE MODEL NUMBER	DESCRIPTION	CEC PTC* RATING	NOTES
BP Solar	TF-80B	80W Thin-film CdTe Laminate w/mounting brackets	71.3	NA
BP Solar	TF-80I	80W Thin-film CdTe Module (Integra frame)	71.2	NA
BP Solar	TF-90B	90W Thin-film CdTe Laminate w/mounting brackets	80.4	NA
BP Solar	TF-90I	90W Thin-film CdTe Module (Integra frame)	80.3	NA
BP Solar	VLX-53	53W Solarex Value-Line poly-Si Module	46.2	NA
BP Solar	VLX-80	80W Solarex Value-Line poly-Si Module	69.7	NA
Dunasolar Inc.	DS-30	30W Unframed 2J a-Si Module	28.8	NA
Dunasolar Inc.	DS-40	40W Unframed 2J a-Si Module	38.4	NA
Energy Photovoltaics, Inc.	EPV-30	30W Unframed 2J a-Si Module	28.8	NA
Energy Photovoltaics, Inc.	EPV-40	40W Unframed 2J a-Si Module	38.4	NA
Evergreen Solar	E-25	25W String Ribbon poly-Si Module	22.2	NA
Evergreen Solar	E-28	28W String Ribbon poly-Si Module	24.9	NA
Evergreen Solar	E-30	30W String Ribbon poly-Si Module	26.7	NA
Evergreen Solar	E-50	50W String Ribbon poly-Si Module	44.4	NA
Evergreen Solar	E-56	56W String Ribbon poly-Si Module	49.8	NA
Evergreen Solar	E-60	60W String Ribbon poly-Si Module	53.4	NA
Evergreen Solar	EC-102	102W Cedar Line Module	91.2	NA
Evergreen Solar	EC-110	110W Cedar Line Module	98.4	NA
Evergreen Solar	EC-115	115W String Ribbon Cedar Line Module	103.1	NA
Evergreen Solar	EC-47	47W Cedar Line Module	41.9	NA
Evergreen Solar	EC-51	51W Cedar Line Module	45.6	NA

Evergreen Solar	EC-55	55W Cedar Line Module	49.2	NA
Evergreen Solar	EC-94	94W Cedar Line Module	83.9	NA
Evergreen Solar	ES-112	112W String Ribbon poly-Si AC Module (with Trace MS100)	99.7	NA
Evergreen Solar	ES-240	240W String Ribbon poly-Si AC Module (with AES MI-250)	213.8	NA
First Solar, LLC	FS-40	40W Thin-Film CdTe Laminate	38	NA
First Solar, LLC	FS-40D	40W Thin-Film CdTe Module w/ D-channel mounting rails	38	NA
First Solar, LLC	FS-45	45W Thin-Film CdTe laminate	42.8	NA
First Solar, LLC	FS-45D	45W Thin-Film CdTe Module with mounting rails	42.8	NA
First Solar, LLC	FS-50	50W/65V Thin-Film CdTe laminate	47.6	NA
First Solar, LLC	FS-50C	50W/65V Thin-Film CdTe laminate w/ C-channel mounting rails	47.6	NA
First Solar, LLC	FS-50D	50W Thin-Film CdTe Module with mounting rails	47.6	NA
First Solar, LLC	FS-50Z	50W/65V Thin-Film CdTe laminate w/ Z-channel mounting rails	47.6	NA
First Solar, LLC	FS-55	55W Thin-Film CdTe Laminate	52.4	NA
First Solar, LLC	FS-55D	55W Thin-Film CdTe Module w/ D-channel mounting rails	52.4	NA
First Solar, LLC	FS-60	60W Thin-Film CdTe laminate	57.2	NA
First Solar, LLC	FS-60D	60W Thin-Film CdTe Module with mounting rails	57.2	NA
GE Energy	GEPV-030-MNA	30W Single Crystal Module w/o connectors	26.5	NA
GE Energy	GEPV-030-MNB	30W Single Crystal Module w/o connectors B	26.5	NA
GE Energy	GEPV-045-MNA	45W Single Crystal Module w/o connectors	39.7	NA

(Continued)

TABLE C.1 LIST OF ELIGIBLE PHOTOVOLTAIC MODULES CALIFORNIA ENERGY COMMISSION EMERGING RENEWABLES PROGRAM (FEBRUARY 2005) (Continued)

MANUFACTURER NAME	MODULE MODEL NUMBER	DESCRIPTION	CEC PTC* RATING	NOTES
GE Energy	GEPV-050-MNA	50W Single Crystal Module w/o connectors	44.2	NA
GE Energy	GEPV-050-MNB	50W Single Crystal Module w/o connectors B	44.2	NA
GE Energy	GEPV-055-GCA	55W Single Crystal Module w/MC connectors	48.9	NA
GE Energy	GEPV-055-GCB	55W Single Crystal Module w/MC connectors B	48.9	NA
GE Energy	GEPV-065-MNA	65W Single Crystal Module w/o connectors	57.7	NA
GE Energy	GEPV-065-MNB	65W Single Crystal Module w/o connectors B	57.7	NA
GE Energy	GEPV-070-MNA	70W Single Crystal Module w/o connectors	62.2	NA
GE Energy	GEPV-070-MNB	70W Single Crystal Module w/o connectors B	62.2	NA
GE Energy	GEPV-100-MCA	100W Single Crystal Module w/MC connectors	88.7	NA
GE Energy	GEPV-100-MCB	100W Single Crystal Module w/MC connectors B	88.7	NA
GE Energy	GEPV-100-MNA	100W Single Crystal Module w/o connectors	88.7	NA
GE Energy	GEPV-100-MNB	100W Single Crystal Module w/o connectors B	88.7	NA
GE Energy	GEPV-110-MCA	110W Single Crystal Module w/MC connectors	97.8	NA
GE Energy	GEPV-110-MCB	110W Single Crystal Module w/MC connectors B	97.8	NA
GE Energy	GEPV-110-MNA	110W Single Crystal Module w/o connectors	97.8	NA
GE Energy	GEPV-110-MNB	110W Single Crystal Module w/o connectors B	97.8	NA
GE Energy	GEPV-165-MCA	165W Single Crystal Module w/MC connectors	146.7	NA
GE Energy	GEPV-165-MCB	165W Single Crystal Module w/MC connectors B	146.7	NA
GE Energy	GEPV-173-MCA	173W Single Crystal Module w/MC connectors	154	NA
GE Energy	GEPV-173-MCB	173W Single Crystal Module w/MC connectors B	154	NA
Isofoton	I-100/12	100W Monocrystalline Rail-Mounted Module X2	89.5	NA

Isofoton	I-100/24	100W Monocrystalline Rail-Mounted Module	89.5	NA
Isofoton	I-106/12	106W Monocrystalline Rail-Mounted Module	95	NA
Isofoton	I-106/24	106W Monocrystalline Rail-Mounted Module X2	95	NA
Isofoton	I-110-24	110W Monocrystalline Rail-Mounted Module X2	98.7	NA
Isofoton	I-110/12	110W Monocrystalline Rail-Mounted Module	98.7	NA
Isofoton	I-130/12	130W Monocrystalline Rail-Mounted Module	116	NA
Isofoton	I-130/24	130W Monocrystalline Rail-Mounted Module X2	116	NA
Isofoton	I-140 R/12	140W Monocrystalline Rail-Mounted Module	125	NA
Isofoton	I-140 R/24	140W Monocrystalline Rail-Mounted Module X3	125	NA
Isofoton	I-140 S/12	140W Monocrystalline Rail-Mounted Module X4	125	NA
Isofoton	I-140 S/24	140W Monocrystalline Rail-Mounted Module X5	125	NA
Isofoton	I-150	150W Monocrystalline Rail-Mounted Module X2	134.3	NA
Isofoton	I-150 S/12	150W Monocrystalline Rail-Mounted Module	134.3	NA
Isofoton	I-150 S/24	150W Monocrystalline Rail-Mounted Module	134.3	NA
Isofoton	I-159	159W Monocrystalline Rai-Mounted Module	142.6	NA
Isofoton	I-165	165W Monocrystalline Rail-Mounted Module	148.1	NA
Isofoton	I-36	36W Monocrystalline Rail-Mounted Module	32.2	NA
Isofoton	I-50	50W Monocrystalline Rail-Mounted Module	44.8	NA
Isofoton	I-53	53W Monocrystalline Rail-Mounted Module	47.5	NA
Isofoton	I-55	55W Monocrystalline Rail-Mounted Module	49.4	NA
Isofoton	I-65	65W Monocrystalline Rail-Mounted Module	58	NA
Isofoton	I-70 R	70W Monocrystalline Rail-Mounted Module	62.5	NA
Isofoton	I-70 S	70W Monocrystalline Rail-Mounted Module X2	62.5	NA
Isofoton	I-75	75W Monocrystalline Rail-Mounted Module	67.1	NA
Isofoton	I-94/12	94W Monocrystalline Rail-Mounted Module X2	84.2	NA

(Continued)

TABLE C.1 LIST OF ELIGIBLE PHOTOVOLTAIC MODULES CALIFORNIA ENERGY COMMISSION EMERGING RENEWABLES PROGRAM (FEBRUARY 2005) (Continued)

MANUFACTURER NAME	MODULE MODEL NUMBER	DESCRIPTION	CEC PTC* RATING	NOTES
Isofoton	I-94/24	94W Monocrystalline Rail-Mounted Module	84.2	NA
Kaneka Corporation	CSA201	58W a-Si Module	54.1	NA
Kaneka Corporation	CSA211	58W a-Si Module (CSA)	54.1	NA
Kaneka Corporation	CSB211	58W a-Si Module (CSB)	54.1	NA
Kaneka Corporation	GSA211	60W a-Si Module	56	NA
Kaneka Corporation	LSU205	58W a-Si Module	54.1	NA
Kaneka Corporation	TSA211	116W a-Si Twin Type Module	108.2	NA
Kaneka Corporation	TSB211	116W a-Si Twin Type Module B	108.2	NA
Kaneka Corporation	TSC211	120W a-Si Twin Type Module (TSC)	111.9	NA
Kaneka Corporation	TSD211	120W a-Si Twin Type Module (TSD)	111.9	NA
Kyocera Solar, Inc.	KC120-1	120W High-Efficiency Multicrystal PV Module	105.7	NA
Kyocera Solar, Inc.	KC125G	125W High-Efficiency Multicrystal PV Module	111.8	NA
Kyocera Solar, Inc.	KC158G	158W High-Efficiency Multicrystal PV Module	139.7	NA
Kyocera Solar, Inc.	KC167G	167W High-Efficiency Multicrystal PV Module	149.6	NA
Kyocera Solar, Inc.	KC187G	187W Multicrystal PV Module, Deep Blue	167.4	NA
Kyocera Solar, Inc.	KC50	50W High-Efficiency Multicrystal PV Module	43.9	NA
Kyocera Solar, Inc.	KC60	60W High-Efficiency Multicrystal PV Module	52.7	NA
Kyocera Solar, Inc.	KC70	70W High-Efficiency Multicrystal PV Module	61.6	NA
Kyocera Solar, Inc.	KC80	80W High-Efficiency Multicrystal PV Module	70.4	NA
Matrix Solar/Photowatt	PW1000-100	100W Large-Scale Dual-Voltage multi-Si Module	90	NA
Matrix Solar/Photowatt	PW1000-105	105W Large-Scale Dual-Voltage multi-Si Module	94.6	NA

Matrix Solar/Photowatt	PW1000-90	90W Large-Scale Dual-Voltage Multi-Si Module	80.9	NA
Matrix Solar/Photowatt	PW1000-95	95W Large-Scale Dual-Voltage Multi-Si Module	85.4	NA
Matrix Solar/Photowatt	PW1250-115	115W Large-Scale Dual-Voltage Multi-Si Module	103.7	NA
Matrix Solar/Photowatt	PW1250-125	125W Large-Scale Dual-Voltage Multi-Si Module	112.9	NA
Matrix Solar/Photowatt	PW1250-135	135W Large-Scale Dual-Voltage Multi-Si Module	122.2	NA
Matrix Solar/Photowatt	PW1650-155	155W Large-Scale Dual-Voltage Multi-Si Module	139.9	NA
Matrix Solar/Photowatt	PW1650-165	165W Large-Scale Dual-Voltage Multi-Si Module	149	NA
Matrix Solar/Photowatt	PW1650-175	175W Large-Scale Dual-Voltage Multi-Si Module	158.3	NA
Matrix Solar/Photowatt	PW750-70	70W Large-Scale Multi-Si Module	62.9	NA
Matrix Solar/Photowatt	PW750-75	75W Large-Scale Multi-Si Module	67.5	NA
Matrix Solar/Photowatt	PW750-80	80W Large-Scale Multi-Si Module	72	NA
Matrix Solar/Photowatt	PW750-90	90W Large-Scale Multi-Si Module	81.2	NA
MC Solar	BP970B	70W Thin-Film CdTe Laminate w/mounting brackets	62.3	NA
MC Solar	BP980B	80W Thin-Film CdTe Laminate w/mounting brackets	71.3	NA
MC Solar	BP990B	90W Thin-Film CdTe Laminate w/mounting brackets	80.4	NA
MC Solar	TF-80B	80W Thin-Film CdTe Laminate w/mounting brackets (now BP980B)	71.3	NA
MC Solar	TF-90B	90W Thin-Film CdTe Laminate w/mounting brackets (now BP990B)	80.4	NA
Midway Labs, Inc.	MLB3416-115	115W Concentrator (335x) Module	105	NA
Mitsubishi Electric Corporation	PV-MF110EC3	110W Polycrystalline Lead-free Solder w/o cable	98.4	NA
Mitsubishi Electric Corporation	PV-MF120EC3	120W Polycrystalline Lead-free Solder w/o cable	107.6	NA

(Continued)

TABLE C.1 LIST OF ELIGIBLE PHOTOVOLTAIC MODULES CALIFORNIA ENERGY COMMISSION EMERGING RENEWABLES PROGRAM (FEBRUARY 2005) (Continued)

MANUFACTURER NAME	MODULE MODEL NUMBER	DESCRIPTION	CEC PTC° RATING	NOTES
Mitsubishi Electric Corporation	PV-MF125E	125W Polycrystalline Module w/multicontact connectors	110.7	NA
Mitsubishi Electric Corporation	PV-MF125EA2LF	125W Polycrystalline Lead-free Solder Module w/MC connector	110.7	NA
Mitsubishi Electric Corporation	PV-MF130E	130W Polycrystalline Module w/multicontact connectors	115.2	NA
Mitsubishi Electric Corporation	PV-MF130EA2LF	130W Polycrystalline Lead-free Solder Module w/MC connector	115.2	NA
Mitsubishi Electric Corporation	PV-MF160EB3	160W Polycrystalline Lead-free Solder Module w/MC connector	142.4	NA
Mitsubishi Electric Corporation	PV-MF165EB3	165W Polycrystalline Lead-free Solder Module w/MC connector	146.9	NA
Mitsubishi Electric Corporation	PV-MF170EB3	170W Polycrystalline Lead-free Solder Module w/MC connector	152.5	NA
Pacific Solar Pty Limited	PP-USA-213-B5	150W BP Solar 2150L SunEmpower Modular Mount	131.8	NA
Pacific Solar Pty Limited	PP-USA-213-L6	160W BP Solar 5160L SunEmpower Modular Mount	140.9	NA
Pacific Solar Pty Limited	PP-USA-213-L7	170W BP Solar 5170L SunEmpower Modular Mount	149.9	NA
Pacific Solar Pty Limited	PP-USA-213-N5	150W BP Solar 4150L SunEmpower Modular Mount	131.8	NA
Pacific Solar Pty Limited	PP-USA-213-N6	160W BP Solar 4160L SunEmpower Modular Mount	140.9	NA

Pacific Solar Pty Limited	PP-USA-213-N7	170W BP Solar 4170L SunEmpower Modular Mount	149.9	NA
Pacific Solar Pty Limited	PP-USA-213-P5	150W BP Solar 3150L SunEmpower Modular Mount	131.8	NA
Pacific Solar Pty Limited	PP-USA-213-P6	160W BP Solar 3160L SunEmpower Modular Mount	140.9	NA
Pacific Solar Pty Limited	PP-USA-213-S5	150W Shell Solar SP-150-PL, -PLC SunEmpower Modular Mount	135	NA
Powerlight Corp.	PL-AP-120L	120W PowerGuard Roof Tile (AstroPower)	104.9	NA
Powerlight Corp.	PL-AP-130	130W PowerGuard Roof Tile (AstroPower AP-130)	115.1	NA
Powerlight Corp.	PL-AP-65	One AstroPower AP-65 laminate mounted on one PowerGuard backerboard	56.6	NA
Powerlight Corp.	PL-AP-65 Double Module	Two AstroPower AP-65 laminates mounted on one PowerGuard backerboard	113.2	NA
Powerlight Corp.	PL-AP-75 Double Module	150W PowerGuard Roof Tile (two AstroPower modules)	131.1	NA
Powerlight Corp.	PL-APx-110-SL	110W PowerGuard Roof Tile (AstroPower)	93.7	NA
Powerlight Corp.	PL-ASE-100	100W PowerGuard Roof Tile (ASE Americas)	88.6	NA
Powerlight Corp.	PL-BP-2150S	150W PowerGuard Roof Tile (BP Solar)	129.2	NA
Powerlight Corp.	PL-BP-3160L	160W PowerGuard Roof Tile (BP Solar)	139.2	NA
Powerlight Corp.	PL-BP-380L Double Module	160W PowerGuard Roof Tile (Two BP-380L modules)	139.1	NA
Powerlight Corp.	PL-BP-485L Double Module	170W Powerguard Roof Tile (Two BP-485L modules)	148	NA
Powerlight Corp.	PL-BP-TF-80L	80W PowerGuard Roof Tile (BP Solar)	70.7	NA
Powerlight Corp.	PL-FS-415-A	50W PowerGuard Roof Tile (First Solar)	47.4	NA
Powerlight Corp.	PL-KYOC-FL120-1B	120W PowerGuard Roof Tile (Kyocera)	103.8	NA

(Continued)

TABLE C.1 LIST OF ELIGIBLE PHOTOVOLTAIC MODULES CALIFORNIA ENERGY COMMISSION EMERGING RENEWABLES PROGRAM (FEBRUARY 2005) (Continued)

MANUFACTURER NAME	MODULE MODEL NUMBER	DESCRIPTION	CEC PTC° RATING	NOTES
Powerlight Corp.	PL-KYOC-FL125	125W PowerGuard Roof Tile (Kyocera)	110.8	NA
Powerlight Corp.	PL-KYOC-FL158	158W PowerGuard Roof Tile (Kyocera)	138.2	NA
Powerlight Corp.	PL-KYOC-FL167	167W PowerGuard Roof Tile (Kyocera)	148	NA
Powerlight Corp.	PL-MST-43	43W PowerGuard Roof Tile (Solarex a-Si)	40.5	NA
Powerlight Corp.	PL-MSX-120	120W PowerGuard Roof Tile (Solarex poly-Si)	103.7	NA
Powerlight Corp.	PL-PW-750	75-80W PowerGuard Roof Tile (Matrix Solar/Photowatt)	69.9	NA
Powerlight Corp.	PL-PW-750 Double Module	150W PowerGuard Roof Tile (two Matrix Solar Photowatt modules)	135.3	NA
Powerlight Corp.	PL-SHAR-ND-N6E1D	146W PowerFuard Roof Tile (Sharp Corp.)	125.1	NA
Powerlight Corp.	PL-SP-135	135W (Pre-2003 Rating) PowerGuard Roof Tile (Siemens)	119.1	NA
Powerlight Corp.	PL-SP-135 (2003+)	135W (2003 Rating) PowerGuard Roof Tile (Siemens)	119.8	NA
Powerlight Corp.	PL-SP-150-24L	150W PowerGuard Roof Tile (Rectangular Siemens)	133.5	NA
Powerlight Corp.	PL-SP-150-CPL	150W PowerGuard Roof Tile (Siemens)	133.6	NA
Powerlight Corp.	PL-SP-70 Double Module	Two Sharp SP-70 modules on one PowerGuard tile	124.8	NA
Powerlight Corp.	PL-SP-75	75W PowerGuard Roof Tile (Siemens)	66.8	NA
Powerlight Corp.	PL-SP-75 Double Module	150W PowerGuard Roof Tile (Two Siemens modules)	133.7	NA
Powerlight Corp.	PL-SQ75-CPL	75W Power Guard Roof Tile (Shell)	65.1	NA

Powerlight Corp.	PL-SQ75-CPL Double Module	150W PowerGuard Roof Tile (Shell)	130.3	NA
Powerlight Corp.	PL-SQ77-CPL Double Module	154W Double Module PowerGuard Roof Tile (Shell Solar)	134.6	NA
Powerlight Corp.	PL-SQ85-P DOUBLE MODULE	170W PowerGuard Roof Tile (Two Shell modules)	152.3	NA
Powerlight Corp.	PL-SY-HIP-190BA2	190W PowerGuard Roof Tile (Sanyo)	177.5	NA
Powerlight Corp.	PL-SY-HIP-190CA2	190W PowerGuard Roof Tile (Sanyo)	177.5	NA
Powerlight Corp.	PL-SY-HIP-H552BA2	175W PowerGuard Roof Tile (Sanyo)	162.2	NA
RWE SCHOTT Solar	ASE-250DGF/17	250W/17V EFG Module, framed	223.7	NA
RWE SCHOTT Solar	ASE-250DGF/50	250W/50V EFG Module, framed	223.7	NA
RWE SCHOTT Solar	ASE-270DGF/17	270W/17V EFG Module, framed	242	NA
RWE SCHOTT Solar	ASE-270DGF/50	270W/50V EFG Module, framed	242	NA
RWE SCHOTT Solar	SAPC-175	175W Monocrystalline Silicon Module	154.4	NA
Sanyo Electric Co. Ltd.	HIP-167BA	167W HIT Hybrid a-Si/c-Si Solar Cell Module	156.7	NA
Sanyo Electric Co. Ltd.	HIP-175BA3	175W HIT Hybrid a-Si/c-Si Solar Cell Module	163.3	NA
Sanyo Electric Co. Ltd.	HIP-175BA5	175W HIT Hybrid a-Si/c-Si Solar Cell Module (5)	163.3	NA
Sanyo Electric Co. Ltd.	HIP-180BA	180W HIT Hybrid a-Si/c-Si Solar Cell Module	169.1	NA
Sanyo Electric Co. Ltd.	HIP-180BA3	180W HIT Hybrid a-Si/c-Si Solar Cell Module (3)	168	NA
Sanyo Electric Co. Ltd.	HIP-180BA5	180W HIT Hybrid a-Si/c-Si Solar Cell Module (5)	168	NA
Sanyo Electric Co. Ltd.	HIP-190BA	190W HIT Hybrid a-Si/c-Si Solar Cell Module	178.7	NA

(Continued)

TABLE C.1 LIST OF ELIGIBLE PHOTOVOLTAIC MODULES CALIFORNIA ENERGY COMMISSION EMERGING RENEWABLES PROGRAM (FEBRUARY 2005) (Continued)

MANUFACTURER NAME	MODULE MODEL NUMBER	DESCRIPTION	CEC PTC* RATING	NOTES
Sanyo Electric Co. Ltd.	HIP-190BA1	190W HIT Hybrid a-Si/c-Si Solar Cell Module (std. j.b.)	178.7	NA
Sanyo Electric Co. Ltd.	HIP-190BA2	190W HIT Hybrid a-Si/c-Si Solar Cell Module (std. j.b. w/addl. wiring)	178.7	NA
Sanyo Electric Co. Ltd.	HIP-190BA3	190W HIT Hybrid a-Si/c-Si Solar Cell Module (3)	178.7	NA
Sanyo Electric Co. Ltd.	HIP-190BA5	190W HIT Hybrid a-Si/c-Si Solar Cell Module (5)	178.7	NA
Sanyo Electric Co. Ltd.	HIP-G751BA1	167W HIT Hybrid a-Si/c-Si Solar Cell Module (std. j.b.)	155.8	NA
Sanyo Electric Co. Ltd.	HIP-G751BA2	167W HIT Hybrid a-Si/c-Si Solar Cell Module (std. j.b. w/addl. wiring)	155.8	NA
Sanyo Electric Co. Ltd.	HIP-H552BA1	175W HIT Hybrid a-Si/c-Si Solar Cell Module (std. j.b.)	163.3	NA
Sanyo Electric Co. Ltd.	HIP-H552BA2	175W HIT Hybrid a-Si/c-Si Solar Cell Module (std. j.b. w/addl. wiring)	163.3	NA
Sanyo Electric Co. Ltd.	HIP-J54BA1	180W HIT Hybrid a-Si/c-Si Solar Cell Module (std. j.b.)	168.1	NA
Sanyo Electric Co. Ltd.	HIP-J54BA2	180W HIT Hybrid a-Si/c-Si Solar Cell Module (std. j.b. w/addl. wiring)	168.1	NA
Schott Applied Power Corp.	SAPC-123	123W Multisilicon Module	107.8	NA
Schott Applied Power Corp.	SAPC-165	165W Multicrystalline Silicon Module	144.8	NA

Schott Applied Power Corp.	SAPC-80	80W Multisilicon Module	70.2	NA
Schuco USA LP	S125-SP	130W Polycrystalline Module w/multicontact connectors	115.2	NA
Schuco USA LP	S158-SP	165W Polycrystalline Module w/multicontact connectors	146.9	NA
Schuco USA LP	S162-SP	170W Polycrystalline Lead-free Solder Module w/MC connector	152.5	NA
Sharp Corporation	ND-160U1Z	160 W, Multicrystalline Silicon Module	140.6	Changed Power temp Coefficient
Sharp Corporation	ND-167U1	167W Multisilicon Module	146.9	Changed Power temp Coefficient
Sharp Corporation	ND-167U3	167W Multisilicon Module (3)	146.9	Changed Power temp Coefficient
Sharp Corporation	ND-70ELU	70W Multicrystalline Silicon Module (left)	61.1	Changed Power temp Coefficient
Sharp Corporation	ND-70ERU	70W Multicrystalline Silicon Module (right)	61.1	Changed Power temp Coefficient
Sharp Corporation	ND-L3E1U	123W Multisilicon Module	108.1	Changed Power temp Coefficient
Sharp Corporation	ND-L3EJE	123W Multisilicon Module (w/junction box)	108.1	Changed Power temp Coefficient
Sharp Corporation	ND-N0ECU	140W Multisilicon Residential Module	123	Changed Power temp Coefficient
Sharp Corporation	ND-N6E1U	146W Multisilicon Module	128.3	Changed Power temp Coefficient
Sharp Corporation	ND-Q0E2U	160W Multisilicon Module	140.6	Changed Power temp Coefficient
Sharp Corporation	NE-165U1	165W Multisilicon Module (flat screw type, same as NE-Q5E2U)	145.2	Changed Power temp Coefficient

(Continued)

TABLE C.1   LIST OF ELIGIBLE PHOTOVOLTAIC MODULES CALIFORNIA ENERGY COMMISSION EMERGING RENEWABLES PROGRAM (FEBRUARY 2005) (Continued)

MANUFACTURER NAME	MODULE MODEL NUMBER	DESCRIPTION	CEC PTC° RATING	NOTES
Sharp Corporation	NE-80E1U	80W Multisilicon Module	70.4	Changed Power temp Coefficient
Sharp Corporation	NE-80EJE	80W Multisilicon Module (w/junction box)	70.4	Changed Power temp Coefficient
Sharp Corporation	NE-K125U1	125W Multisilicon Module (non-flat screw type, black color frame)	110	Changed Power temp Coefficient
Sharp Corporation	NE-K125U2	125W Multisilicon Module (flat screw type)	110.1	Changed Power temp Coefficient
Sharp Corporation	NE-Q5E1U	165W Multisilicon Module (non-flat screw type)	145.2	Changed Power temp Coefficient
Sharp Corporation	NE-Q5E2U	165W Multisilicon Module (flat screw type)	145.2	Changed Power temp Coefficient
Sharp Corporation	NT-175U1	175W Monocrystalline Silicon Module	154.2	Changed Power temp Coefficient
Sharp Corporation	NT-188U1	188W Single Crystal Silicon Module	166	Changed Power temp Coefficient
Sharp Corporation	NT-R5E1U	175 W Multisilicon Module	154.2	Changed Power temp Coefficient
Sharp Corporation	NT-S5E1U	185W Multisilicon Module	163.3	NA
Shell Solar Industries	SM110	110W PowerMax Module	99.2	NA
Shell Solar Industries	SP130-PC	130W PowerMax Monocrystalline Module w/cable assembly	116.6	NA
Shell Solar Industries	SP140-PC	140W PowerMax Monocrystalline Module w/cable assembly	125.8	NA

Manufacturer	Model	Description	Value	
Shell Solar Industries	SP150-PC	150W PowerMax Monocrystalline Module w/cable assembly	134.9	NA
Shell Solar Industries	SQ140-P	140W PowerMax Monocrystalline Module	123.4	NA
Shell Solar Industries	SQ140-PC	140W PowerMax Monocrystalline Module w/multicontact cable assembly	123.4	NA
Shell Solar Industries	SQ150-P	150W PowerMax Monocrystalline Module	132.5	NA
Shell Solar Industries	SQ150-PC	150W PowerMax Monocrystalline Module w/multicontact cable assembly	132.5	NA
Shell Solar Industries	SQ160-P	160W PowerMax Monocrystalline Module	141.5	NA
Shell Solar Industries	SQ160-PC	160W PowerMax Monocrystalline Module w/multicontact cable assembly	141.5	NA
Shell Solar Industries	SQ165-P	165W PowerMax Monocrystalline Module	149.1	NA
Shell Solar Industries	SQ165-PC	165W PowerMax Monocrystalline Module w/multicontact cable assembly	149.1	NA
Shell Solar Industries	SQ175-P	175W PowerMax Monocrystalline Module	158.3	NA
Shell Solar Industries	SQ175-PC	175W PowerMax Monocrystalline Module w/multicontact cable assembly	158.3	NA
Shell Solar Industries	SQ70	75W PowerMax Monocrystalline Module	61.8	NA
Shell Solar Industries	SQ75	75W PowerMax Monocrystalline Module	66.3	NA
Shell Solar Industries	SQ80	80W PowerMax Monocrystalline Module	70.8	NA
Shell Solar Industries	SQ80-P	80W PowerMax Monocrystalline Module	72.3	NA
Shell Solar Industries	SQ85-P	85W PowerMax Monocrystalline Module	76.9	NA
Siemens Solar Industries	SM-110	110W PowerMax Module	99.2	NA
Siemens Solar Industries	SM10	10W PowerMax Module	9	NA
Siemens Solar Industries	SM20	20W PowerMax Module	18	NA
Siemens Solar Industries	SM46	46W PowerMax Module	41.5	NA
Siemens Solar Industries	SM46J	46W PowerMax Module w/conduit-ready J-box	41.5	NA

(Continued)

TABLE C.1   LIST OF ELIGIBLE PHOTOVOLTAIC MODULES CALIFORNIA ENERGY COMMISSION EMERGING RENEWABLES PROGRAM (FEBRUARY 2005) (Continued)

MANUFACTURER NAME	MODULE MODEL NUMBER	DESCRIPTION	CEC PTC* RATING	NOTES
Siemens Solar Industries	SM50	50W PowerMax Module	45	NA
Siemens Solar Industries	SM50-H	50W 33 cell PowerMax Module	45.1	NA
Siemens Solar Industries	SM50-HJ	50W 33 cell PowerMax Module w/conduit-ready J-box	45.1	NA
Siemens Solar Industries	SM50-J	50W PowerMax Module w/conduit-ready J-box	45	NA
Siemens Solar Industries	SM55	55W PowerMax Module	49.6	NA
Siemens Solar Industries	SM55-J	55W PowerMax Module w/conduit-ready J-box	49.6	NA
Siemens Solar Industries	SM6	6 W PowerMax Module	5.4	NA
Siemens Solar Industries	SP130-24P	130W 24V PowerMax Module	116.7	NA
Siemens Solar Industries	SP140-24P	140W 24V PowerMax Module	125.8	NA
Siemens Solar Industries	SP150-24P	150W 24V PowerMax Module	134.9	NA
Siemens Solar Industries	SP18	18W 6V/12V PowerMax Module	16.2	NA
Siemens Solar Industries	SP36	36W 6V/12V PowerMax Module	32.4	NA
Siemens Solar Industries	SP65	65W 6V/12V PowerMax Module	58.4	NA
Siemens Solar Industries	SP70	70W 6V/12V PowerMax Module	62.9	NA
Siemens Solar Industries	SP75	75W 6V/12V PowerMax Module	67.5	NA
Siemens Solar Industries	SR100	100W 6V/12V PowerMax Module	89.9	NA
Siemens Solar Industries	SR50	50W 6V/12V PowerMax Module	44.9	NA
Siemens Solar Industries	SR90	90W 6V/12V PowerMax Module	80.8	NA

Siemens Solar Industries	ST36	36W 12V PowerMax Module	31.7	NA
Siemens Solar Industries	ST40	40W 12V PowerMax Module	35.3	NA
Solar Integrated Technologies	SR2001A	816W Flat Plate Single-Ply Roofing Membrane	771.6	NA
Solar Integrated Technologies	SR2004	1488W Flat Plate Single-Ply Roofing Membrane	1407	NA
Solar Integrated Technologies	SR2004A	744W Flat Plate Single-Ply Roofing Membrane	703.5	NA
Solar Integrated Technologies	SR372	372W Flat Plate Single-Ply Roofing Membrane	351.7	NA
Solec International, Inc.	S-055	55W Framed Crystalline Solar Electric Module	47.7	NA
Solec International, Inc.	S-100D	100W Dual Voltage Crystalline Solar Electric Module	86.5	NA
Solec International, Inc.	SQ-080	80W Framed Crystalline Solar Electric Module	68.9	NA
Solec International, Inc.	SQ-090	90W Framed Crystalline Solar Electric Module	77.8	NA
Spire Solar Chicago	SS75	75W Rail Mounted Monocrystalline Module	66.7	NA
Spire Solar Chicago	SSC 75	75W Rail Mounted Monocrystalline Module x	66.7	NA
Sunpower Corporation	SPR-200	200W Monocrystalline Module	180	NA
Sunpower Corporation	SPR-210	210W Monocrystalline Module	190.9	NA
Sunpower Corporation	SPR-90	90W Monocrystalline Module	81.8	NA
SunWize Technologies, LLC	SW 100	100W Monocrystalline PV Module	87.1	NA
SunWize Technologies, LLC	SW 110	110W Monocrystalline PV Module	96	NA
SunWize Technologies, LLC	SW 115	115W Monocrystalline PV Module	100.5	NA
SunWize Technologies, LLC	SW 120	120W Monocrystalline PV Module	105	NA
SunWize Technologies, LLC	SW 150L	150W Monocrystalline PV Module	130.5	NA
SunWize Technologies, LLC	SW 155L	155W Monocrystalline PV Module	134.9	NA
SunWize Technologies, LLC	SW 160L	160W Monocrystalline PV Module	139.4	NA
SunWize Technologies, LLC	SW 165L	165W Monocrystalline PV Module	143.9	NA
SunWize Technologies, LLC	SW 75	75W Monocrystalline PV Module	65.1	NA
SunWize Technologies, LLC	SW 85	85W Monocrystalline PV Module	73.9	NA

(Continued)

TABLE C-1 LIST OF ELIGIBLE PHOTOVOLTAIC MODULES CALIFORNIA ENERGY COMMISSION EMERGING RENEWABLES PROGRAM (FEBRUARY 2005) (Continued)

MANUFACTURER NAME	MODULE MODEL NUMBER	DESCRIPTION	CEC PTC* RATING	NOTES
SunWize Technologies, LLC	SW 90	90W Monocrystalline PV Module	78.4	NA
SunWize Technologies, LLC	SW 95	95W Monocrystalline PV Module	82.8	NA
United Solar Systems Corp.	ASR-120	120W Arch. Standing Seam 3J a-Si Module	110.9	NA
United Solar Systems Corp.	ASR-128	128W Arch. Standing Seam 3J a-Si Module	118.3	NA
United Solar Systems Corp.	ASR-136	136W Arch. Standing Seam 3J a-Si Module	130	NA
United Solar Systems Corp.	ASR-60	60W Arch. Standing Seam 3J a-Si Module	55.4	NA
United Solar Systems Corp.	ASR-64	64W Arch. Standing Seam 3J a-Si Module	59.1	NA
United Solar Systems Corp.	ASR-68	68W Arch. Standing Seam 3J a-Si Module	65	NA
United Solar Systems Corp.	ES-116	116W a-Si Module with Black Anodized Frame	109.8	NA
United Solar Systems Corp.	ES-124	124W a-Si Module with Black Anodized Frame	117.4	NA
United Solar Systems Corp.	ES-58	58W a-Si Module with Black Anodized Frame	54.9	NA
United Solar Systems Corp.	ES-62T	62W a-Si Module with Black Anodized Frame	58.7	NA
United Solar Systems Corp.	PVL-116(DM)	116W Field Applied 3J a-Si Laminate, Deck-Mounted	107.4	NA
United Solar Systems Corp.	PVL-116(PM)	116W Field Applied 3J a-Si Laminate, Purlin-Mounted	109.9	NA
United Solar Systems Corp.	PVL-124	124W Field Applied 3J a-Si Laminate	118.5	NA
United Solar Systems Corp.	PVL-128(DM)	128W Field Applied 3J a-Si Laminate, Deck-Mounted	118.4	NA
United Solar Systems Corp.	PVL-128(PM)	128W Field Applied 3J a-Si Laminate, Purlin-Mounted	121.2	NA
United Solar Systems Corp.	PVL-136	136W Field Applied 3J a-Si Laminate	130	NA

United Solar Systems Corp.	PVL-29(DM)	29W Field Applied 3J a-Si Laminate, Deck-Mounted	26.9	NA
United Solar Systems Corp.	PVL-29(PM)	29W Field Applied 3J a-Si Laminate, Purlin-Mounted	27.5	NA
United Solar Systems Corp.	PVL-31	31W Field Applied 3J a-Si Laminate	29.6	NA
United Solar Systems Corp.	PVL-58(DM)	58W Field Applied 3J a-Si Laminate, Deck-Mounted	53.7	NA
United Solar Systems Corp.	PVL-58(PM)	58W Field Applied 3J a-Si Laminate, Purlin-Mounted	55	NA
United Solar Systems Corp.	PVL-62	62W Field Applied 3J a-Si Laminate	59.3	NA
United Solar Systems Corp.	PVL-64(DM)	64W Field Applied 3J a-Si Laminate, Deck-Mounted	59.2	NA
United Solar Systems Corp.	PVL-64(PM)	64W Field Applied 3J a-Si Laminate, Purlin-Mounted	60.6	NA
United Solar Systems Corp.	PVL-68	68W Field Applied 3J a-Si Laminate	65	NA
United Solar Systems Corp.	PVL-87(DM)	87W Field Applied 3J a-Si Laminate, Deck-Mounted	80.6	NA
United Solar Systems Corp.	PVL-87(PM)	87W Field Applied 3J a-Si Laminate, Purlin-Mounted	82.5	NA
United Solar Systems Corp.	PVL-93	93W Field Applied 3J a-Si Laminate	88.9	NA
United Solar Systems Corp.	PVR10T	62W a-Si Roof Module	59.3	NA
United Solar Systems Corp.	PVR15T	93W a-Si Roof Module	88.9	NA
United Solar Systems Corp.	PVR20T	124W a-Si Roof Module	118.5	NA
United Solar Systems Corp.	PVR5T	31W a-Si Roof Module	29.6	NA
United Solar Systems Corp.	SFS-11L-10	10-68W Structural Standing Seam 3J a-Si Module	650	NA
United Solar Systems Corp.	SFS-11L-11	11-68W Structural Standing Seam 3J a-Si Module	715	NA

(Continued)

TABLE C.1   LIST OF ELIGIBLE PHOTOVOLTAIC MODULES CALIFORNIA ENERGY COMMISSION EMERGING RENEWABLES PROGRAM (FEBRUARY 2005) (Continued)

MANUFACTURER NAME	MODULE MODEL NUMBER	DESCRIPTION	CEC PTC° RATING	NOTES
United Solar Systems Corp.	SFS-11L-12	12-68W Structural Standing Seam 3J a-Si Module	780	NA
United Solar Systems Corp.	SFS-22L-10	10-136W Structural Standing Seam 3J a-Si Module	1300	NA
United Solar Systems Corp.	SFS-22L-11	11-136W Structural Standing Seam 3J a-Si Module	1430	NA
United Solar Systems Corp.	SFS-22L-12	12-136W Structural Standing Seam 3J a-Si Module	1560	NA
United Solar Systems Corp.	SHR-15	15W Shingle 3J a-Si Module	13.9	NA
United Solar Systems Corp.	SHR-17	17W Shingle 3J a-Si Module	15.7	NA
United Solar Systems Corp.	SSR-120	120W Structural Standing Seam 3J a-Si Module, Purlin-Mounted	113.5	NA
United Solar Systems Corp.	SSR-120(DM)	120W Structural Standing Seam 3J a-Si Module, Deck-Mounted	110.9	NA
United Solar Systems Corp.	SSR-120J	120W Structural Standing Seam 3J a-Si Module w/J-box, Purlin-Mounted	113.5	NA
United Solar Systems Corp.	SSR-120J(DM)	120W Structural Standing Seam 3J a-Si Module w/J-box, Deck-Mounted	110.9	NA
United Solar Systems Corp.	SSR-128	128W Structural Standing Seam 3J a-Si Module, Purlin-Mounted	121.1	NA
United Solar Systems Corp.	SSR-128(DM)	128W Structural Standing Seam 3J a-Si Module, Deck-Mounted	118.3	NA
United Solar Systems Corp.	SSR-128J	128W Structural Standing Seam 3J a-Si Module w/J-box, Purlin-Mounted	121.1	NA

United Solar Systems Corp.	SSR-128J(DM)	128W Structural Standing Seam 3J a-Si Module w/J-box, Deck-Mounted	118.3	NA
United Solar Systems Corp.	SSR-136	136W Structural Standing Seam 3J a-Si Module	130	NA
United Solar Systems Corp.	SSR-60	60W Structural Standing Seam 3J a-Si Module, Purlin-Mounted	56.8	NA
United Solar Systems Corp.	SSR-60(DM)	60W Structural Standing Seam 3J a-Si Module, Deck-Mounted	55.4	NA
United Solar Systems Corp.	SSR-60J	60W Structural Standing Seam 3J a-Si Module w/J-box, Purlin-Mounted	56.8	NA
United Solar Systems Corp.	SSR-60J(DM)	60W Structural Standing Seam 3J a-Si Module w/J-box, Deck-Mounted	55.4	NA
United Solar Systems Corp.	SSR-64	64W Structural Standing Seam 3J a-Si Module, Purlin-Mounted	60.6	NA
United Solar Systems Corp.	SSR-64(DM)	64W Structural Standing Seam 3J a-Si Module, Deck-Mounted	59.1	NA
United Solar Systems Corp.	SSR-64J	64W Structural Standing Seam 3J a-Si Module w/J-box, Purlin-Mounted	60.6	NA
United Solar Systems Corp.	SSR-64J(DM)	64W Structural Standing Seam 3J a-Si Module w/J-box, Deck-Mounted	59.1	NA
United Solar Systems Corp.	SSR-68	68W Structural Standing Seam 3J a-Si Module	65	NA
United Solar Systems Corp.	US-116	116W Framed Triple-Junction a-Si Module	109.9	NA
United Solar Systems Corp.	US-32	32W Framed Triple-Junction a-Si Module	30.3	NA
United Solar Systems Corp.	US-39	39W Framed Triple-Junction a-Si Module	36.9	NA
United Solar Systems Corp.	US-42	42W Framed Triple-Junction a-Si Module	39.8	NA
United Solar Systems Corp.	US-60	60W Framed Triple-Junction a-Si Module	56.8	NA
United Solar Systems Corp.	US-64	64W Framed Triple-Junction a-Si Module	60.6	NA
Webel-SL Energy Systems	W1000-100	100W Monocrystalline PV Module	87.1	NA

(*Continued*)

TABLE C.1 LIST OF ELIGIBLE PHOTOVOLTAIC MODULES CALIFORNIA ENERGY COMMISSION EMERGING RENEWABLES PROGRAM (FEBRUARY 2005) (Continued)

MANUFACTURER NAME	MODULE MODEL NUMBER	DESCRIPTION	CEC PTC* RATING	NOTES
Webel-SL Energy Systems	W1000-110	110W Monocrystalline PV Module	96	NA
Webel-SL Energy Systems	W1000-115	115W Monocrystalline PV Module	100.5	NA
Webel-SL Energy Systems	W1000-120	120W Monocrystalline PV Module	105	NA
Webel-SL Energy Systems	W1600-150	150W Monocrystalline PV Module	130.5	NA
Webel-SL Energy Systems	W1600-155	155W Monocrystalline PV Module	134.9	NA
Webel-SL Energy Systems	W1600-160	160W Monocrystalline PV Module	139.4	NA
Webel-SL Energy Systems	W1600-165	165W Monocrystalline PV Module	143.9	NA
Webel-SL Energy Systems	W900-75	75W Monocrystalline PV Module	65.1	NA
Webel-SL Energy Systems	W900-80	80W Monocrystalline PV Module	69.5	NA
Webel-SL Energy Systems	W900-85	85W Monocrystalline PV Module	73.9	NA
Webel-SL Energy Systems	W900-90	90W Monocrystalline PV Module	78.4	NA

* PTC stands for "PVUSA Test Conditions." The PTC watt rating is based on 1000 W/m^2 solar irradiance, 20°C ambient temperature, and 1-m/s wind speed. The PTC watt rating is lower than the Standard Test Conditions (STC), a watt-rating used by manufacturers.

TABLE C.2 LIST OF ELIGIBLE INVERTERS CALIFORNIA ENERGY COMMISSION EMERGING RENEWABLES PROGRAM (FEBRUARY 2005)

MANUFACTURER NAME	INVERTER MODEL NUMBER	DESCRIPTION	POWER RATING (W)	75% LOAD EFFICIENCY	APPROVED BUILT-IN METER	NOTES
Alpha Technologies, Inc.	Solaris 3500	3.5kW, 240Vac, 96-200Vdc, NEMA-3R, Grid Interactive PV Inverter, LCD, MPPT	3,500	93	Yes	NA
Ballard Power Systems Corporation	EPC-PV-208-30kW	Utility Interactive 208V 30kW PV Power Converter System	30,000	95	No	NA
Ballard Power Systems Corporation	EPC-PV-208-75kW	Utility Interactive 75kW PV Power Converter System	75,000	93	Yes	NA
Ballard Power Systems Corporation	EPC-PV-480-30kW	Utility Interactive 480V 30kW PV Power Converter System	30,000	95	No	NA
Ballard Power Systems Corporation	EPC-PV-480-75kW	Utility Interactive 75kW PV Power Converter System	75,000	93	Yes	NA
Beacon Power Corporation	M5	5kW Power Conversion System	5,000	90	No	NA
Bergey Windpower Co.	Gridtek 10	10kW, 240Vac Split-Phase, Utility Interactive Inverter	10,000	93	No	NA
Fronius USA, LLC	IG 2000	2,000W Grid-tied units with integrated breakers and LCD	2,000	94	Yes	NA

(Continued)

MANUFACTURER NAME	INVERTER MODEL NUMBER	DESCRIPTION	POWER RATING (W)	75% LOAD EFFICIENCY	APPROVED BUILT-IN METER	NOTES
Fronius USA, LLC	IG 2500-LV	2,350W Grid-tied units with integrated breakers and LCD	2,350	94	Yes	NA
Fronius USA, LLC	IG 3000	2,700W Grid-tied units with integrated breakers and LCD	2,700	94	Yes	NA
Fronius USA, LLC	IG 4000	4,000W Grid-tied unit with integrated disconnects and performance meter	4,000	94	Yes	NA
Fronius USA, LLC	IG 4500-LV	4,500W Grid-tied unit with integrated disconnects and performance meter	4,500	93	Yes	NA
Fronius USA, LLC	IG 5100	5,100W Grid-tied unit with integrated disconnects and performance meter	5,100	94	Yes	NA
Magnetek	PVI-3000-I-OUTD-US	3kW, 150-600 VDC Utility Interactive Inverter	3,000	93	Yes	NA
Nextek Power Systems, Inc.	NPS-1000	1,000W Direct Coupling dc Rectifier	1,000	94	No	NA
OutBack Power Systems	GTFX 2524	2,500W Utility Interactive (w/battery backup) 24Vdc Inverter	2,500	91	No	NA
OutBack Power Systems	GTFX 3048	3,000W Utility Interactive (w/battery backup) 48Vdc Inverter	3,000	92	No	NA
OutBack Power Systems	GVFX 3524	3,500W Utility Interactive (w/battery backup) 24Vdc Inverter	3,500	91	No	NA

Manufacturer	Model	Description		Efficiency		
OutBack Power Systems	GVFX 3648	3,600W Utility Interactive (w/ battery backup) 48Vdc Inverter	3,600	92	No	NA
Pacific Solar Pty Limited	SDEIP2-09	240W, 240V Module PV Inverter for the SunEmpower (PP-USA-213)	240	93	No	NA
PV Powered LLC	PVP1100E	1,100W Utility Interactive Inverter	1,100	95	No	NA
PV Powered LLC	PVP1800	1,800W Utility Interactive Inverter	1,800	95	Yes	NA
PV Powered LLC	PVP2800-208	2,800W (208Vac) Utility Interactive Inverter	2,800	97	Yes	NA
PV Powered LLC	PVP2800-240	2,800W (240Vac) Utility Interactive Inverter	2,800	97	Yes	NA
SatCon Power Systems Canada Ltd.	AE-100-60-PV-A	Three-phase 100kW Utility Interactive inverter	100,000	96	No	NA
SatCon Power Systems Canada Ltd.	AE-225-60-PV-A	225kW Three-phase inverter 480Vac	225,000	95	No	NA
SatCon Power Systems Canada Ltd.	AE-30-60-PV-E	30kW Single-phase Utility Interactive Inverter	30,000	93	No	NA
SatCon Power Systems Canada Ltd.	AE-50-60-PV-A	50kW 480Vac Three phase Utility Interactive Inverter	50,000	94	Yes	NA
Sharp Corporation	JH-3500U	Utility Interactive Inverter 240Vac L-L, 3.5kW	3,500	92	Yes	NA
SMA America	SB6000U	Sunny Boy 6,000W Utility Interactive Inverter with performance meter	6,000	94	Yes	NA
SMA America	SC125U	125kW 3-phase 480Vac, 275-600Vdc, Utility Interactive Inverter	125,000	95	Yes	NA

(Continued)

MANUFACTURER NAME	INVERTER MODEL NUMBER	DESCRIPTION	POWER RATING (W)	75% LOAD EFFICIENCY	APPROVED BUILT-IN METER	NOTES
SMA America	SWR1100U	1,100W, 240Vac Sunny Boy String Inverter	1,100	93	No	NA
SMA America	SWR1100U-SBD	1,100W, 240Vac Sunny Boy String Inverter with display	1,100	93	Yes	NA
SMA America	SWR1800U	1.8kW, 120Vac Sunny Boy String Inverter	1,800	93	No	NA
SMA America	SWR1800U-SBD	1.8kW, 120Vac Sunny Boy String Inverter, with display	1,800	93	Yes	NA
SMA America	SWR2500U (208V)	2.1kW, 208Vac Sunny Boy String Inverter	2,100	94	No	NA
SMA America	SWR2500U (240V)	2.5kW, 240Vac Sunny Boy String Inverter	2,500	94	No	NA
SMA America	SWR2500U-SBD (208V)	2.1kW, 208Vac Sunny Boy String Inverter, with display	2,100	94	Yes	NA
SMA America	SWR2500U-SBD (240V)	2.5kW, 240Vac Sunny Boy String Inverter, with display	2,500	94	Yes	NA
SMA America	SWR700U	700W, 120Vac Sunny Boy String Inverter	700	93	No	NA
SMA America	SWR700U-SBD	700W, 120Vac Sunny Boy String Inverter with display	700	93	Yes	NA

Solectria Renewables, LLC	PVI 13kW	13kW 208 and 480Vac Commercial Grid-Tied Solar PV Inverter	13,200	94	No	NA
Xantrex Technology, Inc.	BWT10240	10kW, 240Vac Split-phase, Utility Interactive Inverter	10,000	93	No	NA
Xantrex Technology, Inc.	GT3.0-NA-DS-240	3.0kW, 240Vac, 195-600Vdc Grid-Tied Inverter	3,000	94	Yes	NA
Xantrex Technology, Inc.	PV-100208	100kW 208Vac/3-phase Utility Interactive Inverter	100,000	95	No	NA
Xantrex Technology, Inc.	PV-100S-208	100kW 208Vac, 330-600Vdc Inverter System with automatic transformer disconnect	100,000	95	Yes	NA
Xantrex Technology, Inc.	PV-100S-480	100kW 480Vac, 330-600Vdc Inverter System with automatic transformer disconnect	100,000	95	Yes	NA
Xantrex Technology, Inc.	PV-10208	10kW 208Vac/3-phase Utility Interactive Inverter	10,000	94	Yes	NA
Xantrex Technology, Inc.	PV-15208	15kW 208Vac/3-phase Utility Interactive Inverter	15,000	95	Yes	NA
Xantrex Technology, Inc.	PV-20208	20kW 208Vac/3-phase Utility Interactive Inverter	20,000	94	Yes	NA
Xantrex Technology, Inc.	PV-225208	225kW 208Vac/3-phase Utility Interactive Inverter	225,000	95	No	NA
Xantrex Technology, Inc.	PV-30208	30kW 208Vac/3-phase Utility Interactive Inverter	30,000	94	Yes	NA
Xantrex Technology, Inc.	PV-45208	45kW 208Vac/3-phase Utility Interactive Inverter	45,000	94	Yes	NA

(Continued)

TABLE C.2 LIST OF ELIGIBLE INVERTERS CALIFORNIA ENERGY COMMISSION EMERGING RENEWABLES PROGRAM (FEBRUARY 2005) (Continued)

MANUFACTURER NAME	INVERTER MODEL NUMBER	DESCRIPTION	POWER RATING (W)	75% LOAD EFFICIENCY	APPROVED BUILT-IN METER	NOTES
Xantrex Technology, Inc.	PV-5208	5kW, 208Vac/3-phase, Photovoltaic Utility Interactive Inverter	5,000	93	No	NA
Xantrex Technology, Inc.	STXR1000	1.0kVA, 42-85Vdc, 240Vac, Trace Engr. Sine Wave Inverter (Sunsweep MPPT)	1,000	88	No	NA
Xantrex Technology, Inc.	STXR1500	1.5kVA, 42-85Vdc, 240Vac, Trace Engr. Sine Wave Inverter (Sunsweep MPPT)	1,500	89	No	NA
Xantrex Technology, Inc.	STXR1500 v5.0	1.5kVA, 42-85Vdc, 240Vac, Trace Eng, Sine Wave Inverter (Sunsweep MPPT)	1,500	89	Yes	NA
Xantrex Technology, Inc.	STXR2000	2.0kVA, 42-85Vdc, 240Vac, Trace Engr. Sine Wave Inverter (Sunsweep MPPT)	2,000	90	No	NA
Xantrex Technology, Inc.	STXR2500	2.5kVA, 42-75Vdc, 240Vac, Trace Engr. Sine Wave Inverter (Sunsweep MPPT)	2,500	90	No	NA

Xantrex Technology, Inc.	STXR2500 v5.0	2.5kVA, 42-75Vdc, 240Vac, Trace Eng, Sine Wave Inverter (Sunsweep MPPT)	2,500	90	Yes	NA
Xantrex Technology, Inc.	SW4024 (w GTI)	4.0kVA, 24Vdc, 120Vac, Trace Engr. batt bkp, Sine Wave Inverter	4,000	88	No	NA
Xantrex Technology, Inc.	SW4048 (w GTI)	4.0kVA, 48Vdc, 120Vac, Trace Engr. batt bkp, Sine Wave Inverter	4,000	88	No	NA
Xantrex Technology, Inc.	SW5548 (w GTI)	5.5kVA, 48Vdc, 120Vac, Trace Engr. batt bkp, Sine Wave Inverter	5,500	89	No	NA

TABLE C.3 LIST OF ELIGIBLE SYSTEM PERFORMANCE METERS CALIFORNIA ENERGY COMMISSION EMERGING RENEWABLES PROGRAM (FEBRUARY 2005)

MANUFACTURER NAME	MODEL NUMBER	DISPLAY TYPE	NOTES
ABB/Elster	1S	LCD	NA
ABB/Elster	2S	LCD	NA
ABB/Elster	3S	LCD	NA
ABB/Elster	A3 Alpha	LCD	NA
ABB/Elster	AB1	Cyclometer	NA
ABB/Elster	ABS	Cyclometer	NA
ABB/Elster	Alpha	LCD	NA
ABB/Elster	Alpha Plus	LCD	NA
ABB/Elster	REX	LCD	NA
Astropower	APM2 SunChoice	LCD	NA
BP Solar	HSSM-1	LCD	NA
Brand Electronic	20-1850	LCD	NA
Brand Electronic	20-1850CI	LCD	NA
Brand Electronic	20-CTR	LCD	NA
Brand Electronic	21-1850CI	LCD	NA
Brand Electronic	4-1850	LCD	NA
Brand Electronic	ONE Meter	LCD	NA
Draker Solar Design	PVDAQ Basic	Computer monitor	NA
Draker Solar Design	PVDAQ Commercial	Computer monitor	NA
E-MON	D-MON 208100 KIT	LCD	NA
E-MON	D-MON 208100C KIT	LCD	NA
E-MON	D-MON 208100D KIT	LCD	NA

E-MON	D-MON 2081600 KIT	LCD	NA
E-MON	D-MON 2081600C KIT	LCD	NA
E-MON	D-MON 2081600D KIT	LCD	NA
E-MON	D-MON 208200 KIT	LCD	NA
E-MON	D-MON 208200C KIT	LCD	NA
E-MON	D-MON 208200D KIT	LCD	NA
E-MON	D-MON 20825 KIT	LCD	NA
E-MON	D-MON 20825C KIT	LCD	NA
E-MON	D-MON 20825D KIT	LCD	NA
E-MON	D-MON 2083200 KIT	LCD	NA
E-MON	D-MON 2083200C KIT	LCD	NA
E-MON	D-MON 2083200D KIT	LCD	NA
E-MON	D-MON 208400 KIT	LCD	NA
E-MON	D-MON 208400C KIT	LCD	NA
E-MON	D-MON 208400D KIT	LCD	NA
E-MON	D-MON 20850 KIT	LCD	NA
E-MON	D-MON 20850C KIT	LCD	NA
E-MON	D-MON 20850D KIT	LCD	NA
E-MON	D-MON 208800 KIT	LCD	NA
E-MON	D-MON 208800C KIT	LCD	NA
E-MON	D-MON 208800D KIT	LCD	NA
E-MON	D-MON 480100 KIT	LCD	NA
E-MON	D-MON 480100C KIT	LCD	NA
E-MON	D-MON 480100D KIT	LCD	NA
E-MON	D-MON 4801600 KIT	LCD	NA
E-MON	D-MON 4801600 KIT	LCD	NA

(Continued)

TABLE C.3 LIST OF ELIGIBLE SYSTEM PERFORMANCE METERS CALIFORNIA ENERGY COMMISSION EMERGING RENEWABLES PROGRAM (FEBRUARY 2005) (Continued)

MANUFACTURER NAME	MODEL NUMBER	DISPLAY TYPE	NOTES
E-MON	D-MON 4801600C KIT	LCD	NA
E-MON	D-MON 4801600C KIT	LCD	NA
E-MON	D-MON 4801600D KIT	LCD	NA
E-MON	D-MON 4801600D KIT	LCD	NA
E-MON	D-MON 480200 KIT	LCD	NA
E-MON	D-MON 480200C KIT	LCD	NA
E-MON	D-MON 480200D KIT	LCD	NA
E-MON	D-MON 48025 KIT	LCD	NA
E-MON	D-MON 48025C KIT	LCD	NA
E-MON	D-MON 48025D KIT	LCD	NA
E-MON	D-MON 4803200 KIT	LCD	NA
E-MON	D-MON 4803200C KIT	LCD	NA
E-MON	D-MON 4803200D KIT	LCD	NA
E-MON	D-MON 480400 KIT	LCD	NA
E-MON	D-MON 480400C KIT	LCD	NA
E-MON	D-MON 480400D KIT	LCD	NA
E-MON	D-MON 48050 KIT	LCD	NA
E-MON	D-MON 48050C KIT	LCD	NA
E-MON	D-MON 48050D KIT	LCD	NA
E-MON	D-MON 480800 KIT	LCD	NA
E-MON	D-MON 480800C KIT	LCD	NA
E-MON	D-MON 480800D KIT	LCD	NA

E-MON	E-CON 2120100-SA KIT	LCD	NA
E-MON	E-CON 2120200-SA KIT	LCD	NA
E-MON	E-CON 212025-SA KIT	LCD	NA
E-MON	E-CON 212050-SA KIT	LCD	NA
E-MON	E-CON 2277100-SA KIT	LCD	NA
E-MON	E-CON 2272000-SA KIT	LCD	NA
E-MON	E-CON 227725-SA KIT	LCD	NA
E-MON	E-CON 227750-SA KIT	LCD	NA
E-MON	E-CON 3208100-SA KIT	LCD	NA
E-MON	E-CON 3208200-SA KIT	LCD	NA
E-MON	E-CON 320825-SA KIT	LCD	NA
E-MON	E-CON 320850-SA KIT	LCD	NA
Fat Spaniel Technologies, Inc.	PV2Web	LCD and other digital display types	(PC-based for SMA-America inverters)
General Electric	I70S	Electromechanical or cyclometer register	NA
General Electric	KV	LCD	NA
General Electric	KV2	LCD	NA
Global Power Products	ENER-COMM ECE-100	LCD	NA
Global Power Products	ENER-COMM ECE-200	LCD	NA
Global Power Products	ENER-COMM ECED-100	LCD	NA
Home Energy Systems, Inc.	100 A	LCD	NA
Home Energy Systems, Inc.	25 A	LCD	NA
Home Energy Systems, Inc.	50 A	LCD	NA
Integrated Metering Systems	1101201	LCD	NA

(Continued)

TABLE C.3 LIST OF ELIGIBLE SYSTEM PERFORMANCE METERS CALIFORNIA ENERGY COMMISSION EMERGING RENEWABLES PROGRAM (FEBRUARY 2005) (Continued)

MANUFACTURER NAME	MODEL NUMBER	DISPLAY TYPE	NOTES
Integrated Metering Systems	1101201-T	LCD	NA
Integrated Metering Systems	1101202	LCD	NA
Integrated Metering Systems	1101202-T	LCD	NA
Integrated Metering Systems	1102401	LCD	NA
Integrated Metering Systems	1102401-T	LCD	NA
Integrated Metering Systems	1102402	LCD	NA
Integrated Metering Systems	1102402-T	LCD	NA
Integrated Metering Systems	1102771	LCD	NA
Integrated Metering Systems	1102771-T	LCD	NA
Integrated Metering Systems	1102772	LCD	NA
Integrated Metering Systems	1102772-T	LCD	NA
Integrated Metering Systems	1103471	LCD	NA
Integrated Metering Systems	1103471-T	LCD	NA
Integrated Metering Systems	1103472	LCD	NA
Integrated Metering Systems	1103472-T	LCD	NA
Integrated Metering Systems	1201201	LCD	NA
Integrated Metering Systems	1201201-T	LCD	NA
Integrated Metering Systems	1201202	LCD	NA
Integrated Metering Systems	1201202-T	LCD	NA
Integrated Metering Systems	1202401	LCD	NA

Integrated Metering Systems	1202401-T	LCD	NA
Integrated Metering Systems	1202402	LCD	NA
Integrated Metering Systems	1202402-T	LCD	NA
Integrated Metering Systems	1202771	LCD	NA
Integrated Metering Systems	1202771-T	LCD	NA
Integrated Metering Systems	1202772	LCD	NA
Integrated Metering Systems	1202772-T	LCD	NA
Integrated Metering Systems	1203471	LCD	NA
Integrated Metering Systems	1203471-T	LCD	NA
Integrated Metering Systems	1203472	LCD	NA
Integrated Metering Systems	1203472-T	LCD	NA
Integrated Metering Systems	1301201	LCD	NA
Integrated Metering Systems	1301201-T	LCD	NA
Integrated Metering Systems	1301202	LCD	NA
Integrated Metering Systems	1301202-T	LCD	NA
Integrated Metering Systems	1302401	LCD	NA
Integrated Metering Systems	1302401-T	LCD	NA
Integrated Metering Systems	1302402	LCD	NA
Integrated Metering Systems	1302402-T	LCD	NA
Integrated Metering Systems	1302771	LCD	NA
Integrated Metering Systems	1302771-T	LCD	NA
Integrated Metering Systems	1302772	LCD	NA
Integrated Metering Systems	1302772-T	LCD	NA
Integrated Metering Systems	1303471	LCD	NA
Integrated Metering Systems	1303471-T	LCD	NA

(Continued)

TABLE C.3 LIST OF ELIGIBLE SYSTEM PERFORMANCE METERS CALIFORNIA ENERGY COMMISSION EMERGING RENEWABLES PROGRAM (FEBRUARY 2005) (Continued)

MANUFACTURER NAME	MODEL NUMBER	DISPLAY TYPE	NOTES
Integrated Metering Systems	1303472	LCD	NA
Integrated Metering Systems	1303472-T	LCD	NA
Integrated Metering Systems	2111201	LCD	NA
Integrated Metering Systems	2111201-T	LCD	NA
Integrated Metering Systems	2111202	LCD	NA
Integrated Metering Systems	2111202-T	LCD	NA
Integrated Metering Systems	2112401	LCD	NA
Integrated Metering Systems	2112401-T	LCD	NA
Integrated Metering Systems	2112402	LCD	NA
Integrated Metering Systems	2112402-T	LCD	NA
Integrated Metering Systems	2112771	LCD	NA
Integrated Metering Systems	2112771-T	LCD	NA
Integrated Metering Systems	2112772	LCD	NA
Integrated Metering Systems	2112772-T	LCD	NA
Integrated Metering Systems	2113471	LCD	NA
Integrated Metering Systems	2113471-T	LCD	NA
Integrated Metering Systems	2113472	LCD	NA
Integrated Metering Systems	2113472-T	LCD	NA
Integrated Metering Systems	2221201	LCD	NA
Integrated Metering Systems	2221201-T	LCD	NA
Integrated Metering Systems	2221202	LCD	NA

Integrated Metering Systems	2221202-T	LCD	NA
Integrated Metering Systems	2222401	LCD	NA
Integrated Metering Systems	2222401-T	LCD	NA
Integrated Metering Systems	2222402	LCD	NA
Integrated Metering Systems	2222402-T	LCD	NA
Integrated Metering Systems	2222771	LCD	NA
Integrated Metering Systems	2222771-T	LCD	NA
Integrated Metering Systems	2222772	LCD	NA
Integrated Metering Systems	2222772-T	LCD	NA
Integrated Metering Systems	2223471	LCD	NA
Integrated Metering Systems	2223471-T	LCD	NA
Integrated Metering Systems	2223472	LCD	NA
Integrated Metering Systems	2223472-T	LCD	NA
iSYS Systems	PVM-Net	Computer monitor	NA
Landis + Gyr Inc.	AL Altimus 2S	LCD	NA
OutBack Power Systems	MX60	LCD	NA
Pacific Solar	Sunlogger	LCD	SunEmpower product only
Poobah Industries	SB-001	LCD	Software for PalmOne for Sunny Boy inverters
Power Measurement	ION 6200	LCD	NA

(Continued)

TABLE C.3   LIST OF ELIGIBLE SYSTEM PERFORMANCE METERS CALIFORNIA ENERGY COMMISSION EMERGING RENEWABLES PROGRAM *(FEBRUARY 2005) (Continued)*

MANUFACTURER NAME	MODEL NUMBER	DISPLAY TYPE	NOTES
Righthand Engineering, LLC	WinVerter-Monitor	Computer monitor	PC-based for Xantrex Trace SW series inverters
Schlumberger/Sangamo	Centron C1S	LCD or cyclometer	NA
Schlumberger/Sangamo	J4S	LCD or cyclometer	NA
Schlumberger/Sangamo	J5S	LCD or cyclometer	NA
SMA America	Sunny Boy Control	LCD	NA
SMA America	Sunny Boy Control Light	LCD	NA
SMA America	Sunny Boy Control Plus	LCD	NA
SMA America	Sunny Boy Control Plus-485	LCD	NA
SMA America	Sunny Boy Control-485	LCD	NA
SMA America	Sunny Data	Computer monitor	NA
SMA America	Sunny Data Control	Computer monitor	NA
SMA America	Sunny Data Control	Computer monitor	NA
SMA America	SWR LCD	LCD	NA
SolarQuest	rMeter	LCD or computer monitor	NA
Xantrex	STRM—Remote Meter	LCD	NA

California Energy Commission Forms and Worksheets

CEC-1038 R1, (1-2005)

R1	**RESERVATION APPLICATION FORM** *EMERGING RENEWABLES PROGRAM*	☐ Modify Existing Record # _____ ☐ Affordable Housing Project ☐ New Construction

1. Physical Site of System Installation

Street Address:

City:	State:	Zip

2. Purchaser Name and Mailing Address

Phone: (   )   Fax: (   )

3. Equipment Seller (Must be registered)

Company:

City:	CEC ID (if known)

Phone: (   )   Fax: (   )

4. System Installation (Write "Owner" if not hiring contractor)

Company:

City:	License No.:
Phone:	Fax:

5. Electric Utility (Attach all pages of monthly statement)

☐ PG&E ☐ SCE ☐ SDG&E ☐ BVE	Service ID:
Billing Period:	KWh Used:

Note: If new construction attach building permit. Permit No. _____

Submit complete application to:

California Energy Commission
Emerging Renewables Program (MS-45)
1516 Ninth Street
Sacramento, CA 95814-5512

6. Equipment (PV modules, turbines, inverters, meters)

	Quantity	Manufacturer, Model (see CEC lists)
Generating Equipment		
Inverters, Meters		

Estimated annual energy production _____ kWh/Year

7. Rebate and Other Incentives

System Rated Output	_____ watts
Total System Cost:	$ _____
Expected Rebate:	$ _____
Pay Rebate to:	☐ Purchaser ☐ Seller

Reassign payment?   ☐ Yes   ☐ No
If yes, submit form 1038 R5 with payment request.

Other Incentives: $ _____
Source/Record No.:   _____

8. Declaration

The undersigned parties declare under penalty of perjury that the information in this form and the supporting documentation submitted herewith is true and correct to the best of their knowledge and that the following is true:

1. All system equipment is new and unused and has been purchased within the last 18 months,
2. The generating system is intended primarily to offset Purchaser's electrical needs at the site of installation
3. The Purchaser's intent is to operate the system at the above site of installation for its useful life or the duration of the lease agreement and
4. The generating system will be interconnected with the distribution system of the electric utility identified above.

The undersigned parties further acknowledge that they are aware of the requirements and conditions of receiving funding under the Emerging Renewables Program (ERP) and agree to comply with all such requirements and conditions as provided in the Energy Commission's ERP Guidebook and Overall Program Guidebook as a condition to receiving funding under the ERP. The undersigned Purchaser authorizes the Energy Commission during the term of the ERP to exchange information on this form with the Purchaser's electric utility to verify compliance with the requirements of the ERP.

Purchaser Signature	**Equipment Seller Signature**
Print Name: _____	Print Name: _____
Signature: _____ Date: _____	Signature: _____ Date: _____

Necessary Supporting Documentation.
1. All pages of a monthly electric utility bill.
2. Agreements to purchase and install equipment.
3. Payee Data Record (Form STD-204) if payee identified has not previously been paid by the Energy Commission.
4. If not a standard rebate application, attach other required documentation as specified in the ERP Guidebook.

CEC-1038 R1—Reservation Application Form (all technologies).

CEC- 1038 R2 (1-2005)

R2 | REBATE PAYMENT CLAIM FORM
EMERGING RENEWABLES PROGRAM

RENEWABLE
ENERGY
PROGRAM

CALIFORNIA ENERGY COMMISSION

Mail complete payment claim to:
California Energy Commission
ERP, Payment Claim
1516 Ninth Street (MS-45)
Sacramento, CA 95814-5512

Record Number _____

Payee Number _____

[CEC use only]

[CEC use only]

Tot.Elig.Cost: $_____ Date CFA: _____

SRO watts: _____ Rebate @ _____ = $_____

1. Confirmation of Reservation Amount

_____ has been granted a reservation of $ _____ for a _____ kW renewable energy generating system. The reservation will expire on _____. The system is being installed at _____ and is expected to produce _____ (kWh per year). The payment will be made to the _____.

The generation system must be completed and the claim submitted with the appropriate documentation by the deadline. Claims must be postmarked by the expiration date or the reservation will expire. This reservation is non-transferable. System must be installed at the installation address and sold to the above.

2. System Equipment (Modules, Wind Turbines, Inverters, kWh Meters)

Number	Manufacturer	Model
_____	_____	_____
_____	_____	_____
_____	_____	_____
_____	_____	_____

Total System Price $ _____

Amount paid by purchaser to date: $_____

Orientation: *(Circle One)* W, SW, S, SE, E, Other

Tilt: *(Circle one)* None, 1-15, 15-30, >30 Degrees

Tracking system type: _____

3. Modifications

Has any of the information in section 1 or 2 above changed? ☐ Yes ☐ No
If yes note the changes before claiming payment.

_____
_____
_____

The undersigned parties declare under penalty of perjury that the information in this form and the supporting documentation submitted herewith is true and correct to the best of their knowledge. The parties further declare under penalty of perjury that the following statements are true and correct to the best of their knowledge:

(1) The electrical generating system described above and in any attached documents meets the terms and conditions of the Energy Commission's Emerging Renewables Program and has been installed and is operating satisfactorily as of the date stated below.
(2) The electrical generating system described above and in any attached documents is property interconnected to the utility distribution grid and has or will be issued utility approval to operate the system as interconnected to the distribution grid.
(3) The rated electrical output of the generating system, the physical location of the system, and the equipment identified were installed as stated above.
(4) Except as noted above, there were no changes in the information regarding the seller, installer, purchaser, generating system specifications, installation location, or price from that information provided in the Reservation Request Form originally submitted by the undersigned.

The undersigned parties further acknowledge that they are aware of the requirements and conditions of receiving funding under the Emerging Renewables Program (ERP) and agree to comply with all such requirements and conditions as provided in the Energy Commission's ERP Guidebook and Overall Program Guidebook as a condition to receiving funding under the ERP. As specified in the ERP Guidebook, the undersigned Purchaser authorizes the Energy Commission during the term of the ERP to exchange purchaser information on this form with the Purchaser's electric utility in order to verify compliance with the ERP requirements. If a copy of the utility "letter of authorization to operate" the system *is not* submitted with this payment claim form, the undersigned Purchaser understands that he/she is obligated to submit a copy of this letter to the Energy Commission once it is received.

Purchaser	**Seller**	**Is payment assigned to**
Print Name: _____	Print Name: _____	**another party?** ☐ Yes ☐ No
Signature: _____	Signature: _____	If yes, attach the payment
Date: _____	Date: _____	assignment form (CEC-1038 R5) with original signatures.

IMPORTANT - Necessary Supporting Documentation

1. Final building permit and final inspection signoff; 2. Final invoice(s) confirming the total amount paid for the system equipment and installation; 3. Five-year warranty (CEC-1038 R3 form); 4. Utility letter of authorization to interconnect the system or the Purchaser's authorization form to access Purchaser's utility data; 5. Utility bill or other proof of electrical service and consumption at the site of installation if not previously provided; 6. Payee Data Record (STD-204)

CEC-1038 R2—Rebate Payment Claim Form.

CEC- 1038 R3 (1-2005)

# R3	## MINIMUM WARRANTY FORM *EMERGING RENEWABLES PROGRAM*

System Information

This warranty applies to the following _____ kW renewable energy electric generating system
Description: _____
Located at: _____

What is Covered

This five year warranty is subject to the terms below (check one of the boxes):

☐ **All** components of the generating system **AND** the system's installation. Said warrantor shall bear the full cost of diagnosis, repair and replacement of any system or system component, at no cost to the customer. This warranty also covers the generating equipment against breakdown or degradation in electrical output of more than ten percent from the originally rated output (PTC rating for modules, manufacturers rating for wind turbines); or

☐ **System's installation only.** Said warrantor shall bear the full cost of diagnosis, repair and replacement of any system or system component, exclusive of the manufacturer's coverage. (Copies of five-year warranty certificates for the major system components (i.e., solar modules, wind turbines, etc. and inverter- MUST be provided with this form).

General Terms

This warranty extends to the original purchaser and to any subsequent purchasers or owners at the same location during the warranty period. For the purpose of this warranty, the terms "purchaser," "subsequent owner," and "purchase" include a lessee, assignee of a lease, and a lease transaction. This warranty is effective from _____ (date of completion of the system installation).

Exclusions

This warranty does not apply to:
* Damage, malfunction, or degradation of electrical output caused by failure to properly operate or maintain the system in accordance with the printed instructions provided with the system.
* Damage, malfunction, or degradation of electrical output caused by any repair or replacement using a part or service not provided or authorized in writing by the warrantor.
* Damage malfunction, or degradation of electrical output resulting from purchaser or third party abuse, accident, alteration, improper use, negligence or vandalism, or from earthquake, fire, flood, or other acts of God.

Obtaining Warranty Service

Contact the following warrantor for service or instructions:

Name: _____   Phone: ( )
Company: _____   Fax: ( )
Address: _____

Signature: _____   Date: _____

CEC-1038 R3—Minimum Warranty Form.

CEC- 1038 R4 (1-2005)

R4	**EQUIPMENT SELLER INFORMATION FORM** *EMERGING RENEWABLES PROGRAM*

This information must be submitted before a company can become eligible to participate in the ERP. To remain eligible, a company must resubmit this form annually, by March 31. This annual submittal is required even it the information identified in the company's prior R4 submittal has not changed. In addition, a company must submit an updated R4 form any time its reported information has changed. The updated R4 form must be submitted to the Energy Commission within 30 days of the change of any reported information. Registered companies are listed at [www.consumerenergycenter.org/erprebate/database/]

Business name: Address:	Phone: (　　) Fax : (　　) Email: Web Site:
Owner or principal, Title: Business license number: Reseller's license number: Contractor license number (if applicable):	Select one of the following: ☐ Corporate, LLC, LLP or other that is registered with the California Secretary of State (or appropriate state attached) ☐ Not a corporation, LLC or LLP

The above information applies solely to the business identified above:

Print Name: _____　　Title:_____

Signature: _____

Date: _____

Send this completed form by tele fax to (916) 653-2543 or by mail to:

ERP Seller Registration
California Energy Commission
1516 9th Street, MS-45
Sacramento, CA 95814-5512

Reminder:
This form must be on file with the Energy Commission for a rebate application with the above company to be considered. It must be resubmitted annually by March 31 for sellers to remain eligible from year to year.

CEC-1038 R4—Equipment Seller Information Form.

CEC- 1038 R5 (07-2004)

R5

RESERVATION PAYMENT ASSIGNMENT FORM
EMERGING RENEWABLES PROGRAM

RENEWABLE
ENERGY
PROGRAM
CALIFORNIA ENERGY COMMISSION

Record Number _____
Payee ID Number _____

Reservation Information
Payee Name: _____
Payee Address: _____
_____
Payee Contact: _____
Payee Phone #: _____

Assignment Request

I, _____, the designated payee or authorized representative of the payee, hereby assign the right to receive payment for the above noted reservation under the Emerging Renewables Program to the following individual or entity:

Name: _____
Address: _____
_____
Phone #: _____

I request that payment be forwarded to this individual or entity at the address noted. Upon request proof of payment will be forwarded to me.

Acknowledgement

As the designated payee or authorized representative, I understand that I remain responsible for complying with the requirements of the Emerging Renewables Program and will remain liable for any tax consequences associated with the reservation payment, despite the payment's assignment. I further understand that I may revoke this payment assignment at any time prior to the Energy Commission's processing of the payment by providing written notice to the Energy Commission's Technology Market Development Office. Such notice shall be provided to: Emerging Renewables Program, California Energy Commission, 1516 9th Street, MS-45, Sacramento, CA 95814-5512.

Executed on: _____
Signature: _____
Name: _____
Title: _____

This completed form may be submitted with either the Reservation Request Form (CEC-1038 R1) or the Payment Claim Form (CEC-1038 R2) for standard rebates. This form may not be submitted by telefax, as original signatures are needed to process assignment requests.

CEC-1038 R5—Reservation Payment Assignment Form.

Solar Schools Reservation Request Form

RENEWABLE
ENERGY
PROGRAM

CALIFORNIA ENERGY COMMISSION

The ERP guidebook is available at
www.consumerenergycenter.org/erprebate]
or by calling the Energy Commission's
Call Center at **1-800-555-7794**:

Completed forms must be mailed to:
ERP Schools
California Energy Commission
1516 9th Street, MS-45
Sacramento, CA 95814-5512

Payee Designation:

Reserve rebate for: ☒ School District

1. School District

District Name:
Contact Person:

Mailing Address:

Business: (   )   Fax: (   )
Email:

2. Participating Schools

School Name:

Address:

Business: (   )   Fax: (   )
Email:

Estimated size of photovoltaic system: _____ watts.

School Name:

Address:

Business: (   )   Fax: (   )
Email:

Estimated size of photovoltaic system: _____ watts.

School Name:

Address:

Business: (   )   Fax: (   )
Email:

Estimated size of photovoltaic system: _____ watts.

3. Utility Provider

Electric Utility Provider: ☐ PG&E   ☐ SCE
                             ☐ SDG&E

4. Program Information

How did you hear about our program?
Have you previously applied for funding rebates from the ERP?
        ☐ yes   ☐ no   If yes, reservation # _____

5. Statement of energy efficient measures

For each of the participating schools identified on this form check the
appropriate box:

☐   At least 80% of the classrooms use high efficiency
fluorescent lighting (T8 lamps and electronic ballasts).

☐   Other energy efficient measures with equivalent or greater
energy savings to Item 1 (as determined by the Commission) are in
use (documentation attached).

6. Incentive Requested

Sum of Total watts _____

Total Incentive Requested $_____

☐ Full reservation request (CEC-1038 R1 and supporting
documentation for each site attached) or:

☐ Preliminary 6 month reservation request

 NOTE: Sum of all new systems installed in school district may not
exceed 30 kW.

7. Information for Additional Schools

☐ See attached for information on other schools submitted with
this application (name, address, phone, estimated size of system,
signature of school official).

Declaration: On behalf of the school district, the undersigned declares under penalty of perjury that 1) the information provided in this form and the supporting documentation submitted herewith is true and correct to the best of his/her knowledge, 2) the above described generating system is intended primarily to offset part or all of the school's electrical needs at the site of installation, 3) the site of installation is located within service territory of an eligible electric utility. The undersigned acknowledges that he/she is are aware of the requirements and conditions of receiving funding under the Emerging Renewables Program (ERP) and agrees, on behalf of the school district, to comply with all such requirements and conditions as provided in the Energy Commission's ERP Guidebook and Overall Program Guidebook as a condition to receiving funding under the ERP. As specified in the ERP Guidebook, the undersigned, on behalf of the school district, authorizes the Energy Commission during the term of the ERP to exchange school district information on this form with the school district's electric utility in order to verify compliance with the ERP requirements.

School District representative

Print Name _____

Signature _____

Date _____

CEC Review

IMPORTANT: Attach the following minimum documentation: (1) a signed copy of the board resolution(s) indicating support for the solar project and intent to purchase a photovoltaic system; (2) a copy of a monthly electricity statement; (3) evidence that energy efficient measures equivalent to the use of high efficiency fluorescent lighting in 80 percent of the school's classrooms. This form will not be processed without the required attachments.

Solar Schools Reservation Request Form.

CEC- 1038 R7 (1-2005)

R7

PILOT PERFORMANCE BASED INCENTIVE PROGRAM
PRELIMINARY RESERVATION REQUEST FORM
EMERGING RENEWABLES PROGRAM

1. Physical Site of System Installation

Street Address:

City: | State: | Zip

2. Purchaser Name and Mailing Address

Phone: | Fax:

3. Equipment Seller (Must be registered)

Company:

City: | CEC ID (if known)

Phone: | Fax:

4. System Installation (Write "Owner" if not hiring contractor)

Company:

City: | License No.:

Phone: | Fax:

5. Electric Utility (Attach all pages of monthly statement)

☐ PG&E ☐SCE ☐SDG&E ☐BVE | Service ID:

Billing Period: | KWh Used:

Submit complete application to:

ERP, PBI Preliminary Reservation Request
California Energy Commission
1516 Ninth Street (MS-45)
Sacramento, CA 95814-5512

6. Equipment (PV modules, inverters, meters)

	Quantity	Manufacturer, Model (see CEC lists)
Generating Equipment		
Inverters, DAS/Meters		

7. Incentive Amount

System Rated Output _____ watts

Total System Cost: $ _____

Expected Incentive Rate: $ _____ /kWh

Make Payments to: ☐Purchaser ☐ Seller

Other Incentives: $ _____

Source/Record No.: _____

8. Declaration

The undersigned parties declare under penalty of perjury that the information in this form and the supporting documentation submitted herewith is true and correct to the best of their knowledge and that the following is true:

1. All system equipment is new and unused and has been purchased within the last 18 months,
2. The generating system is intended primarily to offset Purchaser's electrical needs at the site of installation
3. The Purchaser's intent is to operate the system at the above site of installation for its useful life or the duration of the lease agreement and
4. The generating system will be interconnected with the distribution system of the electric utility identified above.

The undersigned parties further acknowledge that they are aware of the requirements and conditions of receiving funding under the Emerging Renewables Program (ERP) and agree to comply with all such requirements and conditions as provided in the Energy Commission's ERP Guidebook and Overall Program Guidebook as a condition to receiving funding under the ERP. The undersigned Purchaser authorizes the Energy Commission during the term of the ERP to exchange information on this form with the Purchaser's electric utility to verify compliance with the requirements of the ERP.

Purchaser Signature	**Equipment Seller Signature**
Print Name: _____	Print Name: _____
Signature: _____ Date: _____	Signature: _____ Date: _____

IMPORTANT - Necessary Supporting Documentation.
1. All pages of a monthly electric utility bill.
2. Agreements to purchase and install equipment.
3. Payee Data Record (Form STD-204) if payee identified has not previously been paid by the Energy Commission in the last 2 years.

CEC-I038 R7—Pilot Performance Program Preliminary Reservation Request Form.

CEC- 1038 R8 (1-2005)

R8

PILOT PERFORMANCE BASED INCENTIVE
PRELIMINARY RESERVATION CONFIRMATION FORM
EMERGING RENEWABLES PROGRAM

RENEWABLE
ENERGY
PROGRAM

CALIFORNIA ENERGY COMMISSION

Mail completed form to:
ERP, PBI Preliminary Reservation Confirmation
California Energy Commission
1516 Ninth Street (MS-45)
Sacramento, CA 95814-5512

Record
*Number*_____

Payee
*Number*_____

[CEC use only]   Eligible Cost: $_____   Date CFA: _____   SRO watts: _____
Date for LTR: _____   Incentive Rate _____$/kWh

1. Confirmation of Preliminary Reservation Amount

_____ has been granted a 12-month preliminary reservation in the amount of $ _____ for a _____ kW photovoltaic generating system to be installed at _____. This generating system is configured and comprised of the system equipment described in Section 2 below. The 12-month preliminary reservation will expire on _____. This Preliminary Reservation Confirmation Form must be completed by the applicant and submitted to the Energy Commission once the system is installed and becomes fully operational. The Preliminary Reservation Confirmation Form must be submitted with the necessary supporting documentation prior to the noted expiration date. In addition, the applicant must provide written authorization allowing the applicant's utility or web-based monitoring system administrator to share applicant data with the Energy Commission for auditing purposes. Upon review and approval of an applicant's Preliminary Reservation Confirmation Form, the Energy Commission will issue the applicant a PBI Final Reservation (CEC-1038 R9), which identifies the applicant, the installed system, the amount of funds reserved for the installed system, and the three-year final reservation funding period during which incentive payments may be claimed by the applicant.

2. System Equipment (PV Modules, Inverters, kWh Meters and DAS)

Number	Manufacturer	Model
_____	_____	_____
_____	_____	_____
_____	_____	_____
_____	_____	_____
_____	_____	_____

Total System Price $ _____

Amount paid by customer to date: $_____

Orientation: (*Circle One*) W, SW, S, SE, E, Other

Tilt: (*Circle One*) None, 1-15, 15-30, >30 Degrees

Tracking system type: _____

3. Modifications

Has any of the information in section 1 or 2 above changed? ☐ Yes ☐ No
If no, skip to the next section. If yes, note the changes below before submitting this form.

_____
_____
_____
_____

4. Reporting Method

Monitoring System Administrator*:	
Reporting method:	

Date of first meter read: _____   Initial reading (kWh): _____

*Monitoring system administrator refers to either the utility or web based third party provider that tracks the electrical production of the system on a monthly basis.

CEC-1038 R8—Pilot Performance Program Reservation Confirmation Form.

CEC- 1038 R9 (1-2005)

R9

PILOT PERFORMANCE BASED INCENTIVE PROGRAM
FINAL RESERVATION CONFIRMATION
EMERGING RENEWABLES PROGRAM

RENEWABLE ENERGY PROGRAM

CALIFORNIA ENERGY COMMISSION

To claim payments mail invoices to:
ERP, PBI Payment Claim
California Energy Commission
1516 Ninth Street (MS-45)
Sacramento, CA 95814-5512

Final Reservation
Number _____

Payee Number _____

1. Final Reservation

_____ (Purchaser) is hereby granted a final reservation in the amount of $_____ at an incentive rate of $0.50 per kilowatt-hour of energy produced. This Final Reservation is issued pursuant to the Energy Commission's Emerging Renewables Program (ERP) Guidebook and based on Purchaser's previously submitted Preliminary Reservation Request and Preliminary Reservation Confirmation applications and the supporting documentation submitted therewith. This Final Reservation is specific to the _____ $kW_{(PTC)}$ photovoltaic generating system (System) installed at _____ and described in Section 2 below, and may not be transferred to another generating system or system location. This Final Reservation expires on _____ and will not be extended under any circumstances. Payments under this Final Reservation will be made to _____ as specified in the Purchaser's preliminary Reservation Request application. Payments under this Final Reservation may be claimed at any time prior to the expiration date by submitting a proper payment claim using Invoice Form (CEC-1038 R10) and attaching the necessary supporting documentation to verify the actual energy produced by the System. Notwithstanding the expiration date of this Final Reservation, payment claims may be submitted for energy produced by the System during the reservation period for a period of 90 days after the expiration date of the reservation. Payment claims submitted after this 90-day period will be denied. Under no circumstance will the incentive payments made under this Final Reservation exceed the amount reserved under the reservation.

Purchaser assumes full responsibility for the operation and funding eligibility of the System. Incentive payments will only be made if the System remains operational and eligible for funding, and the necessary payment claims and supporting documentation are submitted to the Energy Commission. If any equipment is added to or removed from the System during the three-year final reservation period, the Purchaser must notify the Energy Commission in writing. If the System is reduced in size, the amount of funds reserved will be reduced accordingly and the Purchaser will be issued a revised final reservation. If the system is increased in size, the funds reserved will remain the same. In either case, the three-year final reservation period will not change. If, as a result of system changes, the Purchaser or System is no longer eligible for funding, this Final Reservation will be cancelled.

As a condition of receiving incentive payments under the pilot Performance Based Incentive Program, Purchaser agrees to participate in an evaluation process whereby the Energy Commission or its representatives may conduct telephone interviews and/or on-site visits, and analyze system performance data.

2. System Equipment (PV Modules, Inverters, kWh Meters and DAS)

Number	Manufacturer	Model	
_____	_____	_____	Orientation:____
_____	_____	_____	
_____	_____	_____	Tilt:____
_____	_____	_____	
_____	_____	_____	Tracking system type: _____

3. Initial Meter Reading and Data Collection Information

Monitoring System Administrator:	
Reporting Method:	

Date of first meter read: _____   Initial reading (kWh): _____

Please note the Final Reservation Number for future use.
The above information is based on applicant submitted information. Please review this form and promptly report any mistakes to the ERP call center at (800) 555-7794 or fax (916) 653-2543.

CEC-1038 R9—Pilot Performance Program Final Reservation Confirmation Form.

HISTORICAL TIMELINE
OF SOLAR ENERGY

This appendix is an adaptation of the "Solar History Timeline," courtesy of the U.S. Department of Energy.

Seventh Century BC. A magnifying glass is used to concentrate the sun's rays on a fuel and light a fire for light, warmth, and cooking.

Third Century BC. Greeks and Romans use mirrors to light torches for religious purposes.

Second Century BC. As early as 212 BC, the Greek scientist Archimedes makes use of the reflective properties of bronze shields to focus sunlight and set fire to Rome's wooden ships, which were besieging Syracuse. Although there is no proof that this actually happened, the Greek navy re-created the experiment in 1973 and successfully set fire to a wooden boat 50 m away.

AD 20. The Chinese report using mirrors to light torches for religious purposes.

First to Fourth Century. In the first to the fourth centuries, Roman bath houses are built with large, south-facing windows to let in the sun's warmth.

Sixth Century. Sunrooms on houses and public buildings are so common that the Justinian Code establishes "sun rights" to ensure that a building has access to the sun.

Thirteenth Century. In North America, the ancestors of Pueblo people known as *Anasazi* build south-facing cliff dwellings that capture the warmth of the winter sun.

1767. Swiss scientist Horace de Saussure is credited with building the world's first solar collector, later used by Sir John Herschel to cook food during his South African expedition in the 1830s.

1816. On September 27, 1816, Robert Stirling applies for a patent for his *economiser*, a solar thermal electric technology that concentrates the sun's thermal energy to produce electric power.

1839. French scientist Edmond Becquerel discovers the photovoltaic effect while experimenting with an electrolytic cell made up of two metal electrodes placed in an electricity-conducting solution; the electricity generation increases when exposed to light.

1860s. French mathematician August Mouchet proposes an idea for solar-powered steam engines. In the next two decades, he and his assistant, Abel Pifre, will construct the first solar-powered engines for a variety of uses. The engines are the predecessors of modern parabolic dish collectors.

1873. Willoughby Smith discovers the photoconductivity of selenium.

1876. William Grylls Adams and Richard Evans Day discover that selenium produces electricity when exposed to light. Although selenium solar cells fail to convert enough sunlight to power electrical equipment, they prove that a solid material can change light into electricity without heat or moving parts.

1880. Samuel P. Langley invents the bolometer, used to measure light from the faintest stars and the sun's heat rays. It consists of a fine wire connected to an electric circuit. When radiation falls on the wire, it becomes warmer, and this increases the electrical resistance of the wire.

1883. American inventor Charles Fritts describes the first solar cells made of selenium wafers.

1887. Heinrich Hertz discovers that ultraviolet light alters the lowest voltage capable of causing a spark to jump between two metal electrodes.

1891. Baltimore inventor Clarence Kemp patents the first commercial solar water heater.

1904. Wilhelm Hallwachs discovers that a combination of copper and cuprous oxide is photosensitive.

1905. Albert Einstein publishes his paper on the photoelectric effect, along with a paper on his theory of relativity.

1908. William J. Bailley of the Carnegie Steel Company invents a solar collector with copper coils and an insulated box, which is roughly the same collector design used today.

1914. The existence of a barrier layer in photovoltaic devices is noted.

1916. Robert Millikan provides experimental proof of the photoelectric effect.

1918. Polish scientist Jan Czochralski develops a way to grow single-crystal silicon.

1921. Albert Einstein wins the Nobel Prize for his theories explaining the photoelectric effect; for details, see his 1904 technical paper on the subject.

1932. Audobert and Stora discover the photovoltaic effect in cadmium sulfide.

1947. Because energy had become scarce during the long Second World War, passive solar buildings in the United States are in demand; Libbey-Owens-Ford Glass Company publishes a book titled, *Your Solar House*, which profiles 49 of the nation's greatest solar architects.

1953. Dr. Dan Trivich of Wayne State University makes the first theoretical calculations of the efficiencies of various materials of different band-gap widths based on the spectrum of the sun.

1954. Photovoltaic technology is born in the United States when Daryl Chapin, Calvin Fuller, and Gerald Pearson developed the silicon photovoltaic or PV cell at Bell Labs which was the first solar cell capable of generating enough power from the sun to run everyday electrical equipment. Bell Laboratories then produced a solar cell with 6 percent efficiency which was later augmented to 11 percent.

1955. Western Electric begins to sell commercial licenses for silicon photovoltaic technologies. Early successful products include PV-powered dollar changers and devices that decode computer punch cards and tape.

1950s. Architect Frank Bridgers designs the world's first commercial office building featuring solar water heating and design. The solar system has operated continuously since then; the Bridgers-Paxton Building is listed in the National Historic Register as the world's first solar-heated office building.

1956. William Cherry of the U.S. Signal Corps Laboratories approaches RCA Labs' Paul Rappaport and Joseph Loferski about developing photovoltaic cell for proposed Earth-orbiting satellites.

1957. Hoffman Electronics achieves 8 percent efficient photovoltaic cells.

1958. T. Mandelkorn of U.S. Signal Corps Laboratories fabricates n-on-p (negative layer on positive layer) silicon photovoltaic cells, making them more resistant to radiation; this is critically important for cells used in space.

Hoffman Electronics achieves 9 percent solar cell efficiency.

A small array (less than 1 W) on the Vanguard I space satellite powers its radios. Later that year, Explorer III, Vanguard II, and Sputnik-3 will be launched with PV-powered systems on board. Silicon solar cells become the most widely energy source for space applications, and remain so today.

1959. Hoffman Electronics achieves a 10 percent efficient, commercially available cell. Hoffman also learns to use a grid contact, significantly reducing the series resistance.

On August 7, the Explorer VI satellite is launched with a PV array of 9600 solar cells, each measuring 1 cm × 2cm. On October 13 Explorer VII is launched.

1960. Hoffman Electronics achieves 14 percent photovoltaic cells.

Silicon Sensors, Inc., of Dodgeville, Wisconsin, is founded and begins producing selenium photovoltaic cells.

1962. Bell Telephone Laboratories launches Telstar, the first telecommunication satellite; its initial power is 14 W.

1963. Sharp Corporation succeeds in producing silicon PV modules.

Japan installs a 242-W photovoltaic array, the world's largest to date, on a lighthouse.

1965. Peter Glaser conceives the idea of the satellite solar power station.

1966. NASA launches the first orbiting astronomical observatory powered by a 1-kW photovoltaic array; it provides astronomical data in the ultraviolet and x-ray wavelengths filtered out by Earth's atmosphere.

1969. A "solar furnace" is constructed in Odeillo, France; it features an eight-story parabolic mirror.

1970. With help from Exxon corporation. Dr. Elliot Berman designs a significantly less costly solar cell, bringing the price down from $100/W to $20/W. Solar cells begin powering navigation warning lights and horns on offshore gas and oil rigs, lighthouses, and railroad crossings. Domestic solar applications are considered good alternatives in remote areas where utility-grid connections are too expensive.

1972. French workers install a cadmium sulfide photovoltaic system at a village school in Niger.

The Institute of Energy Conversion is established at the University of Delaware to do research and development on thin-film photovoltaic and solar thermal systems, becoming the world's first laboratory dedicated to PV research and development.

1973. The University of Delaware builds "Solar One," a PV/thermal hybrid system. Roof-integrated arrays feed surplus power through a special meter to the utility during the day; power is purchased from the utility at night. In addition to providing electricity, the arrays are like flat-plate thermal collectors; fans blow warm air from over the array to heat storage bins.

1976. The NASA Lewis Research Center starts installing the first of 83 photovoltaic power systems on every continent except Australia. They provide power for vaccine refrigeration room lighting, medical clinic lighting, telecommunications, water pumping, grain milling, and television. The project takes place from 1976 to 1985 and then from 1992 to completion in 1995. David and Christopher Wronski of RCA Laboratories produce the first amorphous silicon photovoltaic cells, which could be less expensive to manufacture than crystalline silicon devices.

1977. In July, the U.S. Energy Research and Development Administration, a predecessor of the U.S. Department of Energy, launches the Solar Energy Research Institute [today's National Renewable Energy Laboratory (NREL)], a federal facility dedicated to energy finding and improving ways to harness and use energy from the sun.

Total photovoltaic manufacturing production exceeds 500 kW; 1 kW is enough power to light about ten 100-W lightbulbs.

1978. NASA's Lewis Research Center installs a 3.5-kW photovoltaic system on the Indian Reservation in southern Arizona—the world's first village system. It provides power for water pumping and residential electricity in 15 homes until 1983, when grid power reaches the village. The PV system is then dedicated to pumping water from a community well.

1980. ARCO Solar becomes the first company to produce more than 1 MW (1000 kW) of photovoltaic modules in 1 year.

At the University of Delaware, the first thin-film solar cell exceeds 10 percent efficiency; it's made of copper sulfide and cadmium sulfide.

1981. Paul MacCready builds the first solar-powered aircraft—the Solar Challenger—and flies it from France to England across the English Channel. The aircraft has more than 16,000 wing-mounted solar cells producing 3000 W of power.

1982. The first megawatt-scale PV power station goes on line in Hisperia, California. The 1-MW-capacity system, developed by ARCO Solar, has modules on 108 dual-axis trackers.

Australian Hans Tholstrup drives the first solar-powered car—the Quiet achiever—almost 2800 mi between Sydney and in 20 days—10 days faster than the first gasoline-powered car to do so.

1983. ARCO Solar dedicates a 6-MW photovoltaic substation in central California. The 120-acre, unmanned facility supplies the Pacific Gas Electric Company's utility grid with enough power for up to 2500 homes. Solar Design Associates completes a home powered by an integrated, stand-alone, 4-kW photovoltaic system in the Hudson River Valley. Worldwide, photovoltaic production exceeds 21.3 MW, and sales top $250 million.

1984. The Sacramento Municipal Utility District commissions its first 1-MW photovoltaic electricity generating facility.

1985. Researchers at the University of South Wales break the 20 percent efficiency barrier for silicon solar cells.

1986. The world's largest solar thermal facility is commissioned in Kramer Junction, California. The solar field contains rows of mirrors that concentrate the sun's energy onto a system of pipes circulating a heat transfer fluid. The heat transfer fluid is used to produce steam, which powers a conventional turbine to generate electricity.

1988. Dr. Alvin Marks receives patents for two solar power technologies: Lepcon and Lumeloid. Lepcon consists of glass panels covered with millions of aluminum or copper strips, each less than a thousandth of a millimeter wide. As sunlight hits the metal strips, light energy is transferred to electrons in the metal, which escape at one end in the form of electricity. Lumeloid is similar but substitutes cheaper, film-like sheets of plastic for the glass panels and covers the plastic with conductive polymers.

1991. President George Bush announces that the U.S. Department of Energy's Solar Energy Research Institute has been designated the National Renewable Energy Laboratory.

1992. Researchers at the University of South Florida develop a 15.9 percent efficient thin-film photovoltaic cell made of cadmium telluride, breaking the 15 percent barrier for this technology.

A 7.5-kW prototype dish system that includes an advanced membrane concentrator begins operating.

1993. Pacific Gas & Electric installs the first grid-supported photovoltaic system in Kerman, California. The 500-kW system is the first "distributed power" PV installation.

The National Renewable Energy Laboratory (formerly the Solar Energy Research Institute) completes construction of its Solar Energy Research Facility; it will be recognized as the most energy-efficient of all U.S. government buildings in the world.

1994. The first solar dish generator to use a free-piston engine is hooked up to a utility grid.

The National Renewable Energy Laboratory develops a solar cell made of gallium indium phosphide and gallium arsenide; it's the first one of its kind to exceed 30 percent conversion efficiency.

1996. The world's most advanced solar-powered airplane, the *Icare*, flies over Germany. Its wings and tail surfaces are covered by 3000 superefficient solar cells, for a total area of 21 m^2. The U.S. Department of Energy and an industry consortium begin operating Solar Two—an upgrade of the Solar One concentrating solar power tower. Until the project's end in 1999, Solar Two demonstrates how solar energy can be stored efficiently and economically so power is produced even when the sun isn't shining; it also spurs commercial interest in power towers.

1998. On August 6, a remote-controlled, solar-powered aircraft, *Pathfinder*, sets an altitude record of 80,000 ft on its thirty-ninth consecutive flight in Mojave, California—higher than any prop-driven aircraft to date.

Subhendu Guha, a scientist noted for pioneering work in amorphous silicon, leads the invention of flexible solar shingles, a roofing material and state-of-the-art technology for converting sunlight to electricity on buildings.

1999. Construction is completed on 4 Times Square in New York, the tallest skyscraper built in the city in the 1990s. It has more energy-efficient features that any other commercial skyscraper and includes building-integrated photovoltaic (BIPV) panels on the thirty-seventh through forty-third floors on the south- and west-facing facades to produce part of the building's power.

Spectrolab, Inc. and the National Renewable Energy Laboratory develop a 32.3 percent efficient solar cell. The high efficiency results from combining three layers of photovoltaic materials into a single cell, which is most efficient and practical in devices with lenses or mirrors to concentrate the sunlight. The concentrator systems are mounted on trackers to keep them pointed toward the sun.

Researchers at the National Renewable Energy Laboratory develop a breaking prototype solar cell that measures 18.8 percent efficient, topping the previous record for thin-film cells by more than 1 percent. Cumulative installed photovoltaic capacity reaches 1000 MW, worldwide.

2000. First Solar begins production at the Perrysburg, Ohio, photovoltaic manufacturing plant, the world's largest at the time; estimates indicate that it can produce enough solar panels each year to generate 100 MW of power. At the International Space Station, astronauts begin installing solar panels on what will be the largest solar power array deployed in space, each wing consisting of an array of 2800 solar cell modules.

Industry Researchers develop a new inverter for solar electric systems that increases safety during power outages. Inverters convert the dc electric output of solar systems to alternating current—the standard for household wiring as well as for power lines to homes.

Two new thin-film solar modules developed by BP Solarex break previous performance records. The company's 0.5-m^2 module has a 10.8 percent conversion efficiency—the highest in the world for similar thin-film modules. Its 0.9-m^2 module achieves 10.6 percent efficiency and a power output of 91.5 W—the highest in the world for a thin-film module.

The 12-kW solar electric system of a Morrison, Colorado, family is the largest residential installation in the United States to be registered with the U.S. Department of Energy's Million Solar Roofs program. The system provides most of the electricity for the family of eight's 6000-ft^2 home.

2001. Home Depot begins selling residential solar power systems in three stores in California. A year later it expands sales to 61 stores nationwide.

NASA's solar-powered aircraft, *Helios*, sets a new world altitude record for non-rocket-powered craft: 96,863 ft (more than 18 mi up).

2002. ATS Automation Tooling Systems, Inc., in Canada begins commercializing spheral solar technology. Employing tiny silicon beads between two sheets of aluminum foil, this solar-cell technology uses much less silicon than conventional multicrystalline silicon solar cells, thus potentially reducing costs. The technology was first championed in the early 1990s by Texas Instruments, but TI later discontinued work on it. For more, see the DOE Photovoltaic Manufacturing Technologies Web site.

The largest solar power facility in the Northwest—the 38.7-kW system White Bluffs Solar Station—goes on line in Richland, Washington.

PowerLight Corporation installs the largest rooftop solar power system in the United States—a 1.18-MW system at Santa Rita Jail, in Dublin, California.

LIST OF SUSTAINABLE ENERGY

EQUIPMENT SUPPLIERS AND

CONSULTANTS

An updated listing of this roster can be accessed through www.pvpower.com/pvinteg. html. The author does not endorse the listed companies or assume responsibility for inadvertent errors. Photovoltaic (PV) design and installation companies who wish to be included in the list could register by e-mail under the Web site.

A & M Energy Solutions

Business type: manufacturer of very high efficiency
  industrial grade mega solar power photovoltaic systems
3425 Fujita Street, Torrance, CA 90505
Tel 1-310-325-8091

A & M Energy Solutions

Business type: solar power contractors
2118 Wilshire Blvd., # 718, Santa Monica, California 90403
Phone: 1-310-445-9888
URL: www.amenenergysolutions.com

ABS Alaskan, Inc.

Business type: systems design, integration
Product types: systems design, integration, PV and other small power systems
2130 Van Horn Rd, Fairbanks, AK 99701
Phone: Fairbanks (907) 451-7145; Anchorage (907) 562-4949
Toll Free: (800) 478-7145
U.S. Toll Free: (800) 235-0689
Fax: Fairbanks (907) 451-1949
E-mail: abs@absak.com
URL: www.absak.com

AES Alternative Energy Systems, Inc.

Business type: PV, wind and micro-hydrosystems design, integration
Product types: systems design, integration, charge control/load centers
Contact: J. Fernando Lamadrid B.
9 E 78th St. New York, NY 10021
Phone: (212) 517-9326
Fax: (212) 517-5326
E-mail: aes@altenergysys.com
URL: www.altenergysys.com

Alternative Power Systems

Business type: systems design, integration, sales
Product types: systems design, integration
Sales contact: James Hart San Diego, CA
Phone: (877) 946-3786 (877-WindSun)
Fax: (760) 434-3407
E-mail: engineering@aapspower.com
URL: www.aapspower.com

Alternatif Enerji Sistemleri Sanayi Ticaret Ltd. Sti.

Business type: systems design, integration product sales
Product types: systems design, integration, product sales
Nispetiye Cad. No: 18/A Blok D.6 1.Levent Istanbul, Turkey
Phone: 90 (212) 283 74 45 pbx
Fax: ~90 (212) 264 00 87
E-mail: info@alternatifenerji.com
URL: www.alternatifenerji.com

Abraham Solar Equipment

Business type: PV systems installation, distribution
Product types: system installation, distribution
124 Creekside Pl, Pagosa Springs, CO 81147
Phone: (800) 222-7242

Advanced Energy Systems, Inc.

Business type: manufacture and distribute PV systems and lighting packages
Product types: PV power packages and lighting systems and
   energy-efficient lighting
9 Cardinal Dr., Longwood, FL 32779
Phone: (407) 333-3325
Fax: (407) 333-4341
E-mail: magicpwr@magicnet.net
URL: www.advancednrg.com

AeroVironment, Inc.

Business type: PV systems design
Product type: system design
222 E Huntington Dr, Monrovia, CA 91016
Phone: (818) 357-9983
Fax: (818) 359-9628
E-mail: avgill@aol.com

Alpha Real, A.G.

Business type: PV systems design, installation
Product types: power electronics and systems engineering
Feldeggstrasse 89, CH-8008 Zurich, Switzerland
Phone: 01-383-02-08
Fax: 01-383-18-95

Altair Energy, LLC

Business type: PV systems design, installation, service
Product types: turnkey PV systems, services, warranty and maintenance
   contracts, financing
600 Corporate Cir, Suite M, Golden, CO 80401
Phone: (303) 277-0025 or (800) 836-8951
Fax: (303) 277-0029
E-mail: info@altairenergy.com
URL: altairenergy.com

ALTEN srl

Business type: PV systems design, engineering, installation, BOS,
  and module manufacture
Product types: system design, engineering, installation, modules, BOS
Via della Tecnica 57/B4, 40068 S. Lazzaro, Bologna, Italy
Phone: 39 051-6258396 or 39 051-6258624
Fax: 39 051-6258398
E-mail: alten@tin.it
URL: www.bo.cna.it/cermac/alt.htm

Alternative-Energie-Technik GmbH

Business type: PV systems design, engineering, installation, distribution
Product types: system design, engineering, installation, distribution
Industriestraße, 12D-66280, Sulzbach-Neuweiler, Germany
Phone: 06897-54337
Fax: 06897-54359
E-mail: info@aet.de
URL: www.aet.de

Alternative Energy Engineering

Business type: PV systems design, installation, distribution
Product types: system design, installation, distribution
P.O. Box 339-PV, Redway, CA 95560
Phone/order line: (800) 777-6609
Phone/techline: (707) 923-7216
URL: www.alt-energy.com

Alternative Energy Store

Business type: on-line component and system sales
Product types: solar and wind
4 Swan St, Lawrence, MA 01841
Fax/voice: (877) 242-6718
Phone orders: (877) 242-6718 or (207) 469-7026
URL: www.AltEnergyStore.com

Alternative Power, Inc.

Business type: PV systems design, installation, distribution
Product types: system design, installation, distribution
160 Fifth Ave, Suite 711, New York, NY 10010-7003
Phone: (212) 206-0022
Fax: (212) 206-0893
E-mail: dbuckner@altpower.com
URL: www.altpower.com

Alternative Solar Products

Business type: PV systems design, installation, distribution
Product types: system design, installation, distribution
Contact: Greg Weidhaas
27420 Jefferson Ave., Suite 104 B, Temecula, CA 92590-2668
Phone: (909) 308-2366
Fax: (909) 308-2388

American Photovoltaic Homes and Farms, Inc.

Business type: PV systems integration
Product type: design and construct PV-integrated homes
5951 Riverdale Ave, Riverdale, NY 10471
Phone: (718) 548-0428

Applied Power Corporation

Business type: PV systems design
Product type: system design
1210 Homann Dr, SE Lacey, WA 98503
Phone: (360) 438-2110
Fax: (360) 438-2115
E-mail: info@appliedpower.com
URL: www.appliedpower.com

Arabian Solar Energy & Technology (ASET)

Business type: PV and DC systems design, manufacture, and sales
Product types: PV systems, and dc systems
11 Sherif St, Cairo, Egypt
Phone: 20 2 393 6463 or 20 2 395 3996
Fax : 20 2 392 9744
E-mail: aset@asetegypt.com
URL: www.asetegypt.com

AriStar Solar Electric

Business type: PV systems sales, integration
Product types: systems sales and integration
3101 W Melinda Ln., Phoenix, AZ 85027
Phone: (623) 879-8085 or (888) 878-6786
Fax: (623) 879-8096
E-mail: aristar@uswest.net
URL: www.azsolar.com

Ascension Technology, Inc.

Business type: PV systems design, integration, BOS manufacturer
Product types: systems design and BOS
PO Box 6314, Lincoln, MA 01773
Phone: (781) 890-8844
Fax: (781) 890-2050
URL: www.ascensiontech.com

Atlantic Solar Products, Inc.

Business type: PV systems design and integration
Product types: systems design and integration
PO Box 70060, Baltimore, MD 21237
Phone: (410) 686-2500
Fax: (410) 686-6221
E-mail: mail@atlanticsolar.com
URL: www.atlanticsolar.com

Atlantis Energy Systems

Business type: PV system designer, manufacturer
Product types: PV systems, modules
9275 Beatty Dr., Sacramento, CA 95820
Phone: (916) 438-2930
E-mail: jomo13@atlantisenergy.com

B.C. Solar

Business type: PV system design, installation, training
Product types: systems design, installation, and training
PO Box 1102, Post Falls, ID 82854
Phone: (208) 667-9608
Phone: (208) 263-4290

Big Frog Mountain Corp.

Business type: PV and wind energy systems design, sales, integration
Product types: systems design, installation, distribution
Contact: Thomas Tripp
100 Cherokee Blvd, Suite 2109, Chattanooga, TN 37405
Phone: (423) 265-0307
Fax: (423) 265-9030
E-mail: sales@bigfrogmountain.com
URL: www.bigfrogmountain.com

Burdick Technologies Unlimited (BTU)

Business type: PV systems design, installation
Product types: systems design, installation; roofing systems
701 Harlan St, #64, Lakewood, CO 80214
Phone: (303) 274-4358

C-RAN Corporation

Business type: PV systems design, packaging
Product types: systems design (water purification, lighting, security)
666 4th St, Largo, FL 34640
Phone: (813) 585-3850
Fax: (813) 586-1777

CEM Design

Business type: PV systems design
Product types: systems design, architecture
520 Anderson Ave, Rockville, MD 20850
Phone: (301) 294-0682
Fax: (301) 762-3128

CI Solar Supplies Co.

Business type: PV systems
Product type: systems
PO Box 2805, Chino, CA 91710
Phone: (800) 276-5278 (800-2SOLAR8)
E-mail: jclothi@ibm.net
URL: www.cisolar.com

California Solar

Business type: PV systems design
Product types: systems design
627 Greenwich Dr., Thousand Oaks, CA 91360
Phone: (805) 379-3113
Fax: (805) 379-3027

CANROM Photovoltaics Inc.

Business type: systems integrator
Product types: systems design, installation
108 Aikman Ave, Hamilton, ON, Canada L8M 1P9
Phone: (905) 526-7634
Fax: (905) 526-9341
URL: www.canrom.com

Colorado Solar Electric

Business type: systems integrator
Product type: retail sales
6501 County Rd 313, New Castle, CO 81647
Phone: (970) 618-839
URL: www.cosolar.com

Creative Energy Technologies

Business type: PV systems and products
Product types: systems and efficient appliances
10 Main St., Summit, NY 12175
Phone: (888) 305-0278
Fax: (518) 287-1459
E-mail: info@cetsolar.com
URL: www.cetsolar.com

Currin Corporation

Business type: PV systems design, installation
Product types: systems design, installation
PO Box 1191, Midland, MI 48641-1191
Phone: (517) 835-7387
Fax: (517) 835-7395

Design Engineering Consultants

Business type: sustainable energy systems electromechanical consultants
Contact: Dr. Peter Gevorkian
10850 Riverside Dr., Suite 509, Toluca Lake, CA 91602
Tel: 1-818-980-7583
Fax: 1-818-475-5079
E-mail: Peter@decleed.com
URL: www.decleed.com

DCFX Solar Systems P/L

Business type: PV systems design, integration
Product types: systems design, integration, pyramid power system,
    transportable hybrid package systems
Mt Darragh Rd., PO Box 264, Pambula 2549 N.S.W., Australia
Phone: 612.64956922
Fax: 612.64956922
E-mail: dcfx@acr.net.au

Dankoff Solar Products, Inc.

Business type: PV systems design, installation, distribution
Product types: systems design, installation, distribution
2810 Industrial Rd., Santa Fe, NM 87505
Phone: (505) 473-3800
Fax: (505) 473-3830
E-mail: pumps@danksolar.com

Delivered Solutions

Business type: PV product distributor
Product types: PV products
PO Box 891240, Temecula, CA 92589
Phone: (800) 429-7650 (24 hours every day)
Phone: (909) 694-3820
Fax: (909) 699-6215

Direct Gain, LLC

Business type: PV systems design, installation
Product types: system design, installation
23 Coxing Rd, Cottekill, NY 12419
Phone: (914) 687-2406
Fax: (914) 687-2408
E-mail: RLewand@Worldnet.att.net

Direct Power and Water Corporation

Business type: PV systems design, installation
Product types: systems design, installation
3455-A Princeton NE, Albuquerque, NM 87107
Phone: (505) 889-3585
Fax: (505) 889-3548
E-mail: dirpowdd@directpower.com
URL: www.dirpowdd@directpower.com

Diversified Technologies

Business type: PV systems design, installation
Product types: systems design, installation
35 Wiggins Ave., Bedford, MA 01730-2345
Phone: (617) 466-9444

Eclectic Electric

Business type: PV systems design, installation
Product types: systems design, installation, training
127 Avenida del Monte Sandia Park, NM 87047
Phone: (505) 281-9538

EcoEnergies, Inc.

Business type: PV systems design, integration, distribution
Product types: PV and other renewables 171 Commercial St
Sunnyvale, CA 94086
Contact: Thomas Alexander
Phone: (408) 731-1228
Fax: (408) 746-3890
URL: http://www.ecoenergies.com

Ecotech (HK) Ltd.

Business type: PV systems and DHW design, integration, and distribution
Product types: PV and DHW Room 608, 6/F.
Yue Fung Industrial Building 35-45 Chai Wan Kok St., Tsuen Wan, N.T.,
   Hong Kong
Phone: (852) 2833 1252
Phone: (852) 2405 2252
Fax: (852) 2405 3252
E-mail: inquiry@ecotech.com.hk
URL: www.ecotech.com.hk

ECS Solar Energy Systems

Business type: PV systems packaging
Product types: system packaging modular power stations
6120 SW 13th St. Gainesville, FL 32608
Phone: (904) 377-8866 Phone: (904) 338-0056

Ehlert Electric and Construction

Business type: PV systems design, installation
Product types: pumping systems design, installation
HCR 62, Box 70, Cotulla, TX 78014-9708
Phone: (210) 879-2205
Fax: (210) 965-3010

Electro Solar Products, Inc.

Business type: PV systems design, installation
Product types: system design, manufacture (traffic control, lights, pumping)
502 Ives Pl, Pensacola, FL 32514
Phone: (904) 479-2191
Fax: (904) 857-0070

Electron Connection

Business type: PV systems design, installation, distribution
Product types: system design, installation, distribution
PO Box 203, Hornbrook, CA 96044
Phone/Fax: (916) 475-3401
Phone: (800) 945-7587
E-mail: econnect@snowcrest.net
URL: www.snowcrest.net/econnect

Electronics Trade and Technology Development Corp. Ltd.

Business type: PV distribution and export
Product types: PV distribution and export
Contact: M. H. Rao, General Manager
3001 Redhill Ave, Bldg. 5-103, Costa Mesa, CA 92626
Phone: (714) 557-2703
Fax: (714) 545-2723
E-mail: mhrao@pacbell.net

EMI

Business type: PV systems distribution
Product types: system and products distribution
Phone: (888) 677-6527

Energy Outfitters

Business type: PV systems design, integration
Product types: systems design, integration
136 S Redwood Hwy, PO Box 1888, Cave Junction, OR 87523
Phone: (800) 467-6527 (800-GO-SOLAR)
Office phone: (541) 592-6903
Fax: (503) 592-6747
E-mail: nrgoutfit@cdsnet.net

Energy Products and Services, Inc.

Business type: PV systems design, integration
Product types: systems design, integration, training
321 Little Grove LN., Fort Myers, FL 33917-3928
Phone: (941) 997-7669
Fax: (941) 997-8828

Enertron Consultants

Business type: PV systems design, integration
Product types: systems design, building integration
418 Benvenue Ave, Los Altos, CA 94024
Phone: (415) 949-5719
Fax: (415) 948-3442

Enn Cee Enterprises

Business type: PV systems design, thermal systems, integration
Product types: solar lanterns, indoor lighting, street lighting, garden lighting,
    water heating, dryers
Contact: K.S. Chaugule, Managing Director; Vipul K. Chaugule, Director
#542, First Stage, C M H Rd, Indiranagar, Bangalore, Karnataka, India
Phone: 91 (080) 525 9858 (Time Zone GMT 5:30)
Fax: 91 (080) 525 9858

EnviroTech Financial, Inc.

Business type: equipment trade and finance
Contact: Mr. Gene Beck
Orange, California
Tel: 1-714-532-2731
www.etfinancial.com

EV Solar Products, Inc.

Business type: PV systems design, installation
Product types: systems design, installation
Contact: Ben Mancini
2655 N Hwy 89, Chino Valley, AZ 86323
Phone: (520) 636-2201
Fax: (520) 636-1664
E-mail: evsolar@primenet.com
URL: www.evsolar.com

Feather River Solar Electric

Business type: PV systems design, integration
Product types: systems design, integration
4291 Nelson St, Taylorsville, CA 95983
Phone: (916) 284-7849

Flack & Kurtz Consulting Engineers

Business type: PV consulting engineers
Product types: systems engineering
Contact: Daniel H. Nall, AIA, PE
475 Fifth Ave., New York, NY 10017
Phone: (212) 951-2691
Fax: (212) 689-7489
E-mail: nall@ny.fk.com

Fran-Mar

Business type: PV systems design, integration
Product types: systems design, integration
9245 Babcock Rd, Camden, NY 13316
Phone: (315) 245-3916
Fax: (315) 245-3916

Gebrüder Laumans GmbH & Co. KG

Business type: tile company with installation license for bmc PV tiles in Germany
   and Benelux
Product types: roof and facade PV tile and slate installation
Stiegstrasse 88, D-41379 Brüggen, Germany
Phone: (0 21 57) 14 13 30
Fax: (0 21 57) 14 13 39

Generation Solar Renewable Energy Systems Inc.

Business type: PV/wind systems design, integration, installation, distribution
Product types: systems design, integration, installation, distribution
Contact: Richard Heslett
340 George St N, Suite 405, Peterborough, ON K9H 7E8
Canada phone: (705) 741-1700
E-mail: gensolar@nexicom.net
URL: www.generationsolar.com

GenSun, Incorporated

Business type: integrated PV system builder
Product types: unitized, self-contained, portable, zero installation
10760 Kendall Rd., PO Box 2000, Lucerne Valley, CA 92356
Phone: (760) 248-2689; (800) 429-3777 Fax: (760) 248-2424
E-mail: solar@gensun.com
URL: www.gensun.com

Geosolar Energy Systems, Inc.

Business type: PV systems design, integration
Product types: systems design, integration
3401 N Federal Hwy, Boca Raton, FL 33431 USA
Phone: (407) 393-7127
Fax: (407) 393-7165

Glidden Construction

Business type: PV systems design, integration
Product types: systems design, integration
3727-4 Greggory Way, Santa Barbara, CA 93105
Phone: (805) 966-5555
Fax: (805) 563-1878

Global Resource Options, LLP

Business type: PV systems design, manufacture, sales and consulting
Product type: commercial- and residential scale PV systems, consulting,
   design, installation
P.O. Box 51 Strafford, VT 05072
Phone: 802-765-4632
Fax: 802-765-9983
E-mail: global@sover.net
URL: www.GlobalResourceOptions.com

GO Solar Company

Business type: PV systems sales and integration
Product types: systems, components, integration
12439 Magnolia Blvd 132, North Hollywood, CA 91607
Phone: (818) 566-6870
Fax: (818) 566-6879
E-mail: solarexpert@solarexpert.com
URL: www.solarexpert.com

Grant Electric

Business type: Solar power contractor
Contact: Bruce Grant
16461 Shermanway Suite 175, Van Nuys, California 91406
Phone: 1-818-375-1977

Great Northern Solar

Business type: PV systems design, integration, distribution
Product types: Systems design, integration, distributor
Rte 1, Box 71, Port Wing, WI 54865
Phone: (715) 774-3374

Great Plains Power

Business type: PV systems design, integration
Product types: systems design, integration
1221 Welch St, Golden, CO 80401
Phone: (303) 239-9963
FAX: (303) 233-0410
E-mail: solar@bewellnet.com
URL: http://server2.hypermart.net/solar/public html

Green Dragon Energy

Business type: systems integrator
Product type: PV and wind systems
2 Llwynglas, Bont-Dolgadfan, Llanbrynmair, Powys SY19 7AR, Wales, UK
Phone: 44 (0) 1650 521 589
Mobile: 0780 386 0003
E-mail: dragonrg@globalnet.co.uk

Heinz Solar

Business type: PV lighting systems
Product types: lighting systems design, integration
16575 Via Corto East, Desert Hot Springs, CA 92240
Phone: (619) 251-6886
Fax: (619) 251-6886

Henzhen Topway Solar Co., Ltd/Shenzhen BMEC

Business type: assembles and manufactures components and packages
Product types: lanterns, lamps, systems, BOS
RM8-202, Hualian Huayuan, Nanshan Dadao, Nanshan, Shenzhen, P.R. China, Post
  Code: 518052
Phone: ~86 755 6402765; 6647045; 6650787
Fax: ~86 755 6402722
E-mail: info@bangtai.com
URL: www.bangtai.com

High Resolution Solar

Business type: PV system design, integration, distribution
Product types: systems design, integration, distributor
Contact: Jim Mixan
7209 S 39th St., Omaha, NE 68147
Phone: (402) 738-1538
E-mail: jmixansolar@worldnet.att.net

Hitney Solar Products, Inc.

Business type: PV systems design, integration
Product types: systems design, integration
2655 N Hwy 89, Chino Valley, AZ 86323
Phone: (520) 636-1001
Fax: (520) 636-1664

Horizon Industries

Business type: PV systems and product distribution, service
Product types: systems and product distributor, service
2120 LW Mission Rd, Escondido, CA 92029
Phone: (888) 765-2766 (888-SOLAR NOW)
Fax: (619) 480-8322

Hutton Communications

Business type: PV systems design, integration
Product types: systems design, integration
5470 Oakbrook Pkwy, #G, Norcross, GA 30093
Phone: (770) 729-9413
Fax: (770) 729-9567

I.E.I., Intercon Enterprises, Inc.

Business type: North American distributor, Helios Technology Srl
Product type: PV modules
Contact: Gilbert Stepanian
12140 Hidden Brook Terr, N Potomac, MD 20878
Phone: (301) 926-6097
Fax: (301) 926-9367
E-mail: gilberts@erols.com

Independent Power and Light

Business type: PV systems design, integration, distribution
Product types: systems design, integration, distributor
RR 1, Box 3054, Hyde Park, VT 05655
Phone: (802) 888-7194

Innovative Design

Business type: architecture and design services
Product types: systems design and integration
850 West Morgan St, Raleigh, NC 27603
Phone: (919) 832-6303
Fax: (919) 832-3339
E-mail: innovativedesign@mindspring.com
URL: www.innovativedesign.net
Nevada Office: 8275 S Eastern Suite 220, Las Vegas, NV 89123
Phone: (702) 990-8413
Fax: (702) 938-1017

Integrated Power Corporation

Business type: PV systems design, integration
Product types: systems design, integration
7618 Hayward Rd, Frederick, MD 21702
Phone: (301) 663-8279
Fax: (301) 631-5199
E-mail: sales@integrated-power.com

Integrated Solar, Ltd.

Business type: PV system design, integration, distribution, installation, service
Product type: design, integrator, distributor, catalog
1331 Conant St., Suite 107, Maumee, OH 43537
Phone: (419) 893-8565
Fax: (419) 893-0006
E-mail: ISL 11@ix.netcom.com
URL: www.tpusa.com/isolar

Inter-Island Solar Supply

Business type: PV systems design, integration, distribution
Product types: systems design, integration, distribution
761 Ahua St. Honolulu, HI 96819
Phone: (808) 523-0711
Fax: (808) 536-5586
URL: www.solarsupply.com

ITALCOEL s.r.l., Electronic & Energy Control Systems

Business type: system integrator and BOS manufacturer
Product types: PV systems, PV inverters, design
66, Loc. Crognaleto, I-65010 Villanova (PE), Italy EU
Phone: 39.85.4440.1
Fax: 39.85.4440.240
E-mail: dayafter@iol.it

Jade Mountain

Business type: catalog sales
Product types: PV system components and loads
PO Box 4616, Boulder, CO 80306-4616
Phone: (800) 442-1972; (303) 449-6601
Fax: 303-449-8266
E-mail: jade-mtn@indra.com
URL: www.jademountain.com

Johnson Electric Ltd.

Business type: PV systems design, integration, distribution
Product types: systems design, integration, distributor
2210 Industrial Dr., PO Box 673, Montrose, CO 81402
Phone: (970) 249-0840
Fax: (970) 249-1248

Kyocera Solar, Inc.

Business type: manufacturer and distributor
Product types: PV modules and systems
7812 E Acoma, Scottsdale, AZ 85260
Phone: (800) 223-9580; (480) 948-8003
Fax: (480) 483-2986
E-mail: info@kyocerasolar.com
URL: www.kyocerasolar.com

L and P Enterprise Solar Systems

Business type: PV systems design, integration
Product types: systems design, integration
PO Box 305, Lihue, HI 96766
Phone: (808) 246-9111
Fax: (808) 246-3450

Light Energy Systems

Business type: PV systems design, contracting, consulting
Product types: systems design, integration
965 D Detroit Ave, Concord, CA 94518
Phone: (510) 680-4343
E-mail: solar@lightenergysystems.com
URL: www.lightenergysystems.com

Lotus Energy Pvt. Ltd.

Business type: PV systems design, integration; BOS manufacture; training
Product types: systems design, integration; BOS; training
Contact: Jeevan Goff, Managing Director
PO Box 9219, Kathmandu, Nepal
Phone: 977 (1) 418 203 (Time Zone GMT 5:45)
Fax: 977 (1) 412 924
E-mail: Jeevan@lotusnrg.com.np
URL: www.southasia.com/Nepaliug/lotus

Moonlight Solar

Business type: PV systems design, integration
Product types: design, contracting, repair
3451 Cameo LN. Blacksburg, VA 24060
Phone/Fax: (540) 953-1046
E-mail: moonlightsolar@moonlightsolar.com
E-mail: URL: http://www.moonlightsolar.com

Mytron Systems Ind.

Business type: PV systems
Product type: solar cookers, lantern and PV systems
161, Vidyut Nagar B, Ajmer Rd. Jaipur 302021, India
Phone/Fax: 91-141-351434
E-mail: yogeshc@jp1.dot.net.in

Nekolux: Solar, Wind & Water Systems

Business type: PV, wind and micro-hydrosystems design and installation
Product types: systems design, integration, distributor
Contact: Vladimir Nekola
1433 W. Chicago Ave, Chicago, IL 60622
Phone: 312-738-3776
E-mail: vladimir@nekolux.com
URL: www.nekolux.com

New England Solar Electric (formerly Fowler Solar Electric)

Business type: PV systems design, integration
Product types: systems design, integration, book
401 Huntington Rd, PO Box 435, Worthington, MA 01098
Phone: (800) 914-4131
URL: www.newenglandsolar.com

Nextek Power Systems, Inc.

Business type: lighting integration
Product types: DC lighting for commercial applications using fluorescent
   or HID lighting
992 S Second St., Ronkonkoma, NY 11779
Phone: (631) 585-1005
Fax (631) 585-8643
E-mail: davem@nextekpower.com
URL: www.nextekpower.com
West Coast Office: 921 Eleventh St Suite 501, Sacramento, CA 95814
Phone: (916) 492-2445
Fax (916) 492-2176
E-mail: patrickm@nextekpower.com

Ning Tong High-Tech

Business type: BOS manufacturer
Product types: solar garden light, solar traffic light, batteries, solar tracker,
   solar modules, portable solar systems, solar simulator and tester
Room 404, 383 Panyu Rd., Shanghai, P.R. China 200052
Phone: 86 21 62803172
Fax: 86 21 62803172
E-mail: songchao38@21cn.com
URL: www.ningtong-tech.com

North Coast Power

Business Type: PV systems dealer
Product Type: PV systems dealer PO Box 151
Cazadero, CA 95421
Phone: (800) 799-1122
Fax: (877) 393-3955
E-mail: mmiller@utilityfree.com
URL: www.utilityfree.com

Northern Arizona Wind & Sun

Business type: PV systems integration, products distribution
Product types: systems and products integrator and distributor
PO Box 125, Tolleson, AZ 85353
Phone: (888) 881-6464; (623) 877-2317
Fax: (623) 872-9215
E-mail: Windsun@Windsun.com
URL: www.solar-electric.com, www.windsun.com
Flagstaff Office: 2725 E Lakin Dr, #2, Flagstaff, AZ 86004
Phone: (800) 383-0195; (928) 526-8017
Fax: (928) 527-0729

Northern Power Systems

Business type: power systems design, integration, installation
Product types: controllers, systems design, integration, installation
182 Mad River Park, Waitsfield, VT 05401
Phone: (802) 496-2955, x266
Fax: (802) 879-8600
E-mail: rmack@northernpower.com
URL: www.northernpower.com

Northwest Energy Storage

Business type: PV systems design, integration, distribution
Product types: systems design, integration, distributor
10418 Hwy 95 N, Sandpoint, ID
Phone: (800) 718-8816; (208) 263-6142

Occidental Power

Business type: PV systems design, integration, installation
Product types: systems design, integration, installation
3629 Taraval St., San Francisco, CA 94116
Phone: (415) 681-8861
Fax: (415) 681-9911
E-mail: solar@oxypower.com
URL: www.oxypower.com

Off Line Independent Energy Systems

Business type: PV systems design, integration
Product types: systems design, integration
PO Box 231, North Fork, CA 93643
Phone: (209) 877-7080
E-mail: ofln@aol.com

Oman Solar Systems Company, L.L.C. (Division of AJAY Group of Companies)

Business type: systems, design, integration, installation, consulting
Product types: PV systems, wind generators, water pumps, solar hot water systems
Contact: N.R. Rao, PO Box 1922, RUWI 112, Oman
Phone: 00968 - 592807, 595756, 591692
Fax: 00968 - 591122, 7715490
E-mail: oss.marketing@ajaygroup.com

Phasor Energy Company

Business type: PV systems design, integration
Product types: systems design, integration
4202 E Evans Dr, Phoenix, AZ 85032-5469
Phone: (602) 788-7619
Fax: (602) 404-1765

Photovoltaic Services Network, LLC (PSN)

Business type: PV systems design, integration, package grid-tied systems
Product types: systems design, integration
215 Union Blvd., Suite 620, Lakewood, CO 80228
Phone: (303) 985-0717; (800) 836-8951
Fax: (303) 980-1030
E-mail: tschuyler@neosdenver.com

Planetary Systems

Business type: PV systems design, integration, distributor
Product types: systems design, integration, distributor
PO Box 9876, 2400 Shooting Iron Ranch Rd, Jackson, WY 83001
Phone/Fax: (307) 734-8947

Positive Energy, Inc.

Business type: PV systems design, integration
Product types: systems design, integration
3900 Paseo del Sol #201, Santa Fe, NM 87505
Phone: (505) 424-1112
Fax: (505) 424-1113
E-mail: info@positivenergy.com
URL: www.positivenergy.com

PowerPod Corp.

Business type: PV systems design, integration
Product types: modular PV systems for village electrification
PO Box 321, Placerville, CO 81430
Phone: (970) 728-3159
Fax: (970) 728-3159
E-mail: solar@rmi.com
URL: www.powerpod.com

Rainbow Power Company Ltd.

Business type: system design, integration, maintenance, repair
Product types: systems integration, systems design, maintenance and repair
1 Alternative Way, PO Box 240, Nimbin, NSW, Australia, 2480
Phone: (066) 89 1430
Fax: (066) 89 1109
International phone: 61 66 89 1088
International fax: 61 66 89 1109
E-mail: rpcltd@nor.com.au
URL: www.rpc.com.au

Real Goods Trading Company

Business type: PV systems design, integration, distribution, catalog
Product types: systems design, integration, distributor, catalog
966 Mazzoni St., Ukiah, CA 95482
Phone: (800) 762-7325
E-mail: realgood@realgoods.com
URL: www.realgoods.com

Remote Power, Inc.

Business type: PV systems design, integration
Product types: systems design, integration
12301 N Grant St, #230 Denver, CO 80241-3130
Phone: (800) 284-6978
Fax: (303) 452-9519
E-mail: RPILen@aol.com

Renewable Energy Concepts, Inc.

Business type: PV system design, installation, sales
Product types: PV panels, wind turbines, inverters, batteries
1545 Higuera St, San Luis Obispo, CA 93401
Phone: (805) 545-9700; (800) 549-7053
Fax: (805) 547-0496
E-mail: info@reconcepts.com
URL: www.reconcepts.com

Renewable Energy Services, Inc., of Hawaii

Business type: PV systems design, integration
Product types: systems design, integration
PO Box 278, Paauilo, HI 96776
Phone: (808) 775-8052
Fax: (808) 7775-0852

Resources & Protection Technology

Business type: PV / BIPV system design, integration, distribution
Product types: PV, solar thermal, heat pump
4A, Block 2, Dragon Centre, 25 Wun Sha St. Tai Hang, Hong Kong
Phone: (852) 8207 0801
Fax: (852) 8207 0802
E-mail: info@rpt.com.hk
URL: www.rpt.com.hk

RGA, Inc.

Business type: PV lighting systems
Product types: lighting systems
454 Southlake Blvd Richmond, VA 233236
Phone: (804) 794-1592
Fax: (804) 3779-1016

RMS Electric

Business type: PV systems design, integration
Product types: systems design, integration
2560 28th St., Boulder, CO 80301
Phone: (303) 444-5909
Fax: (303) 444-1615
E-mail: info@rmse.com
URL: www.rmse.com

Roger Preston Partners

Business type: system design, integration, engineering
Product types: systems integration, systems design, energy engineering
1050 Crown Point Pkwy, Suite 1100, Atlanta, GA 30338
Phone: (770) 394-7175
Fax: (770) 394-0733
E-mail: rpreston@atl.mindspring.com

Roseville Solar Electric

Business type: PV systems design, integration, distribution, and installation
Product type: grid-tie, battery backup, residential, commercial
Contact: Kevin Hahner
PO Box 38590, Sacramento, CO 38590
Phone: (916) 772-6977; (916) 240-6977
E-mail: khahner@juno.com

SBT Designs

Business type: system sales and installation
Product types: system sales and installation
25840 IH-10 West #1, Boerne, Texas 78006
Phone: (210) 698-7109
Fax: (210) 698-7147
E-mail: sbtdesigns@bigplanet.com
URL: www.sbtdesigns.com

S C Solar

Business type: PV and solar thermal systems sales, integration
Product types: systems design, sales, and distribution
7073 Henry Harris Rd, Lancaster, SC 29720
Phone/Fax: (803) 802-5522
E-mail: dwhigham@scsolar.com
URL: www.scsolar.com

SEPCO—Solar Electric Power Co.

Business type: manufacturer of PV lighting systems and OEM PV systems
Product types: PV lighting and power systems
Contact: Steven Robbins
7984 Jack James Dr., Stuart, FL 34997
Phone: (561) 220 6615
Fax: (561) 220 8616
E-mail: sepco@tcol.net

Siam Solar & Electronics Co., Ltd.

Business type: Solarex distributor
Product types: laminator of custom-size PV modules, sine wave inverters,
    12-V dc ballasts
Contact: Mr.VIWAT SRI-ON (Managing Director)
62/16-25 Krungthep-Nontaburi Rd, Nontaburi, 11000, Thailand
Phone: 66-2-5260578
Fax: 66-2-5260579
E-mail: sattaya@loxinfo.co.th

Sierra Solar Systems

Business type: PV systems design, integration
Product types: systems design, integration
109 Argall Way, Nevada City, CA 95959
Phone: (800) 517-6527
Fax: (916) 265-6151
E-mail: solarjon@oro.net
URL: www.sierrasolar.com

Solar Age Namibia Pty. Ltd.

Business type: PV systems design, integration
Product types: systems design, integration, village lighting
PO Box 9987 Windhoek, Namibia
Phone: 264-61-215809
Fax: 264-61-215793
E-mail: solarage@iafrica.com.na
URL: www.iafrica.com.na/solar

Solar Century

Business type: PV systems design, integration
Product types: systems design and installation
91-94 Lower Marsh, London SE 1 7AB, UK
Phone: 44 (0)207 803 0100
Fax: 44 (0)207 803 0101
URL: www.solarcentury.com

Solar Creations

Business type: PV systems design, integration
Product types: systems design, integration
2189 SR 511S Perrysville, OH 44864
Phone: (419) 368-4252

Solar Depot

Business type: PV systems design, integration
Product types: systems design, integration
61 Paul Dr., San Rafael, CA 94903
Phone: (415) 499-1333
Fax: (415) 499-0316
URL: www.solardepot.com

Solar Design Associates

Business type: PV systems design, building integration, architecture
Product types: systems design, building integration, architecture
PO Box 242, Harvard, MA 01451
Phone: (978) 456-6855
Fax: (978) 456-3030
E-mail: sda@solardesign.com
URL: www.solardesign.com

Solar Dynamics, Inc.

Business type: manufacture portable PV system package
Product type: portable PV system
152 Simsbury Rd, Building 9, Avon, CT 06001
Phone: (877) 527-6461 (877-JASMINI); (860) 409-2500
Fax: (860) 409-9144
E-mail: info@solar-dynamics.com
URL: www.solar-dynamics.com

Solar Electric, Inc.

Business type: PV systems design, integration, distribution
Product types: systems design, integration, distributor
5555 Santa Fe St., #J San Diego, CA 92109
Phone: (800) 842-5678; (619) 581-0051
Fax: (619) 581-6440
E-mail: solar@cts.com
URL: www.solarelectricinc.com

Solar Electric Engineering, Inc.

Business type: PV systems design, integration, distribution
Product types: systems design, integration, distributor
116 4th St., Santa Rosa, CA 95401
Phone: (800) 832-1986

Solar Electric Light Fund

Business type: PV systems design, integration
Product types: systems design, integration
1734 20th St, NW., Washington, DC 20009
Phone: (202) 234-7265
Fax: (202) 328-9512
URL:www.self.org

Solar Electric Light Company (SELCO)

Business type: PV systems design, integration
Product types: systems design, integration
35 Wisconsin Cir., Chevy Chase, MD 20815
Phone: (301) 657-1161
Fax: (301) 657-1165
URL: www.selco-intl.com
India URL: www.selco-india.com
Vietnam URL: www.selco-vietnam.com
Sri Lanka URL: www.selco-srilanka.com

Solar Electric Specialties Co.

Business type: PV systems design, integration
Product types: systems design, integration
PO Box 537, Willits, CA 95490
Phone: (800) 344-2003
Fax: (707) 459-5132
E-mail: seswillits@aol.com
URL: www.solarelectric.com

Solar Electric Systems of Kansas City

Business type: PV lighting systems
Product types: lighting systems
13700 W 108th St, Lenexa, KS 66215
Phone: (913) 338-1939
Fax: (913) 469-5522
E-mail: solarelectric@compuserve.com
URL: www.solarbeacon.com

Solar Electrical Systems

Business type: PV systems design, integration, distribution
Product types: systems design, integration, distributor
2746 W Appalachian Ct., Westlake Village, CA 91362
Phone: (805) 373-9433, (310) 202-7882
Fax: (805) 497-7121, (310) 202-1399
E-mail: ses@pacificnet.net

Solar Energy Systems of Jacksonville

Business type: PV systems design, integration
Product types: systems design, integration
4533 Sunbeam Rd, #302, Jacksonville, FL 32257
Phone: (904) 731-2549
Fax: (904) 731-1847

Solar Energy Systems Ltd.

Business type: PV systems design, integration
Product types: systems design, integration
Unit 3, 81 Guthrie St., Osborne Park, Western Australia 6017
Phone: ~61 (0)8.9204 1521
Fax: ~61 (0)8.9204 1519
E-mail: amaslin@sesltd.com.au
URL: www.sesltd.com.au

Solar Engineering and Contracting

Business type: PV systems design, integration
Product types: systems design, integration
PO Box 690, Lawai, HI 96765
Phone: (808) 332-8890
Fax: (808) 332-8629

The Solar Exchange

Business type: PV systems design, integration
Product types: water pumping and home, systems
PO Box 1338, Taylor, AZ 85939
Phone: (520) 536-2029; (520) 521-0929
E-mail: solarexchange@cybertrails.com
URL: http://skybusiness.com/thesolarexchange

Solar Grid

Business type: catalog sales
Product types: PV system components
2965 Staunton Rd, Huntington, WV 25702
Order line: (800) 697-4295
Tech line: (304) 697-1477
Fax: (304) 697-2531
E-mail: sales@solarg.com
URL: www.solarg.com

Solar Integrated Technologies

Business: manufacturer of flexible solar power mats
1837 E. Martin Luther King Jr. Blvd, Los Angeles, CA 90058
Phone: 323-231-0411
Fax: 323-231-0517

Solar Online Australia

Business type: PV and wind products, design, supply, integration
Product types: components, systems, design, integration
48 Hilldale Dr., Cameron Park NSW 2285, Australia
Phone: ~61 2 4958 6771
E-mail: info@solaronline.com.au
URL: www.solaronline.com.au

Solar Outdoor Lighting, Inc. (SOL)

Business type: PV street lighting systems
Product types: street lighting systems
3131 SE Waaler St, Stuart, FL 34997
Phone: (407) 286-9461
Fax: (407) 286-9616
E-mail: lightsolar@aol.com
URL: www.solarlighting.com

Solar Quest, Becker Electric

Business type: PV systems design, integration, distribution
Product types: systems design, integration, distributor
28706 New School Rd., Nevada City, CA 95959
Phone: (800) 959-6354; (916) 292-1725
Fax: (916) 292-1321

Solar Sales Pty. Ltd.

Business type: PV systems design, integration
Product types: systems design, integration
97 Kew St, PO Box 190, Welshpool 6986, Western Australia
Phone: 618.03622111
Fax: 618.94721965
E-mail: solar@ois.com.au
URL: www.solarsales.oz.nf

Solar Sense

Business type: PV systems integration
Product types: small portable solar power systems and battery chargers
Contact: Lindsay Hardie
7725 Lougheed Hwy, Burnaby, BC, Canada V5A 4V8
Phone: (800) 648-8110; (604) 656-2132
Fax: (604) 420-1591
E-mail: info@solarsense.com
URL: www.solarsense.com

Solar-Tec Systems

Business type: PV systems sales, integration
Product types: systems sales, integration
33971-A Silver Lantern, Dana Point, CA 92629
Phone: (949) 248-9728
Fax: (949) 248-9729
URL: solar-tec.com

Solartronic

Business type: PV systems design, integration, sales
Product types: systems design, integration; product distributor
Morelos Sur No. 90 62070 Col. Chipitlán Cuernavaca, Mor., Mexico
Phone: 52 (73)18-9714
Fax: 52 (73)18-8609
E-mail: info@solartronic.com
URL: www.solartronic.com

Solartrope Supply Corporation

Business type: wholesale supply house
Product types: systems components
Phone: (800) 515-1617

Solar Utility Company, Inc.

Business type: PV systems design, integration
Product types: systems design, integration
Contact: Steve McKenery
6160 Bristol Pkwy, Culver City, CA 90230
Phone: (310) 410-3934
Fax: (310) 410-4185

Solar Village Institute, Inc.

Business type: PV systems design, integration
Product types: systems design, integration
PO Box 14, Saxapawhaw, NC 27340
Phone: (910) 376-9530

Solar Works!

Business type: PV systems design, integration
Product types: systems design, integration
Contact: Daniel S. Durgin
PO Box 6264, 525 Lotus Blossom Ln, Ocean View, HI 96737
Phone: (808) 929-9820
Fax: (808) 929-9831
E-mail: ddurgin@aloha.net
URL: solarworks.com

Solar Works, Inc.

Business type: PV systems design, integration, distribution
Product types: systems design, integration, distributor
64 Main St., Montpelier, VT 05602
Phone: (802) 223-7804
E-mail: LSeddon@solar-works.com
URL: www.solar-works.com

Sollatek

Business type: systems design, installation, BOS manufacturer
Product types: systems design and installation
Unit 4/5, Trident Industrial Estate, Blackthorne Rd, Poyle Slough, SL3 0AX,
   United Kingdom
Phone: 44 1753 6883000
Fax: 44 1753 685306
E-mail: sollatek@msn.com

Soler Energie S.A. (Total Energie Group)

Business type: PV systems design, integration
Product types: systems design, integration
BP 4100, 98713 Papeete, French Polynesia
Phone: 689 43 02 00
Fax: 689 43 46 00
E-mail: soler@mail.pf
URL: www.total-energie.fr

Solo Power

Business type: PV systems design, integration
Product types: systems design, integration
1011-B Sawmill Rd, NW Albuquerque, NM 87104
Phone: (505) 242-8340
Fax: (505) 243-5187

Soltek Solar Energy Ltd.

Business type: PV systems design, integration, distribution, catalog
Product types: systems design, integration
2-745 Vanalman Ave., Victoria, BC V8Z 3B6 Canada
E-mail: soltek@pinc.com
URL: http://vvv.com/~soltek

SOLutions in Solar Electricity

Business type: PV systems design, sales, installation, consulting, training
Product types: system design, installation, consulting, training
Contact: Joel Davidson
PO Box 5089, Culver City, CA 90231
Phone: (310) 202-7882
Fax: (310) 202-1399
E-mail: joeldavidson@earthlink.net
URL: www.solarsolar.com

Soluz, Inc.

Business type: PV systems design, integration
Product types: international systems, design and distribution
Contact: Steve Cunningham
55 Middlesex St, Suite 221, North Chelmsford, MA 01863-1561
Phone: (508) 251-5290
Fax: (508) 251-5291
E-mail: soluz@igc.apc.org

Southwest Photovoltaic Systems, Inc.

Business type: PV systems design, integration
Product types: systems design, integration
212 E Main St., Tomball, TX 77375
Phone: (713) 351-0031
Fax: (713) 351-8356
E-mail: SWPV@aol.com

Sovran Energy, Inc.

Business type: PV systems design, integration, distribution
Product types: systems design, integration, distributor
13187 Trewhitt Rd., Oyama, BC, Canada V4V 2B17
Phone: (250) 548-3642
Fax: (250) 548-3610
E-mail: sovran@sovran.ca
URL: www.sovran.ca/sovran

Star Power International Limited

Business type: PV systems integration
Product types: systems design, integration
912 Worldwide Industrial Center, 43 Shan Mei St, Fotan, Hong Kong
Phone: (852) 26885555
Fax: (852) 26056466
E-mail: starpwr@hkstar.com

Stellar Sun

Business type: PV systems integration
Product types: systems design, integration
2121 Watt St., Little Rock, AR 72227
Phone: (501) 225-0700
Fax: (501) 225-2920
E-mail: bill@stellarsun.com
URL: http://stellarsun.com

Strong Plant & Supplies FZE

Business type: PV system design, integration, distribution, consulting services
Product types: PV systems, modules, charge controllers, power centers, inverters, lighting, and pumping products
Contact: Toufic E. Kadri
PO Box 61017, Dubai, United Arab Emirates
Phone: 971 4 835 531
Fax: 971 4 835 914
E-mail: strongtk@emirates.net.ae

Sudimara Solar/PT Sudimara Energi Surya

Business type: PV systems design, integration, distribution
Product types: systems design, integration, distributor
JI. Banyumas No. 4, Jakarta 10310 Indonesia
Phone: 3904071-3
Fax: 361639

Sun, Wind and Fire

Business type: PV systems design, integration
Product types: systems design, integration
7637 SW 33rd Ave, Portland, OR 97219-1860
Phone: (503) 245-2661
Fax: (503) 245-0414

SunAmp Power Company

Business type: PV systems design, integration, distribution
Product types: systems design, integration, distributor
7825 E Evans, #400, Scottsdale, AZ 85260
Phone: (800) 677-6527 (800-MR SOLAR)
E-mail: sunamp@sunamp.com
URL: www.sunamp.com

Sundance Solar Designs

Business type: PV systems design, integration
Product types: systems design, integration
PO Box 321, Placerville, CO 81430
Phone: (970) 728-3159
Fax: (970) 728-3159
E-mail: solar@rmi.com

Sunelco

Business type: PV systems design, integration
Product types: systems design, integration
PO Box 1499, 100 Skeels St, Hamilton, MT 59840
Phone: (800) 338-6844; (406) 363-6924
Fax: (406) 363-6046
E-mail: sunelco@montana.com
URL: www.sunelco.com

Sunergy Systems

Business type: PV equipment and systems
Product types: equipment and systems
PO Box 70, Cremona, AB T0M 0R0 Canada
Phone: (403) 637-3973

Sunmotor International Ltd.

Business type: manufacturer and systems installation for PV water pumping
Product type: solar water pumping systems
104, 5037 - 50th St., Olds, AB T4H 1R8 Canada
Phone: (403) 556-8755
Fax: (403) 556-7799
URL: www.sunpump.com

Sunnyside Solar, Inc.

Business type: PV systems design, integration, distribution
Product types: systems design, integration, distributor, lighting
RD 4, Box 808, Green River Rd., Brattleboro, VT 05301
Phone: (802) 257-1482

Sunpower Co.

Business type: PV systems design, integration, distribution
Product types: systems design, integration, distributor, pumping
Contact: Leigh and Pat Westwell
RR3, Tweed, ON K0K 3J0 Canada
Phone: (613) 478-5555
E-mail: sunpower@blvl.igs.net
URL: www.mazinaw.on.ca/sunpower.html

SunWize Technologies, Inc.

Business type: PV systems design, integration, distribution, manufacture
  of portable PV systems
Product types: systems design, integration, distributor, portable PV systems
1155 Flatbush Rd., Kingston, NY 12401
Contact: Bruce Gould, VP, Sales
Phone: (800) 817-6527; (845) 336-0146
Fax: (845) 336-0457
E-mail: sunwize@besicorp.com
URL: www.sunwize.com

Superior Solar Systems, Inc.

Business type: PV systems design, integration
Product types: systems design, integration
1302 Bennett Dr., Longwood, FL 32750
Phone: (800) 478-7656 (800-4PVsolar)
Fax: (407) 331-0305

Talmage Solar Engineering, Inc.

Business type: PV systems design, integration
Product types: systems design, integration
18 Stone Rd., Kennebunkport, ME 04046
Phone: (888) 967-5945
Fax: (207) 967-5754
E-mail: tse@talmagesolar.com
URL: www.talmagesolar.com

Technical Supplies Center Ltd. (TSC)

Business type: BOS distributor and system integrator
Product type: PV, wind, batteries, charge controllers, inverters
South 60th St, East Awqaff Complex, PO Box 7186, Sana'a, Republic of Yemen
Phone: 967 1 269 500
Fax : 967 1 267 067
E-mail: ZABARAH@y.net.ye

Thomas Solarworks

Business type: PV systems design, integration
Product types: systems design, integration
PO Box 171, Wilmington, IL 60481
Phone: (815) 476-9208
Fax: (815) 476-2689

Total Energie

Business type: PV systems design, integration
Product types: systems design, integration
7, chemin du Plateau, 69570 Lyon-Dardilly, France
Phone: 33 (0)4 72 52 13 20
Fax: 33 (0)4 78 64 91 00
E-mail: infos@total-energie.fr
URL: www.total-energie.fr

Utility Power Group

Business type: PV manufacture, systems design, integration
Product types: manufacture, systems design, integration
9410-G DeSoto Ave Chatsworth, CA 91311
Phone: (818) 700-1995
Fax: (818) 700-2518
E-mail: 71263.444@compuserve.com

Vector Delta Design Group, Inc.

Product types: turnkey electric and solar power design and integration
Contact: Dr. Peter Gevorkian
2325 Bonita Dr., Glendale, CA 91208
Phone: (818) 241-7479
Fax: (818) 243-5223
E-mail: vectordeltadesign@charter.net
URL: www.vectordelta.com

Vermont Solar Engineering

Business type: PV systems design, integration, distribution
Product types: systems design, integration, distributor
PO Box 697, Burlington, VT 05402
Phone: (800) 286-1252; (802) 863-1202
Fax: (802) 863-7908
E-mail: vtsolar1@together.net
URL: www.vtsolar.com

Whole Builders Cooperative

Business type: PV systems design, integration
Product types: systems design, integration
2928 Fifth Ave, S. Minneapolis, MN 55408-2412
Phone: (612) 824-6567
Fax: (612) 824-9387

Wind and Sun

Business type: PV systems design, integration, distribution
Product types: systems design, integration, distributor
The Howe, Watlington, Oxford OX9 5EX UK
Phone: (44) 1491-613859
Fax: (44) 1491-614164

WINSUND (Division of Hugh Jennings Ltd.)

Business type: PV and wind systems design, installation, distribution
Product types: systems design, installation, distribution
Tatham St, Sunderland SR1 2AG, England, UK
Phone: 44 191 514 7050
Fax: 44 191 564 1096
E-mail: info@winsund.com
URL: www.winsund.com

Woodland Energy

Business type: portable PV systems design, manufacture, and sales
Product type: portable PV systems
PO Box 247, Ashburnham, MA 01430
Phone: (978) 827-3311
E-mail: info@woodland-energy.com
URL: www.woodland-energy.com

WorldWater & Power Corporation

Type of business: solar power irrigation and water pumping
Pennington Business Park, 55 Route 31 South, Pennington, NJ 08534
Tel: 1-609-818-0700
E-mail: pump@waterworld.com

Zot's Watts

Business type: PV systems design, integration, distribution
Product types: systems design, integration, distributor
Contact: Zot Szurgot
1701 NE 75th St., Gainesville, FL 32641
Phone: (352) 373-1944
E-mail: roselle@gnv.fdt.net

GLOSSARY

Renewable Energy Power Systems

All those technical terms can make renewable energy systems difficult for many people to understand. This glossary aims to cover all the most commonly used terms, as well as a few of the more specific terms.

alternating current (ac): Electric current that continually reverses direction. The frequency at which it reverses is measured in cycles per second, or hertz (Hz). The magnitude of the current itself is measured in amps (A).

alternator: A device for producing ac electricity. Usually driven by a motor, but can also be driven by other means, including water and wind power.

ammeter: An electric or electronic device used to measure current flowing in a circuit.

amorphous silicone: A noncrystalline form of silicon used to make photovoltaic modules (commonly referred to as solar panels).

ampere (A): The unit of measurement of electric current.

ampere-hour (Ah): A measurement of electric charge. One ampere-hour of charge would be removed from a battery if a current of 1 A flowed out of it for 1 hour. The ampere-hour rating of a battery is the maximum charge that it can hold.

anemometer: A device used to measure wind speed.

anode: The positive electrode in a battery, diode, or other electric device.

axial flow turbine: A turbine in which the flow of water is in the same direction as the axis of the turbine.

battery: A device, made up of a collection of cells, used for storing electricity, which can be either rechargeable or nonrechargeable. Batteries come in many forms, and include flooded cell, sealed, and dry cell.

battery charger: A device used to charge a battery by converting (usually) ac alternating voltage and current to a dc voltage and current suitable for the battery. Chargers often incorporate some form of a regulator to prevent overcharging and damage to the battery.

beta limit: The maximum power (theoretically) that can be captured by a wind turbine from the wind, which equals 59.3 percent of the wind energy.

blade: The part of a turbine that water or air reacts against to cause the turbine to spin, which is sometimes incorrectly referred to as the propeller. Most electricity-producing wind turbines will have two or three blades, whereas water-pumping wind turbines will usually have up to 20 or more.

capacitor: An electronic component used for the temporary storage of electricity, as well as for removing unwanted noise in circuits. A capacitor will block direct current but will pass alternating current.

cathode: The negative electrode in a battery, diode, or other electric device.

cell: The most basic, self-contained unit that contains the appropriate materials, such as plates and electrolyte, to produce electricity.

circuit breaker: An electric device used to interrupt an electric supply in the event of excess current flow. Can be activated either magnetically, thermally, or by a combination of both, and can be manually reset.

compact fluorescent lamp: A form of fluorescent lighting that has its tube "folded" into a "U" or other more compact shape, so as to reduce the space required for the tube.

conductor: A material used to transfer or conduct electricity, often in the form of wires.

conduit: A pipe or elongated box used to house and protect electric cables.

converter: An electronic device that converts electricity from one dc voltage level to another.

cross-flow turbine: A turbine where the flow of water is at right angles to the axis of rotation of the turbine.

current: The rate of flow of electricity, measured in amperes. Analogous to the rate of flow of water measured in liters per second, which is also measured in amperes.

Darrius rotor: A form of vertical-axis wind turbine that uses thin blades.

diode: A semiconductor device that allows current to flow in one direction, while blocking it in the other.

direct current (dc): Electric current that flows in one direction only, although it may vary in magnitude.

dry cell battery: A battery that uses a solid paste for an electrolyte. Common usage refers to these as small cylindrical "torch" cells.

earth (or ground): Refers to physically connecting a part of an electric system to the ground, done as a safety measure, by means of a conductor embedded in suitable soil.

earth-leakage circuit breaker (KLCB): A device used to prevent electrical shock hazards in mains voltage power systems, includes independent power systems, which are also known as residual current devices (RCDs).

electricity: The movement of electrons (a subatomic particle) produced by a voltage through a conductor.

electrode: An electrically conductive material, forming part of an electric device, often used to lead current into or out of a liquid or gas. In a battery, the electrodes are also known as plates.

electrolysis: A chemical reaction caused by the passage of electricity from one electrode to another.

electrolyte: The connecting medium, often a fluid, which allows electrolysis to occur. All common batteries contain an electrolyte, such as the sulfuric acid used in lead-acid batteries.

energy: The abstract notion that makes things happen or that has the potential or ability to do work. It can be stored and converted between many different forms, such as heat, light, electricity, and motion. It is never created or destroyed but does become unavailable to us when it ends up as low-temperature heat. It is measured in joules (J) or watt-hours (Wh) but more usually megajoules (MJ) or kilowatt-hours (kWh).

equalizing charge: A flooded lead-acid battery will normally be charged in boost mode until the battery reaches 2.45 to 2.5 V per cell, at which time the connected regulator should switch into "float" mode, where the battery will be maintained at 2.3 to 2.4 V per cell. During an equalizing charge, the cells are overcharged at 2.5 to 2.6 V per cell to ensure that all cells have an equal (full charge). This is normally achieved by charging from a battery charger, though some regulators will perform this charge when energy use is low, such as when the users are not at home.

float charge: A way of charging a battery by varying the charging current, so that its terminal voltage, the voltage measured directly across its terminals, "floats" at a specific voltage level.

flooded cell battery: A form of rechargeable battery where the plates are completely immersed in a liquid electrolyte. The starter battery in most cars is of the flooded cell type. Flooded cell batteries are the most commonly used type for independent and remote area power supplies.

fluorescent light: A form of lighting that uses long thin tubes of glass that contain mercury vapor and various phosphor powders (chemicals based on phosphorus) to produce white light that is generally considered to be the most efficient form of home lighting. See also *compact fluorescent lamp.*

furling: A method of preventing damage to horizontal-axis wind turbines by automatically turning them out of the wind using a spring-loaded tail or other device.

fuse: An electric device used to interrupt an electric supply in the event of excess current flow. Often consists of a wire that melts when excess current flows through it.

gel-cell battery: A form of lead-acid battery where the electrolyte is in the form of a gel or paste. Usually used for mobile installations and when batteries will be subject to high levels of shock or vibration.

generator: A mechanical device used to produce dc electricity. Coils of wire passing through magnetic fields inside the generator produce power. See also *alternator.* Most ac-generating sets are also referred to as generators.

gigawatt (GW): A measurement of power equal to a thousand million watts.

gigawatt-hour (GWh): A measurement of energy. One gigawatt-hour is equal to 1 GW being used for a period of 1 hour or 1 MW being used for 1000 hours.

halogen lamp: A special type of incandescent globe made of quartz glass and a tungsten filament, which also contains a small amount of a halogen gas (hence the name), enabling it to run at a much higher temperature than a conventional incandescent globe. Efficiency is better than a normal incandescent, but not as good as a fluorescent light.

head: The vertical distance that water will fall from the inlet of the collection pipe to the water turbine in a hydropower system.

hertz (Hz): Unit of measurement for frequency. Is equivalent to cycles per second (refer to *alternating current*). Common household mains power is normally 60 Hz.

horizontal-axis wind turbine: The most common form of wind turbine, consisting of two or three airfoil-style blades attached to a central hub, which drives a generator. The axis or main shaft of the machine is horizontal or parallel to the earth's surface.

incandescent globe: This is the most common form of light globe in the home. It usually consists of a glass globe inside which is a wire filament that glows when electricity is passed through it. They are the least efficient of all electric lighting systems.

independent power system: A power generation system that is independent of the tile mains grid.

insolation: The level of intensity of energy from the sun that strikes the earth. Usually given as watts per square meter (W/m^2). A common level in Australia in summer is about 1000 W/m^2.

insulation: A material used to prevent the flow of electricity used on electric wires in order to prevent electric shock. Typical materials used include plastics, such as PVC and polypropylene, ceramics, and minerals, such as mica.

inverter: An electronic device used to convert dc electricity into ac electricity, usually with an increase in voltage. There are several different basic types of inverters, including sine wave and square-wave inverters.

junction box: An insulating box, usually made from plastics, such as PVC, used to protect the connection point of two or more cables.

kilowatt (kW): A measurement of power equal to 1000 W.

kilowatt-hour (kWh): A measurement of energy. One kilowatt-hour is equal to 1 kW being used for a period of 1 hour.

lead-acid battery: A type of battery that consists of plates made of lead and lead oxide, surrounded by a sulfuric acid electrolyte. The most common type of battery used in RAPS systems.

light emitting diode (LED): A semiconductor device, which produces light of a single color or very narrow band of colors. Light-emitting diodes are used for indicator lights, as well as for low-level lighting, readily available in red, green, blue, yellow, and amber. The lights have a minimum life of 100,000 hours of use.

load: The collective appliances and other devices connected to a power source. When used with a shunt regulator, a "dummy" load is often used to absorb any excess power being generated.

megawatt (MW): A measurement of power equal to 1 million W.

megawatt-hour (MWh): A measurement of power with respect to time (energy). One megawatt-hour is equal to 1 MW being used for a period of 1 hour, or 1 kW being used for 1000 hours.

meters per second (m/s): A speed measurement system often used to measure wind speed. One meter per second is equal to 2.2 mi/h or 3.6 km/h.

micro-hydrosystem: A generation system that uses water to produce electricity. Types of water turbines include Pelton, Turgo, cross flow, overshot, and undershot waterwheels.

modified square-wave: A type of waveform produced by some inverters. This type of waveform is better than a square wave but not as suitable for some appliances as a sine wave.

monocrystalline solar cell: A form of solar cell made from a thin slice of a single large crystal of silicon.

nacelle: That part of a wind generator that houses the generator, gearbox, and so forth at the top of the tower.

nickel-cadmium battery (NICAD): A form of rechargeable battery, having higher storage densities than that of lead-acid batteries, NICADs use a mixture of nickel hydroxide and nickel oxide for the anode and cadmium metal for the cathode. The electrolyte is potassium hydroxide. Very common in small rechargeable appliances, but rarely found in independent power systems, due to their high initial cost.

noise: Unwanted electric signals produced by electric motors and other machines that can cause circuits and appliances to malfunction.

ohm: The unit of measurement of electrical resistance. The symbol used is the uppercase Greek letter omega. A resistance of 1 ohm will allow 1 A of current to pass through it at a voltage drop of 1 V.

Ohm's law: A simple mathematical formula that allows voltage, current, or resistance to be calculated when the other two values are known. The formula is

$$V = IR$$

where V is the voltage, I is the current, and R is the resistance.

Pelton wheel: A water turbine in which specially shaped buckets attached to the periphery of a wheel are struck by a jet of water from a narrow nozzle.

photovoltaic effect: The effect that causes a voltage to be developed across the junction of two different materials when they are exposed to light.

pitch: Loosely defined as the angle of the blades of a wind or water turbine with respect to the flow of the wind or water.

plates: The electrodes in a battery. Usually take the form of flat metal plates. The plates often participate in the chemical reaction of a battery, but sometimes just provide a surface for the migration of electrons through the electrolyte.

polycrystalline silicon: Silicon used to manufacture photovoltaic panels, which is made up of multiple crystals clumped together to form a solid mass.

power: The rate of doing work or, more generally, the rate of converting energy from one form to another (measured in watts).

PVC (polyvinyl chloride): A plastic used as an insulator on electric cables, as well as for conduits. Contains highly toxic chemicals and is slowly being replaced with safer alternatives.

quasi sine wave: A description of the type of waveform produced by some inverters. See *modified square-wave.*

ram pump: A water pumping device that is powered by falling water. These devices work by using the energy of a large amount of water falling a small height to lift a small amount of water to a much greater height. In this way, water from a spring or stream in a valley can be pumped to a village or irrigation scheme on a hillside. Wherever a fall of water can be sustained, the ram pump can be used as a comparatively cheap, simple, and reliable means of raising water to considerable heights.

RAPS (remote area power supply): A power generation system used to provide electricity to remote and rural homes, usually incorporating power generated from renewable sources such as solar panels and wind generators, as well as nonrenewable sources, such as petroleum-powered generators.

rechargeable battery: A type of battery that uses a reversible chemical reaction to produce electricity, allowing it to be reused many times. Forcing electricity through the battery in the opposite direction to normal discharge reverses the chemical reaction.

regulator: A device used to limit the current and voltage in a circuit, normally to allow the correct charging of batteries from power sources, such as photovoltaic arrays and wind generators.

renewable energy: Energy that is produced from a renewable source, such as sunlight.

residual current device (RCD): See *earth-leakage circuit breaker*.

resistance: A material's ability to restrict the flow of electric current through itself (measured in ohms).

resistor: An electronic component used to restrict the flow of current in a circuit, also used specifically to produce heat, such as in a water heater element.

sealed lead-acid battery: A form of lead-acid battery where the electrolyte, is contained in an absorbent fiber separator or gel between the batteries plates. The battery is sealed so that no electrolyte can escape, and thus can be used in any position, even inverted.

semiconductor: A material that only partially conducts electricity that is neither an insulator nor a true conductor. Transistors and other electronic devices are made from semiconducting materials and are often called semiconductors.

shunt: A low-value resistance, connected in series with a conductor that allows measurements of currents flowing in the conductor by measurement of voltage across the shunt, which is often used with larger devices, such as inverters to allow monitoring of the power used.

sine wave: A sinusoidal-shaped electrical waveform. Mains power is a sine wave, as is the power produced by some inverters. The sine wave is the most ideal form of electricity for running more sensitive appliances, such as radios, TVs, and computers.

solar cell: A single photovoltaic circuit usually made of silicon that converts light into electricity.

solar module: A device used to convert light from the sun directly into dc electricity by using the photovoltaic effect. Usually made of multiple silicon solar cells bonded between glass and a backing material.

solar power: Electricity generated by conversion of sunlight, either directly through the use of photovoltaic panels, or indirectly through solar-thermal processes.

solar thermal: A form of power generation using concentrated sunlight to heat water or other fluid that is then used to drive a motor or turbine.

square wave: A type of waveform produced by some inverters. The square wave is the least desirable form of electricity for running most appliances. Simple resistors, such as incandescent globes and heating elements, work well on a square wave.

storage density: The capacity of a battery compared to its weight (measured in watt-hours per kilogram).

surge: An unexpected flow of excessive current, usually caused by a high voltage that can damage appliances and other electric equipment. Also, an excessive amount of power drawn by an appliance when it is first switched on.

switch mode: A form of converting one form of electricity to another by rapidly switching it on and off and feeding it through a transformer to effect a voltage change.

tip-speed ratio: The ratio of blade tip speed to wind speed for a wind turbine.

transformer: A device consisting of two or more insulated coils of wire wound around a magnetic material, such as iron, used to convert one ac voltage to another or to electrically isolate the individual circuits.

transistor: A semiconducting device used to switch or otherwise control the flow of electricity.

turbulence: Airflow that rapidly and violently varies in speed and direction that can cause damage to wind turbines. It is often caused by objects, such as trees or buildings.

vertical-axis wind turbine: A wind turbine with the axis or main shaft mounted vertically, or perpendicular to the earth's surface. This type of turbine does not have to be turned to face the wind—it always does.

voltage: The electric pressure that can force an electric current to flow through a closed circuit (measured in volts).

voltage drop: The voltage lost along a length of wire or conductor due to the resistance of that conductor. This also applies to resistors. The voltage drop is calculated using Ohm's law.

voltmeter: An electric or electronic device used to measure voltage.

water turbine: A device that converts the motion of the flow of water into rotational motion, which is often used to drive generators or pumps. See *micro-hydrosystem.*

waterwheel: A simple water turbine, often consisting of a series of paddles or boards attached to a central wheel or hub that is connected to a generator to produce electricity or a pump to move water.

watt (W): A measurement of power commonly used to define the rate of electricity consumption of an appliance.

watt-hour (Wh): A measurement of power with respect to time (energy). One watt-hour is equal to 1 W being used for a period of 1 hour.

wind farm: A group of wind generators that usually feed power into the mains grid.

wind generator: A mechanical device used to produce electricity from the wind. Typically a form of wind turbine connected to a generator.

wind turbine: A device that converts the motion of the wind into rotational motion used to drive generators or pumps. Wind generator, wind turbine, windmill, and other terms are commonly used interchangeably to describe complete wind-powered electricity generating machines.

yaw: The orientation of a horizontal-axis wind turbine.

zener diode: A diode often used for voltage regulation or protection of other components.

Meteorological Terms

altitude: The angle up from the horizon.

angle of incidence: The angle between the normal to a surface and the direction of the sun. Therefore, the sun will be perpendicular to the surface if the angle of incidence is zero.

azimuth: The angle from north measured on the horizon, in the order of N, E, S, and W. Thus, north is 0 degrees, and east is 90 degrees.

civil twilight: Defined as beginning in the morning and ending in the evening when the center of the sun is geometrically 6 degrees below the horizon.

horizon: The apparent intersection of the sky with the earth's surface. For rise and set computations, the observer is assumed to be at sea level, so that the horizon is geometrically 90 degrees from the local vertical direction. The inclination surface tilt is expressed as an angle to the horizontal plane. Horizontal is 0 degrees; vertical is 90 degrees.

horizontal shadow angle (HSA): The angle between the orientation of a surface and the solar azimuth.

local civil time (LCT): It is a locally agreed upon time scale. The time given out on the radio or television, and the time by which we usually set our clocks. Local civil time depends on the time of year and your position on Earth. It can be defined as the time at the Greenwich meridian plus the time zone and the daylight savings corrections.

orientation: The angle of a structure or surface plane relative to north in the order of N, E, S, and W. Thus, north is 0 degrees, and east is 90 degrees.

shadow angles: Shadow angles refer to the azimuth and altitude of the sun, taken relative to the orientation of a particular surface.

solar noon: The time when the sun crosses the observer's meridian. The sun has its greatest elevation at solar noon.

sunrise and sunset: Times when the upper edge of the disk of the sun is on the horizon. It is assumed that the observer is at sea level and that there are no obstructions to the horizon.

twilight: The intervals of time before sunrise and after sunset when there is natural light provided by the upper atmosphere.

vertical shadow angle (VSA): The angle between the HSA and the solar altitude, measured as a normal to the surface plane.

INDEX